Complexation
Chromatography

CHROMATOGRAPHIC SCIENCE

A Series of Monographs

Editor: JACK CAZES
Sanki Laboratories, Inc.
Sharon Hill, Pennsylvania

Complexation Chromatography

edited by
D. Cagniant

University of Metz
Metz, France

CRC Press
Taylor & Francis Group
Boca Raton London New York

CRC Press is an imprint of the
Taylor & Francis Group, an **informa** business

CRC Press
Taylor & Francis Group
6000 Broken Sound Parkway NW, Suite 300
Boca Raton, FL 33487-2742

First issued in paperback 2019

© 1992 by Taylor & Francis Group, LLC
CRC Press is an imprint of Taylor & Francis Group, an Informa business

ISBN-13: 978-0-8247-8577-2 (hbk)
ISBN-13: 978-0-367-40290-7 (pbk)

Library of Congress Cataloging-in-Publication Data

Complexation chromatography/edited by D. Cagniant.
    p. cm. -- (Chromatographic science: v. 57)
    Includes index.
    ISBN 0-8247-8577-0
    1. Ligand exchange chromatography. I. Cagniant, D.
II. Series.
QD79.C4537C66   1991
543' .089--dc20   91-41540
                                  CIP

Visit the Taylor & Francis Web site at
http://www.taylorandfrancis.com

and the CRC Press Web site at
http://www.crcpress.com

Preface

The interaction of solute molecules with liquid phase and/or solid phase is of prime importance in chromatographic separations. Among the possible interaction mechanisms, those involving the transfer of electrons between pairs of molecules, each acting as either donor or acceptor (EDA interactions), are the basis of a great many chromatographic methods.

According to the nature of the interacting pairs, several types of donor–acceptor interactions can be distinguished. Their applications in chromatography produce a variety of results, and this fact can be somewhat troublesome. For example, metals can act as electron acceptors and can form molecular complexes with organic compounds capable of donating electron pairs to the vacant orbitals of the metal. As early as 1955, Bradford applied silver nitrate dissolved in polyethylene glycol as the stationary phase in the separation of saturated and unsaturated hydrocarbons using gas chromatography. This event marked the beginning of *argentation chromatography*, which has found many applications in the separation of olefinic compounds in mixtures of varying complexity.

In 1961, Helfferich described the substitution of metal-ion-coordinated ammonia molecules in a resin phase for organic diamine molecules. He was the first to propose the term "ligand exchange," thus giving rise to *ligand-exchange chromatography* (LEC). Since this pioneering work, various types of ligand-exchange processes have been developed and are well covered in the literature, particularly by one of the contributors to this book (V. A. Davankov).

When π donor–acceptor interactions occur between π-electron donating and withdrawing aromatic or heterocyclic compounds, charge-transfer complexes are obtained, as is well documented in classic organic chemistry and biochemistry. The fundamentals of these interactions were covered in the literature as early as the 1960s by several authors, notably Mulliken (1969). Their utilization in chromatography gave rise to *charge-transfer adsorption chromatography*.

These three fundamental aspects of molecular interactions have been brought together in this book under the general title *complexation chromatography*. A survey of the basic factors involved in complex formation is presented in the first chapter, as complex formation governs the retention mechanism and selectivity in EDA chromatography. The purpose of the contributing authors was to cover the principal applications of complexation chromatography, as reflected in its three fundamental aspects, taking into account recent developments in chromatographic processes (e.g., new detectors and new phases). It is noteworthy that molecular complexes found applications in all types of chromatographic methods (thin-layer, gas, liquid, and high-performance liquid chromatographies).

It was also useful to make a general survey of packings used in all types of EDA chromatographic methods. Independent of Chapter 2 which is devoted to this subject, many examples are given in other chapters. Interestingly, very important classes of compounds can be submitted to complexation chromatography in the fields of either organic compounds (saturated, olefinic, polycyclic aromatic, and heterocyclic compounds) or biochemical derivatives (amino acids, peptides, proteins, fatty acids, lipids, prostaglandins, nucleic acids, drugs, etc.). All these topics are considered as well as more specific cases in the field of coal and petroleum products analysis. Special care was taken to avoid redundancy in areas already well reviewed. For ligand-exchange chromatography, only selected aspects are discussed, focusing on chiral compounds and the more recent literature.

D. Cagniant

Contents

Contributors

D. Cagniant   Laboratory of Organic Chemistry, University of Metz, Metz, France

V. A. Davankov   Institute of Organo-Element Compounds, USSR Academy of Sciences, Moscow, USSR

Guy Felix   Laboratory of Organic and Organometallic Chemistry, Bordeaux I University, Talence, France

L. Nondek   Water Research Analytical Department, Water Research Institute, Praha, Czechoslovakia

Complexation
Chromatography

1
Complexation in Chromatography

L. Nondek
Water Research Institute, Praha, Czechoslovakia

I. GENERAL CONSIDERATIONS

The interactions of analyte molecules with liquid phase (stationary phase in gas–liquid or liquid–liquid chromatography, mobile phase in liquid chromatography) and/or solid phase (stationary phase in gas–solid chromatography or liquid–solid chromatography) are of prime importance from the viewpoint of chromatographic separations. In liquids, two independent interaction mechanisms are operating on the molecular level: nonspecific van der Waals forces and specific interactions involving the transfer of electrons (charge) between interacting molecules. In principle, the nature of the adsorption forces between adsorbate molecules and solid surface is the same.

The formation of covalent bonds or interactions between strong acids and bases, which yields very stable products, is of no importance in the chromatographic process. Only relatively weak interactions with energy in the range of several kcal/mol (Fig. 1) are suitable for this purpose.

All the specific interactions taking place either in the liquid phase or on solid surfaces can be explained on the basis of a modernized concept of acid–base interactions [2,3]. Bases (electron donors) are species that can donate a pair of electrons shared by acids (electron acceptors). In terms of molecular orbital theory, electron

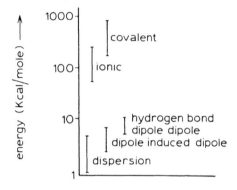

FIGURE 1   Energetics of interactions. (From Ref. 1.)

donor–acceptor (EDA) interactions can be classified according to the interacting molecular orbitals; see Table 1, where n are nonbonding orbitals (e.g., unshared electron pairs, σ and π are bonding (occupied) orbitals, and σ^* and π^* are antibonding (empty) orbitals. The underlined interactions in Table 1 play an important role in chromatographic separations of organic compounds. Hydrogen bonding pertains primarily to n-σ^* interactions. Electron donors and acceptors are classified in Table 2.

In this review of the application of EDA interactions in chromatography, we distinguish between organic complexes (e.g., π-π^* or n-π^*), and coordination complexes (π-n) of some metal cations with organic ligands. The donors (bases) and acceptors (acids) can be classified as *soft* or *hard*. In Klopman's quantum chemical approach [4], the energy change (ΔE) in a chemical interaction is evaluated in terms of a charge–charge interaction (electrostatic forces), orbital–orbital interactions, and solvation:

$$\Delta E = \frac{-q_r q_s e^2}{d\epsilon} + \sum_m \sum_n \frac{2(c_r^m)^2 (c_s^n)^2 \beta^2}{E_m^* - E_n^*} + \Delta_{solv} \qquad (1)$$

where q_r and q_s are charges of reactants R and S, respectively, d the distance between these sites, ε the dielectric constant, c_r^m and c_s^n the coefficients of the atomic orbitals m (occupied) and n (unoccupied) of energy E_m^* and E_n^*, and Δ_{solv} the solvation energy.

When the difference between frontier orbitals ($E_m^* - E_n^*$) is large, the first term of Eq. (1) is dominating and the interaction is said to be *charge controlled*. This is the case of interactions between hard bases and hard acids—species of relatively

TABLE 1 Classification of EDA Interactions

Donor orbital	Acceptor orbital		
	n	σ^*	π^*
n	n-n	$\underline{n\text{-}\sigma^*}$	$\underline{n\text{-}\pi^*}$
σ	σ-n	σ-σ^*	σ-π^*
π	$\underline{\pi\text{-}n}$	π-σ^*	$\underline{\pi\text{-}\pi^*}$

TABLE 2 Examples of Electron Donors and Acceptors

Class of species	Examples
n-Donors	Aliphatic amines, pyridine, sulfides, ketones, ethers, and all neutral organics with nonbonding electron pairs
σ-Donors	Alkanes, peralkylpolysiloxanes
π-Donors	Olefines, aromatics, azaarenes
n-Acceptors	Monatomic cations (e.g., Pd^{2+}, Ag^+, Cu^{2+}, etc.)
σ-Acceptors	Halogenes, halogenated alkanes (e.g., CH_2Cl_2, $CHCl_3$)
π-Acceptors	Alkynes, alkenes, or aromatics with electron-withdrawing substituents (e.g., CN, Cl, etc.)

TABLE 3   Classification of Acids and Bases

Hard bases	RNH_2, ROH, R_2O, AcO^-
Borderline bases	$ArNH_2$, pyridine
Soft bases	Olefins, arenes, RCN, R_3P, RSH, R_2S
Hard acids	Ca^{2+}, Al^{3+}, Fe^{3+}, Cr^{3+}, Ti^{4+}
Borderline acids	Fe^{2+}, Co^{2+}, Cu^{2+}, Zn^{2+}, Sn^{2+}, Pb^{2+}
Soft acids	Cu^+, Ag^+, Cd^{2+}, Hg^{2+}, metal ions, and bulk metals

small size and high effective charge. On the other hand, frontier-controlled interactions are favored by large, highly polarizable species of low or zero charge. These species are called soft acids and bases. Organic compounds and metal ions are classified as hard and soft in Table 3.

The ease of interaction between various bases and acids is given in Table 4 with respect to Klopman's theory. The so called hard–soft acid–base (HSAB) principle does not always predict in an unequivocal manner whether or not a given interaction will be facile, since it does not involve other factors (e.g., steric). The kinetics of complexation is also very important, since the ability to break down a complex readily and to re-form it is a condition necessary for its utilization in a separation process.

The formation of labile complexes has often been utilized in gas or liquid chromatography. According to the nature of these complexes, one can speak of *charge-transfer chromatography* [6,7], *ligand-exchange chromatography* [8,9], or *argentation chromatography* [10]. The term *molecular complexation chromatography* was coined in an early paper of Klemm and Reed (11). Despite the fact that

TABLE 4   Interactions Between Hard and Soft Species

Acid	Base	$E_m{}^* - E_n{}^*$	e^2/d	β	Interaction
Hard	Hard	Large	Large	Small	Easily
Soft	Soft	Small	Very small	Large	Easily
Hard	Soft	Medium	Small	Very small	Difficult
Soft	Hard	Medium	Small	Very small	Difficult

Source: Ref. 5.

energy of complex formation is also in units of kcal/mol, and although the resulting separation mechanism is usually a combination of several types of molecular interaction (e.g., electron donor–acceptor complexation, hydrogen bonding, and steric interactions), one can distinguish cases where complexation is the process governing the retention of analytes.

The molecular complexes are substances with well-defined stoichiometry and geometry and are formed by the interaction of two or more component molecules or ions. Since the complex formation is highly selective, depending on the structure of the complexing species as well as on the temperture or polarity of environment, its utilization in chromatography has resulted in many selective separations of species of similar chemical structure, such as constitutional, configurational, and isotopic isomers. The application of complexation chromatography offers an advantage, especially in cases where reversed-phase liquid chromatography (RPLC), because of being based on the hydrophobic interactions of low selectivities, fails.

In organic chemistry, various types of labile molecular complexes have been described and used in gas or liquid chromatography: electron donor–acceptor (EDA) complexes, organometallic complexes, coordination complexes, inclusion complexes, or clathrates. One must, however, distinguish between the complexation chromatography and the chromatography of complexes [12]; stable complexes can be treated as another stable molecular species. Many chelates (e.g., acetylacetonates) have been separated by gas chromatography (GC) or high-performance liquid chromatography (HPLC). In this case the complexation taking place prior to the chromatographic separation is practically irreversible, not being a part of the separation mechanism.

The first attempt to use the formation of organic complexes in chromatography was made by Godlewicz [13] in 1949. The aromatic fraction of lubricating oils was separated as color zones using a silica gel column impregnated with trinitrobenzene. Since the pioneer work of Godlewicz, the application of complexation processes in gas and liquid chromatography, as well as in the selective handling of analytes prior to chromatographic separation, has been studied in many laboratories.

II. ORGANIC EDA COMPLEXES

Organic electron donor–acceptor (EDA) complexes [14–16], also described as charge-transfer complexes, are in addition to inclusion complexes, widely used in chromatographic separation. EDA complexes (AD) result from a weak interaction of electron donors (D) with electron acceptors (A):

$$A + D \xrightleftharpoons{\quad K_{eq} \quad} AD$$

The enthalpy of EDA complexation is usually on the order of a few kcal/mol, and the rates of formation and dissociation of AD are very high. As the stability of EDA complexes depends not only on the structure of components but also on the temperature and polarity of solvent used, their formation has been utilized in GC and LC for several decades [14–16].

In organic molecules, the electron density is not distributed regularly. Relatively electopositive regions of molecules exist as well as relatively electronegative regions. These inequalities can lead to electron sharing between two interacting molecules A and D. Electron donors are defined as molecules capable of giving up an electron, and their ionization potential, I, is a measure of the donation ability. Electron acceptors are able to accept an electron, and this is related to their electron affinity (EA) or reduction potential.

In many instances, the same molecule can act as an electron donor or acceptor, depending on the circumstances. Whereas polycyclic aromatic hydrocarbons will donate charge to picric acid or trinitrobenzene, they accept charge from, for example, N-dimethylaniline [17], behaving as charge acceptors. Conversely, carbonyl compounds are acceptors to aromatic hydrocarbons but are donors to bromine [15]. In large molecules such as pharmaceuticals, biologically active compounds, and synthetic dyestuffs, several independent electron-accepting and electron-donating parts of the molecule can exist [14–16,18].

Small unsaturated or aromatic hydrocarbons are usually weak donors or very weak acceptors. Their donating or accepting capability increases with the increase in the number of C=C double bonds or aromatic rings. Polynuclear aromatic hydrocarbons (PAHs) and azaarenes are therefore efficient donors of π-electrons. Replacement of a hydrogen atom in the parent molecule of PAHs with an electron-releasing substituent such as an alkyl, alkoxy, or amino group increases the capability of a molecule to donate π-electrons.

On the other hand, aromatic or unsaturated compounds containing several electron-withdrawing substitutents , such as NO_2, Cl, or CN, are efficient electron acceptors. 1,3,5,-Trinitrobenzene (1), tetracyanoethylene (2), and similar molecular structures are efficient π-acceptors. The formation of EDA complexes is assumed to be the main interaction mechanism governing chromatographic behavior in charge-transfer or EDA liquid or gas chromatography.

(1)   (2)

A. Retention Mechanism in EDA Chromatography

In GC or LC separation based on EDA complexation, the chromatographic retention depends on the stability of EDA complexes formed between the stationary phase and the solute. The formation of these complexes, which is assumed to be the governing retention mechanism, is influenced by factors identical to those in a homogeneous phase.

Ideas about the formation of EDA complexes in the gas phase or dilute solutions are based on Mulliken's theory of charge-transfer complexes [19,20]. In this theory, the wave function of the ground state of a 1:1 complex, ψ_N, is described by the expression

$$\psi_N(DA) = a\psi_0(DA) + b\psi_1(D^+A^-) \tag{3}$$

Two electronic states are assumed here: nonbonding (DA) and dative (D^+A^-). The ψ_0 term represents a no-bond wave function of A and D in close proximity with no charge transfer between them. The ψ_1 term is a dative wave function corresponding to the total transfer of an electron from D to A. Although the ratio b^2/a^2 may vary from zero to infinite, it is generally small in the molecular complexes in the ground state (where $a \gg b$).

The main factor influencing the stability of EDA complexes is the electronic structure of D and A. Mulliken [19,20] has stressed the orientation character of charge-transfer interactions as opposed to the dispersion forces, which depend primarily on the polarizabilities of interacting molecules. The orientational character, which along with the electron transfer process, restricts the combination of A and D to simple integral ratios, is predicted by the symmetry of molecular wave functions. According to the theory, a parallel-plane configuration for complexes of planar aromatic donors and acceptors is supposed.

On the basis of crystallographic studies of solid-state complexes [15,21], the donor and acceptor molecules are arranged in parallel planes, with little distortion within A and D molecules. The quantum chemical calculations indicates that only a limited number of alternative structures lead to maximum overlap of π-orbitals in π-π^* complexes (Fig. 2). The interplanar distance between D and A is often significantly less than van der Waals contact distance (3.4 Å in graphite).

The structure of complexes formed from π-acceptors and n-donors is different. The overlap between the n-orbital of the donor with the unoccupied orbital of the π-acceptor will result in structures where the plane of the n-donor (e.g., pyridine) is perpendicular to the plane of the π-acceptor.

The formation of hydrogen-bonded complexes can also be explained as a special case of EDA interaction [22,23]. It can be demonstrated, for example, on the interactions between substituted phenols (acids or σ-acceptors) and triethylamine (base of n-donor). Thus, according to charge-transfer theory, a fraction of the n-electron from nitrogen is transferred to the antibonding σ^*-orbital of OH. This

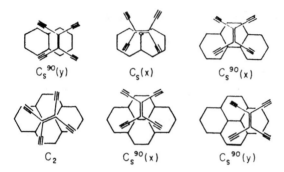

$C_s^{90}(y)$ $C_s(x)$ $C_s^{90}(x)$

C_2 $C_s^{90}(x)$ $C_s^{90}(y)$

FIGURE 2 Structural alternatives discussed in the text for DA complexes between tetracyanoethylene and three donor molecules: naphthalene, phenanthrene, and pyrene. (From Ref. 21.)

causes a weakening and elongation of the O—H bond and the enhancement of the polarity of the O—H $\cdots$ N system. Thus the OH bond is much weaker if the phenol is a stronger σ^*-acceptor. In the case of a low I_D value of the donor and a low pK_a value of the acceptor, complete transfer of the proton is a consequence of n-σ^* EDA interaction.

However, the main difference between complex formation in chromatographic stationary-phase solutions and complexation on a sorbent surface is given by two additional factors:

1. A large surface concentration of complexing species may permit the formation of complexes with "unusual" geometry or stoichiometry.
2. The limited motion of adsorbed or immobilized acceptor or donor species causes steric hindrances for complexation.

For these reasons the equilibrium constant K_{eq} measured in dilute solutions may not always correlate with retention times t_r or capacity factors k'.

As stated above, the stability of EDA complexes depends primarily on the structure of both participants: the solute molecules and the bonded ligands. To separate the contribution of various experimental factors, such as temperatue and solvent effects or varying concentration of "active ligands," the structure–retention relationships will be discussed in terms of chromatographic selectivity α. This approach makes it possible to compare different experiments, evaluate various sorbents, and examine retention mechanism.

B. Chromatographic Selectivity

Chromatographic selectivity $\alpha_{i,1}$ is defined as the ratio of capacity factors k_i'/k_1' for a given pair of solutes. In a series of structurally related compounds, the k_i' values are usually related to the capacity factor k_1' of the compound having the simplest structure. As the individual capacity factors are proportional to the equilibrium constant K_{eq},

$$\log k_i' \simeq \log(K_{eq})_i = -\frac{\Delta G_i^\circ}{RT} \tag{4}$$

Thus $\log \alpha_{i,1}$ can be expressed as

$$\log \alpha_{i,1} = (-\Delta G^\circ)_i - (-\Delta G^\circ)_1 \tag{5}$$

and in this form correlated with other structurally related molecular properties. This type of relationship is known [24] as a *linear free-energy relationship* (LFER).

In the concept of LFER, the additivity of independent interaction mechanisms is assumed. In chromatography, $\log k_{rel}'$ or $\log \alpha_{i,1}$ can be correlated with Taft or Hammett substituent constants, with rate of equilibrium constants of chemical reactions involving studied solutes, with $\log k_{rel}'$ measured in different chromatographic systems, or with other properties that can be expressed in terms of free-energy changes [25].

The most valuable feature of LFERs is that they make it possible to recognize regular patterns of chemicophysical behavior and readily to observe deviations from these patterns. A quantitative expression of the "normal" response of a chromatographic system to variations in either the structure of solutes and bonded ligands or mobile-phase composition is obtained [24].

C. EDA Complexation in Gas Chromatography

After the discovery of organic EDA complexes by Benesi and Hildebrand [26], chromatographers have also attempted to utilize EDA complexation in GC. In 1958, Norman [27] reported the use of 2,4,7-trinitro-9-fluorenone (TNF) as a stationary phase for the separation of nitrotoluenes. Di-*n*-alkyl tetrachlorophtalates were used as a selective stationary phase for GC of aromatic hydrocarbons [28]. This electron acceptor was also employed for the separation of aromatic hydrocarbons and amines [29,30]. Nitroaromatic acceptors were usually used as solutions in a proper stationary phase (e.g., in polyethylene glycol). Their low thermal stability and relatively high volatility restricted their practical utilization.

The GC technique has, however, been utilized for the determination of equilibrium formation constants [31–33], which correlate not only with the stability constants determined by the spectroscopic technique but also with vertical ionization potentials and electron affinities. These physicochemical studies were reviewed by Laub and Pecsok [34].

The differences observed between GC and spectroscopically determined K_{eq} are ascribed in varying degrees to steric effects, hydrogen bonding ,and solvent effects. Thus the separation of EDA interactions from "physical" interactions has been identified as the main problem with this technique [35]. There are therefore several ways to calculate K_{eq} from the primary retention data.

The simplest method, proposed by Gil-Av and Herling [31], is suitable for calculation of the concentration equilibrium constant, K_f^c, if 1:1 complexes are formed. K_f^c is defined as

$$K_c^f = K_{eq} \frac{\gamma_D^c \gamma_A^c}{\gamma_{AD}^c} \tag{6}$$

where γ_i^c are the concentration activity coefficients. K_f^c is calculated from the linear plot [Eq. (8)] of experimentally measured distribution coefficient K_L versus the concentration of acceptor in a stationary phase [A]:

$$K_L = \frac{\text{solute concentration in stationary phase}}{\text{solute concentration in mobile phase}} \tag{7}$$

$$K_L = K_L^o + K_L^o K_f^c[A] \tag{8}$$

K_L^o is the solute (donor) distribution constant observed in the absence of acceptor in the stationary phase.

More sophisticated approaches take into account the differing stoichiometry of the complexes formed and the specific solvation of A, D, and AD species [33,36, 37].

D. EDA Complexes in Liquid Chromatography

In EDA-LC, the complexation can take place in either the mobile phase or the stationary phase; the latter alternative, utilizing stationary phases with electron-accepting or electron-donating ability, is more frequent. The separation of PAHs on silica impregnated with 2,4,6-trinitrobenzene or another organic acceptor was reported several decades ago [13,19,38]. Nevertheless, the adsorbed stationary phases are easily washed out, and this not only causes a stepwise loss of retention but also interferes with the ultraviolet (UV) detection often used in HPLC.

The charge-transfer complexation in mobile phase has generally been used in combination with conventional RPLC for the purpose of separation of enantiomers. Lochmuller and Jensen [39] used N-(2,4-dinitrophenyl)-L-alanine-n-

dodecyl ester as a nonionic chiral mobile phase additive for the separation 1-azahexahelicenes by RPLC. The resolution obtained was found to be a function of the mobile-phase polarity and the concentration of the chiral additive. The alternative to the use of complexing mobile phases is to bound the complexing species to solid surfaces [40]. Two classes of sorbents have been prepared: polymers and chemically modified silicas.

1. Chemically Bonded Donors and Acceptors

Organic electron acceptors bound to the surface of polymers have been studied in LC for over 20 years, since the first such sorbent was prepared by Ayres and Mann [41] in 1964. Divinylbenzene-styrene copolymer modified with benzyl groups was nitrated. The resulting resin (3) permitted complete separation of anthracene and pyrene despite the very low efficiency of the column used. Polymers modified by electron acceptors were also reported by Smets et al. [42].

(3)   (4)

Hydrophilic gels such as polydextrans modified with various donors or acceptors have been prepared and tested by Porath and co-workers [7,43,44]; for example, Sephandex G-25 was modified with pentachlorothiophenyl ligands (4). In this case, separation of donors and acceptors of biological importance was achieved in aqueous mobile phases. Smidl and Pecka [45,46] prepared methacrylate copolymers modified with 2,4-dinitrophenoxy (5) and 3,5-dinitrobenzoyl (6) ligands.

Natural polymers have been also chemically modified with electron donors and acceptors. Glucose esterified with 3-(9-phenanthryl)propionic acid has been used for the thin-layer chromatography of nitrotoluenes and quinones [47]. Riboflavin has been covalently bonded to cellulose for the separation of biologically active electron donors [48].

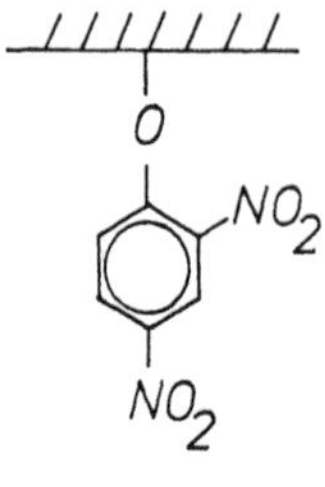

(5) 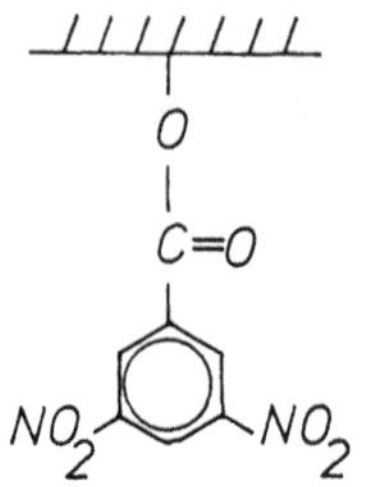   (6)

The advent of chemically bonded stationary phases for HPLC based on silica made it possible to prepare more selective and more efficient electron-accepting or electron-donating sorbents. In practice, the organic donors or acceptors are immobilized on silica surface via a suitable silanization reaction. The first silica modified with chemically bonded acceptor ligands was prepared via the reaction of surface silanols with p-nitrophenyl isocyanate by Ray and Frei [49]. The resulting sorbent (7) has been found unstable, being easily decomposed by moisture and light.

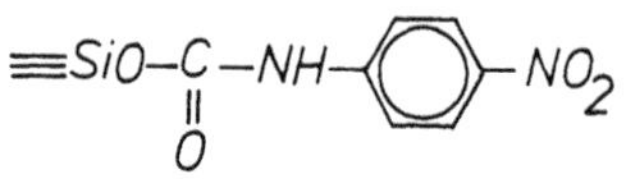

(7)

Since then, several acceptor-modified silicas have been prepared and tested [3,40,50-58]. Most of them are based on nitroaromatic ligands such as 2,4,5,7-tetranitrofluorenonoxime [TENF (8)], 2,4-dinitroaniline [DNA (9)], or 2,4,6-trinitroaniline (TNA) groups bonded to the silica surface by means of a short aliphatic chain. Di- and trinitrophenylmercaptopropyl ligands were bound to silica by Welch and Hoffman [59].

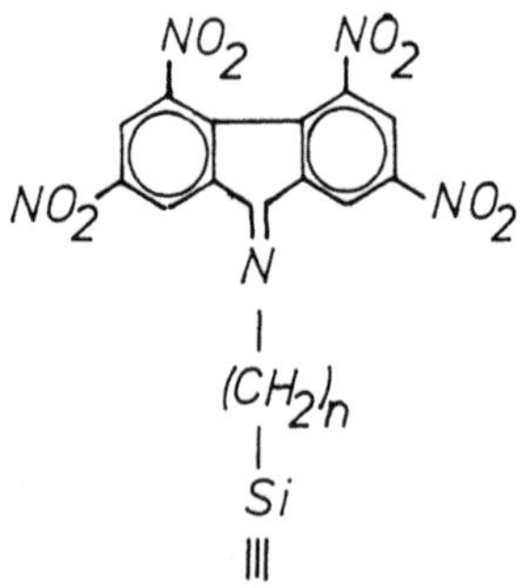

(8) 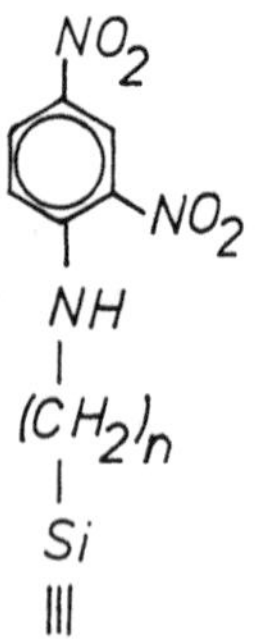   (9)

In addition to nitroaromatic ligands, tetrachlorophtalimidopropyl [TCI (10)] silica [60,61], pentafluorbenzamidopropyl, and pentafluorophenyl ligands have also been prepared [62–64]. 3-2,2,2-trichloroacetamide)propyl and 3-(2,2,2-trichloro-ethoxy)propyl ligands have been bound to silica as nonaromatic electron acceptors [65].

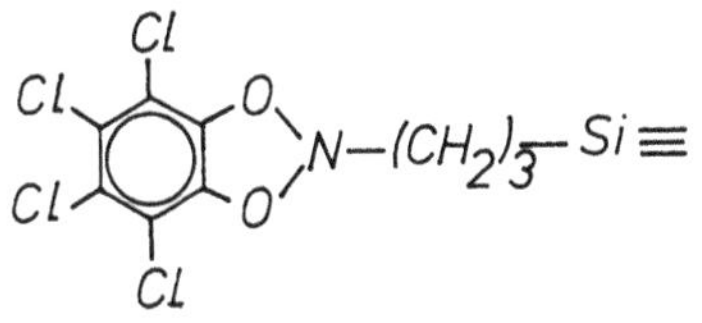

$$(10)$$

Chemically bonded PAHs, azaaromatics, and alkoxybenzenes are expected to be efficient electron donors. Phenoxy [66], pyrene [67], or anthracene (68) ligands have been bonded to the silica surface via corresponding silanes as well as safrol [69] ligands (*11*), which have been found to act as strong π-donating species.

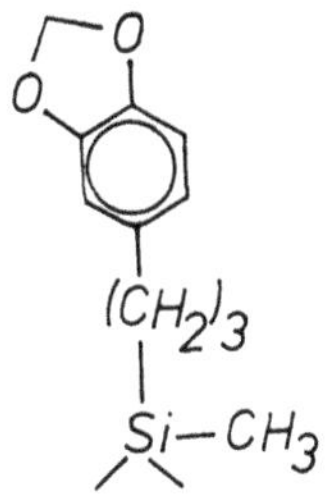

$$(11)$$

Several examples of the application of chemically bonded donors and acceptors in LC and HPLC in the filed of PAHs, aromatic amines, azaarenes, and bilogical compounds are provided in Chapter 3.

2. Quantum-Chemical Model of Selectivity in EDA Chromatography

In recent years, a large number of chemically bonded electron-acceptor and electron-donor stationary phases has been prepared [40,70]. Since the chromatographic selectivity in EDA–LC depends on the ability of the stationary phase to form complexes with solutes, it may be useful to use LFER for an examination of this ability. The chromatographic selectivity $\alpha_{i,1}$ in EDA–LC can be approximated as

$$\log \alpha_{i,1} \doteq (\Delta H^\circ)_i - (\Delta H^\circ)_1 \tag{9}$$

where $i = 1, 2, 3, \ldots n$ in a series of n structurally related solutes (ΔS° = const.). In EDA chromatography it is possible to substitute the enthalpic terms in Eq. (9) with an interaction energy, ΔE, which is a measure of the stability of the complex. Nondek and Ponec [58] attempted to predict $\log \alpha_{i,1}$ using a simple quantum-chemical model in which only π-electrons are transferred between HOMO and LUMO frontier orbitals of A and D. Thus ΔE in Eq. (1) may be approximated by the simplified equation of Klopman and Salem [4,71,72]:

$$- \Delta E = \frac{2 c_{\mu,\text{HOMO}}^2 c_{v,\text{LUMO}}^2}{E_{\text{HOMO}} - E_{\text{LUMO}}} \beta_{\mu,v}^2 \tag{10}$$

where $c_{\mu,\text{HOMO}}$ and $c_{v,\text{LUMO}}$ are frontier orbital coefficients, E_{HOMO} and E_{LUMO} are energies of these orbitals, and $\beta_{\mu,v}$ is the corresponding resonance integral. Assuming that the numerator in Eq. (10) is roughly constant in a series of structurally related solutes, the following relationship between the energy of frontier orbitals and the chromatographic selectivity or relative stability of EDA complexes has been found [58]:

$$\log \alpha_{i,1} = \text{const.} \frac{(\Delta E_{\text{HOMO}})_i}{(E_{\text{HOMO}} - E_{\text{LUMO}})_1} \tag{11}$$

where $(\Delta E_{\text{HOMO}})_i = (E_{\text{HOMO}})_i - (E_{\text{HOMO}})_1$.

The model above can serve as a rational basis for the synthesis of more selective electon–acceptor aromatic ligands. The selectivity depends not only upon the number and mutual position of electron-accepting substituents (NO_2, Cl, CN, etc.) attached to the aromatic skeleton, but also on the nature of the spacer connecting the ligand with the silica surface. The dependence of $\alpha_{i,1}$ on the structure of bonded ligands is also shown in the separation of methylcholanthrenes with nitrofluorenone ligands studied by Lochmuller et al. [51]. The selectivity increases with increase in the number of nitro groups attached to the fluoreniminopropyl skeleton (Fig. 3).

The experimental results of Smidl [45] are also in quantitative agreement with the quantum-chemical model, as 3,5-dinitrobenzoyl ligands attached to a gel matrix show greater selectivity than do 2,4-dinitrophenoxy ligands. As mentioned earlier [40], the quantum-chemical model is only of limited value for numerical calculations of $\alpha_{i,1}$. However, it demonstrates that $\log \alpha_{i,1}$ is a combination of two independent factors [73]: (a) the structural difference of solutes: term $\delta = (\Delta E_{\text{HOMO}})_i$, and (b) the ability of bonded ligands to form EDA complexes with the given class of compounds: $\kappa = 1/(E_{\text{HOMO}} - E_{\text{LUMO}})_1$.

Thus the selectivity can be expressed as

$$\log \alpha_{i,1} = \kappa \delta \tag{12}$$

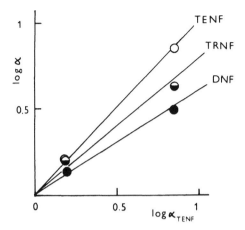

FIGURE 3 Replotted selectivities of 2,4,5,7-tertranitro- (TENF), 2,4,7-trinitro- (TRNF), and 2,6-dinitrofluorenimino (DNF) ligands in the separation of 1-,2-, and 3-methylcholanthrene. (From Ref. 58.)

3. Characterization of Chemically Bonded Acceptors

The LFER above [Eq. (12)] has been used for the quantitative evaluation of various chemically bonded aromatic acceptors by means of a series of structurally related solutes. As PAHs are frequently used in testing acceptor-modified sorbents, the set of δ-constants has been calculated statistically from retention of PAHs on 2,4-dinitroanilinopropyl (DNA) silica (Table 5). The δ constants serve for calculation of the $\mathcal{H}$ parameter characterizing the stationary phases.

Using the $\mathcal{H}$ parameter of Eq. (9), Nondek (40) has recently surveyed various acceptor-modified silicas. The most efficient sorbent is 2,4,7,9-tetranitro-fluorenoneoxime (TENF) silica, prepared by Hemetsberger et al. [53] possessing $\mathcal{H} \doteq 1.5$, and the weakest is Nucleosil-NO2 ($\kappa \doteq 0.5$), which is assumed to be p-nitropropylbenzene silica [57,74].

Welch and Hoffman [59] synthesized and tested nitroaromatics bonded to the silica surface via mercaptoproyl spacers. Nevertheless, the phases prepared have properties comparable with those of 2,4-dinitroanilinopropyl silica despite the fact that the mercaptopropyl group is assumed to be less electron donating than are spacers containing —NH— or —O— substituents. Also, perfluorinated aromatic ligands do not possess better chromatographic properties than those of nitroaromatics ($\kappa \doteq 0.5$).

The relative retention of PAHs on trichloroacetamide and trichloroethoxy bonded phases [75] does not correlate well with the δ-values, which reflect π-π^*

TABLE 5 Retention Constants (σ) of PAHs

PAH	σ
Naphthalene[a]	0.00
Phenanthrene	0.66
Anthracene	0.64
Fluoranthene	1.01
Pyrene	1.04
Chrysene	1.32
Benzo[a]pyrene	1.69
Perylene	1.79
Picene	1.95

Source: Ref. 73 (DNA silica used as reference sorbent).
[a]Used as referenece solute.

interaction. For the interaction of PAHs with nonplanar, nonaromatic electron acceptors, the steric and electronic effects are quite different.

The chromatographic selectivity of the sorbents expressed by $\mathcal{H}$ is also influenced by the density of bonded acceptor ligands, the temperature, and the polarity of the mobile phase. Sorbents with low coverages show poor chromatographic performance, probably due to the nonhomogeneity of the surface layer [40].

4. Structure and Retention of Aromatic π-Donors

The set of δ-constants generalizes retention within the series of fused-ring PAHs observed on 2,4-dinitroanilinopropyl silica together with the other chemically bonded acceptors. The set has been correlated [73] with different structural parameters of PAHs, as, for example, vertical ionization potential (I_D^v), energy of HOMO (E_{HOMO}), delocalization energy calculated by the Hückel method (DE_{HMO}), average molecular polarizability (α_{exp}), molecular connectivity (X), or Snyder's molecular area (A_s). All these parameters are, however, linear dependent on the molecular size expressed for example, by the number of carbon atoms, n_c [73], or the number of π-electrons [59].

The δ-values have been calculated from k' obtained on DNA silica [40]. The reasonable linear correlations of δ with log α measured with the various aromatic acceptors indicate that the steric effects given by various structure of ligands or character of spacer play a minor role [40]. The set of δ-constants thus represents a

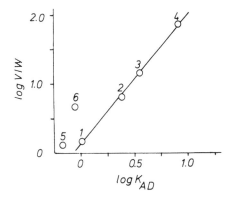

FIGURE 4   The correlation between the retention of PAHs on DNA–silica (Ref. 81) and the stability constants (Ref. 76) measured in solution: 1-naphthalene, 2-phenanthrene, 3-pyrene, 4-benzo[a]pyrene, 5-biphenyl, and 6-terphenyl.

useful tool for quantification of the complexing ability of bonded aromatic acceptors.

A different situation is observed with more flexible aromatic structures, such as biphenyl, p-terphenyl, and *cis*- or *trans*-stilbene. As is evident from the correlation of retention volumes with the K_{eq} value of tetracyanoethylene complexes [76], biphenyl and p-terphenyl seem to be retained on DNA silica more like more-aromatic structures than they are in homogeneous solutions (Fig. 4).

The values of K_{AD} for stilbenes measured in solution are 2.30 L/mol for *trans*-stilbene and 1.25 L/mol for *cis*-stilbene [77], but on TENF a much greater difference has been found for the capacity factors, k': 11.6 (*trans*) and 0.68 (*cis*) with hexane as the mobile phase [53]. Also in this case, *trans*-stilbene in EDA-LC is relatively more aromatic than in the homogeneous phase.

Also, the retention of alkylbenzenes and alkylnaphthalenes [53,75,78] does not follow the pattern found in homogeneous solutions. The methyl group generally exhibits a relatively larger enhancement of retention than do other alkyls. Large n-alkyls, such as butyl, isobutyl, or amyl, have nearly no effect compared to retention of the parent aromatic structures. The correlation of the substitutent effect of alkyls in terms of the Hammett or Taft equation is unsuccessful. The limited applicability of this type of LFER has also been observed for complexation in homogeneous solutions [79,80].

Aromatics with polar electron withdrawing substituents such as —NO_2 are retained on bonded acceptors more than are parent hydrocarbons [81,82]; for example, nitrobenzene is retained more than benzene. The polar acceptor ligands therefore probably act as the other polar stationary phases (CN, NH_2, etc.).

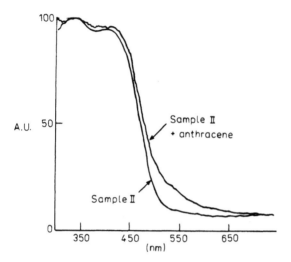

FIGURE 5 Photacoustic spectra of pure picramidopropyl silica (sample II) and the same sorbent with adsorbed anthracene. (From Ref. 83.)

5. Surface Complexes in EDA-LC

It must be pointed out that the existence of well-defined EDA complexes between solutes and surface-bound ligands is deduced primarily by analogy with the complexation taking place in homogeneous solutions. Visual or spectral observations of these complexes, which usually have a very intense color in homogeneous solutions, have been reported for sorbents bearing bound or adsorbed strong and weakly colored acceptors [13,42]. With an intensively colored acceptor ligand such as DNA or TENF, such color changes have not been observed.

The only exception is a report by Nondek and co-workers [83] in which they used photoacoustic spectroscopy to detect a surface EDA complex between bonded picramidopropyl groups and adsorbed anthracene. The slight change in the photoacoustic spectum of intensively yellow picramidopropyl silica after the sorption of anthracene in the region 450 to 550 nm (Fig. 5) is in accordance with the relative weakness of the charge-transfer interaction between the two species.

In other cases, various correlations between the retention of solutes and their stability constants K_{eq} observed in dilute solutions (e.g., Fig. 4), ionization potentials I, electron affinities EA, and so on, have been used as indirect evidence of the existence of surface EDA complexes [40,59,73,78].

In connection with the discussion of surface EDA complexes, the structure and properties of the ligand layer are important. Individual polar ligands can interact either mutually or with the silica surface [84]. The aggregation depends on temperature and solvation, as shown by Hammers et al. [81,85]. The self-association of ligands occurs to some extent in the presence of weakly polar solvents.

If the association equilibrium is perturbed by a temperature change, a new state of the ligand layer is established after a few hours. In nonpolar solvents, adsorption of solute molecules on the top of the ligand layer is supposed to prevail, whereas in more polar solvents the solute molecules are assumed to penetrate the swollen layer. Jadaud et al. [86] proposed a *slot model* of adsorption that requires unhindered penetration of solute into the spaces between acceptor ligands.

6. Temperature and Solvent Effects in EDA-LC

Obviously, the retention of PAHs on bonded acceptors decreases with increasing temperature and solvent polarity. The same holds for chromatographic selectivity, $\alpha_{i,1}$ [40]. As the enthalpy of complexation, $-\Delta H$, decreases with increasing complex stability, the retention of stronger donors will be decreased more with increase in temperature. Consequently, a lower selectivity is observed at higher temperatures.

Nonpolar mobile phases.   As for the composition of nonpolar mobile phases, both retention and selectivity decrease with increasing polarity. It is interesting that similar trends in the relative stability of EDA complexes have been observed in solutions; for example, the relative K_{eq} values are decreased in a series of several donors with fluoranil or 1,4-dicyano-2,3,5,6-tetrafluorbenzene as acceptors if tetrachloromethane is replaced with the more polar chloroform [87].

It must be pointed out that the solvent effects on EDA complexation can be characterized as either nonspecific or specific solvation of A or D with solvent S. The latter can be described as competing equilibria:

$$A + D \overset{K_{eq}}{=\!=\!=} AD \tag{13}$$

$$D + S \overset{K_{solv}^{D}}{=\!=\!=} DS \tag{14}$$

DS is a weak EDA complex of S with solute D. The interaction of S with bonded ligands A is neglected. In the absence of S (e.g., in pure *n*-hexane or other nonpolar, inert mobile phase), the equilibrium constant K_{eq} is expressed as follows:

$$K_{eq} = \frac{[DA]}{[D^\circ - DA][A^\circ - DA]} \tag{15}$$

where $[D^\circ]$ and $[A^\circ]$ are the initial concentrations of A and D. In the presence of S, the equilibrium constants K_{eq}^{solv} and K_{solv}^{D} are written

$$K_{eq}^{solv} = \frac{[DA]}{[D^\circ - DA - DS][A^\circ - DA]} \qquad (16)$$

$$K_{solv}^{D} = \frac{[DS]}{[D^\circ - DA - DS][A^\circ - DA]} \qquad (17)$$

assuming that $[S] >> [DS]$ and $[A] >> [DA]$. These conditions are perfectly fulfilled under LC conditions and one can simply derive (88):

$$K_{eq} = K_{eq}^{solv}(1 + K_{solv}^{D}[S^\circ]) \qquad (18)$$

Since $k'_n = K_{eq}[A^\circ]$ and $k'_s = K_{eq}^{solv}[A^\circ]$, a relationship between k'_n, k'_s, and $[S^\circ]$ can be obtained:

$$k'_n = k'_s(1 + K_{solv}^{D}[S^\circ]) \qquad (19)$$

where k'_s is capacity factor observed at $[S^\circ]$ and k'_n in the absence of S.

Plotting experimental $1/k'_s$ values against $[S]$ for binary mobile phase n-hexane/CH_2Cl_2, one can verify the validity of the solvation model [Eq. (16)] for several PAHs. The K_{solv}^{D} calculated from the linearized dependences are 0.8 L/mol for naphthalene, 1.3. L/mol for phenanthrene and anthracene, and 2.9 L/mol for pyrene [40].

PAHs usually form 1:1 complexes with polynitrobenzenes, even if several aromatic rings are available for complexation [14,15]. Only with very large π-donors such as 1,2,5,6-dibenzanthracene are some π-acceptors reported to form 1:2 complexes. There is however, a number of reports concerning the formation of 1:2 or 1:3 complexes of aromatics with halogenated δ-acceptors such as CH_2Cl_2 and CCl_4 [14]. Assuming an interaction of several molecules of S with one molecule of D or A, one can derive [89] a relationship similar to Eq. (16):

$$k'_n = k'_s(1 + K_{solv}^{D}[S]^n)(1 + K_{solv}^{A}[S]^m) \quad m, n = 1, 2, \text{or } 3 \qquad (20)$$

Since Jadaud et al. [61] obtained a linear plot of $1/\sqrt{k'_s}$ versus $[S]$, two molecules of CH_2Cl_2 probably interact with one molecule of fluoranthene. Nonlinear plots of $1/k'_s$ versus $[S]$ have been also found in case of LC of PAHs on trinitroanilinopropyl ligands using dioxane or isopropanol as polar additives [90].

It should also be stressed that these simple models neglect the change of polarity of mobile phase, which is known to influence K_{eq}. Also, the nonspecific solvation of polar bonded ligands that causes swelling of the ligand layer is not involved. The models above are therefore restricted to relative low concentrations of nonpolar modifiers such as CH_2Cl_2, $ChCl_3$, or CCl_4 added to n-alkanes.

Using a similar approach, Hemetsberger et al. [53] found that the bonded TENF ligands form A2S complexes with several polar solvents. They assumed that D competes with n molecules of S for A. In this way a "sandwich" structure of swollen ligands is assumed to be formed. The complexing strengths of the solvent increase in the order isopropyl chloride < dichloromethane < tetrahydrofuran < acetone < ethyl acetate. The specific solvation of bonded ligands with n-donors (e.g., alcohols, esters, or ethers) can be assumed. However, chlorinated hydrocarbons such as tetrachloromethane of chloroform with no electron-donor ability are not known to interact specifically with organic electron acceptors.

The effect of mobile-phase polarity on the retention of PAHs over TENF silica has been thoroughly discussed by Hemetsberger et al. [53]. They used a model derived by Filakov and Borovikov [91] under the assumption that the stability of EDA complexes is determined by dipole–dipole electrostatic interactions. Assuming that the dipole moments of D and A are unchanged by the complexation, a linear relationship [Eq. (21)] has been derived [91]:

$$\log k' = a + \frac{b}{\varepsilon} \tag{21}$$

The plots of $\log k'$ versus $1/\varepsilon$ are not exactly linear, however, for the experimental data of Hemetsberger et al. [53].

All the considerations above hold for PAHs as model solutes. Aromatics with polar substituents such as OH, NO_2, or NH_2 substituents seem to interact with bonded nitroaromatic lignads not only via simple π-π interactions, and their chromatographic behavior therefore does not correspond with the retention order expected.

Jadaud et al. [61,86] measured retention of naphthalene, 1-nitronaphthalene, and 11,3-dinitronaphthalene on TCE silica using isooctane/methyl *tert*-butyl ether as mobile phase. In pure isooctane, the retention order of the solutes (naphthalene > nitronaphthalene > dinitronaphthalene) is in agreement with the expected stability of EDA complexes formed between the solutes and bonded acceptor ligands. The addition of polar modifier have reversed the retention order above, which, together with nonlinear plots of $1/k'_s$ versus [S] for these solutes, indicates a more complicated nature of mutual interaction among solute, ligands, and methyl *tert*-butyl ether than one can expect from simple models based on EDA interactions only.

Aqueous mobile phases.   As has been mentioned elsewhere [14], there is no simple correlation between K_{eq} and bulk polarity parameters such as dielectric constant, ε, although a general trend to lower K_{eq} with increasing ε is observed. For strong complexes, the reverse trend occurs, however [14–16]. These complexes with significantly larger dipoles gain an extra stabilization from the inductive interactions with polar solvents.

In very polar solvents, the ionic states of both components are stabilized. The stabilization energy will depend on the dielectric constant of the solvent according to Born's equation [7,16]:

$$-\Delta H_{solv} \; \simeq \; (1 - \frac{1}{\varepsilon}) \, (\frac{1}{R_A^-} + \frac{1}{R_D^+}) \tag{22}$$

where R_A^- and R_D^+ are ionic radii of ions A^- and D^+, respectively. Thus the solvation energy calculated with Eq. (22) may amount to about 100 kcal/mol in water [16]. EDA-LC in reversed-phase systems therefore seems to be an interesting alternative to normal-phase separations discussed above.

The first attempt in this direction was made by Hunt et al. [92], who used bonded phtalimide for the separation of PAHs. The retention order of these solutes is very close to that for the C18 phase used for the comparison. Porath [7] also pointed out that EDA complexation might be enhanced due to solvent effects in water-mediated EDA-LC over Sephadex gels. It was clearly shown by Mourey and Siggia [66] that EDA complexation can operate along with the solvatophobic effect obvious in reversed-phase LC. They studied the retention of nitrobenzenes on bonded phenoxy groups acting as an acceptor. The elution order of nitrobenzenes is completely changed by EDA complexation compared with conventional C18 phase.

Hemmetsberger and Ricken [93] studied systematically the chromatographic behavior of bonded TENF ligands in the reversed-phase LC of PAHs. They have found that the EDA complexation acted together with the solvophobic effect. The heats of adsorption, $-\Delta H_{ads}$, are much higher for the nitro than for the C18 phases as a result of strong EDA interactions: the $-\Delta H_{ads}$ values vary to a greater extent with the structure of the solutes, indicating that more specific solute–ligand interactions are involved. In Table 6, the chromatogaphic selectivities of two pairs of structurally related solutes with the same number of carbon atoms are given. Under con-

TABLE 6 Selectivity in EDA-RP LC of Alkylaromatics on 2,4,5,7-Tetranitrofluorenoneoxime (TENF) Silica in Comparison with C18 Stationary Phase[a]

Solute	Chromatographic selectivity, $\alpha_{i,1}$	
	TENF	C18
1-Methylnaphthalene	1.00	1.00
2-Methylnaphthalene	1.22	1.00
Acenaphtene	1.58	1.25
Acenaphtylene	2.41	1.12

Source: Ref. 93.
[a]Mobile phase, methanol-water 95:5.

TABLE 7 Comparison of Retention Times t_r of Aromatics and Nitrotoluenes on Bonded Safrol, Phenyl, and Octadecyl Ligands

Solute	Retention times, t_r (min)		
	Safrol	Phenylalkyl	C_{18}
Benzene	7	7	16
Naphthalene	11.5	10.4	27
Anthracene	19	18	27
2-Nitrotoluene	10	9	9
2,6-Dinitrotoluene	13	9.7	8
2,4,6-Tinitrotoluene	21	10.4	6

Source: Ref. 69.
[a]Mobile phase, methanol-water 73:27.

ventional RPLC conditions, selectivity on a C_{18} phase is lower than that on the nitro phase; for example, acenaphtylene, forming relatively stronger EDA complexes with acceptors than acenaphtene, possesses a considerably enhanced retention in EDA reversed-phase LC.

Lee et al [69] have compared retention of benzene, naphthalene, anthracene, and nitrotoluenes in EDA-RPLC on bonded safrol with phenylalkyl and C_{18} phase (Table 7). Also in this case, EDA interactions with stationary phase strongly change the retention order given by hydrophobicity of solutes in aqueous mobile phase.

N-Propylaniline bonded phase has been studied by Murphy et al. [94] in both normal-and reversed-phase LC. It possesses enhanced selectivity to nitroaromatics.The retention mechanism in RPLC mode is supposed to be a combination of solvophobic effect with EDA complexation. Nitrobenzens elute in both systems in the order benzene < nitrobenzene < *p*-dinitrobenzene < *o*-dinitrobenzene < 1,3,5-trinitrobenzene.

III. METAL COMPLEXES

Electron-deficient species, such as many metal cations, have at least one empty valence orbital available for extra coordination.One of the characteristics of the coordinative unsaturated metal cations is their tendency toward complexation with proper electron-donating species. A great number of various metal complexes is known, and some of them play an important role in chromatographic separation.

The factors influencing the stability of a complex in respect to a selected ligand A (analyte) are:

1. Valence, electronic structure, and radius of central metal ion
2. Spatial arrangement of overlaping orbitals of central ion and ligands
3. Basicity of ligands
4. "Internal" electric effects of ligands transmitted through the central ion
5. Steric effects due to direct contact between the atoms of different ligands
6. "External" effects due to changes in the outer coordination sphere (e.g., solvent effect)

Thus the mutual interplay of all the effects above results in a high sensitivity of complexation processes to the changes in analyte A structure.

From the point of view of the complexation mechanism and stoichiometry, we can distinguish among several cases: (a) change in the number of ligands coordinated in the inner coordination sphere [Eq. (23)], (b) displacement of a ligand L coordinated in the inner coordination sphere [Eq. (24)], and (c) displacement of a ligand S coordinated in the outer coordination sphere [Eq. (25)]:

$$(ML_n) + A \rightleftharpoons (ML_nA) \tag{23}$$

$$(ML_n) + A \rightleftharpoons (ML_{n-1}A) + L \tag{24}$$

$$(ML_n)S_m + A \rightleftharpoons (ML_n)S_{m-1}A) + S \tag{25}$$

The last scheme (Eq.25) involves interactions between the outer coordination sphere of the complex and a molecule of analyte A. The outer sphere is densely packed with the highly organized solvent molecules S.

The general remarks above will be demonstrated briefly with two examples. Argentation chromatography and ligand-exchange chromatography are discussed in more detail in Chapter 4.

A. Argentation Chromatography

Under the term *argentation chromatography* one can discuss utilization of the complexes of transition metals with compounds having C—C double or triple bonds. As olefines and acetylenes are soft bases, they easily complex with many soft acids (see Table 3) (e.g. Cu^+ or Ag^+). From the analytical point of view, the complexes with Ag^+, especially, have been widely utilized in GC and also thoroughly investigated [95-97]. Distribution and solubility measurement data from aqueous systems [98,99] are in good correspondence with the retention data obtained from glycol solutions of $AgNO_3$ used as GC stationary phases [31,100-104]; only a few definite exceptions have been noted [102,103].

These studies indicate that:

1. Usually, one olefin molecule coordinates at a single silver ion, forming a planar complex with triangular structure.
2. Alkyl substitution at the double bond usually decreases the stability of complexes despite the basicity of olefin is enhanced. The steric effects are predominating for *trans*-olefins.
3 With *endo*-cycloolefins, the stability constant increases with increasing ring strain of olefin, which is released via complexation.
4. the most stable diene complexes are formed by 1,5-diene systems.
5. The deuteration of a $C{=}C$ bond increases its basicity and consequently the stability of the complex [95,98].

The effect of double-bond substitution is a combination of electronic and steric effects [98]. The latter can be eliminated in some cases; for example, for a series of *m*- and *p*-substituted styrenes [105] and a series of unsaturated ethers, esters, and ketones [106], the log K_{eq} increases linearly with increasing Hammett σ and σ_m constants, respectively.

The stability of silver–olefin complexes also depends on the anion of silver salt ($BF_4^- > ClO_4^- \gg NO_3^-$ in concentrated solutions; see Refs. 99 and 107). The complexation between Ag^+ and olefins has been used for the separation of olefines and unsaturated carboxylic acids (see also Chapter 4).

The nature of the Ag^+–olefin coordination bonding has been a subject of much discussion; σ-π bonding has been suggested for Ag^+ as well as for most of the transition metals (95). This bonding involves an overlap of the occupied bonding π-orbital of the olefin with a vacant metal orbital (σ-component) and an overlap of the vacant antibonding $\pi*$ orbital of the olefin with an occupied metal d-orbital (π component). Whether for a given metal the σ or π component of the coordinative bond predominates will depend on the energy levels of metal acceptor and donor orbitals relative to those of the olefin [95].

In addition to unsaturated hydrocarbons, other soft bases containing N, O, or S heteroatoms (*n*-donors) have been separated on complexing stationary phases in GC (e.g, dicarbonyl Rh-trifluoroacetyl-*d*-camphorate [108], and dimeric 3-tri-fluoracetyl- or 3- heptafluorobutyryl-(IR)-camphorates of Co^{2+}, Ni^{2+}, or Cu^{2+} [109,110]). Also, various complexing metal ion compounds containing, for example, Hg^{2+}, Cu^{2+}, Pd^{2+}, or Pt^{2+} have been investigated [97,111-113].

In RPLC, the principle of argentation chromatography can also be utilized. The formation of Ag^+ complexes with unsaturated compounds taking place in the mobile phase results in a decrease in retention due to an increase in the hydrophilicity of the complexes in comparison with the parent unsaturated compounds. A linear relationship between log k' and [Ag^+] in the mobile phase has therefore been found (e.g., for RPLC of retinyl esters[114]).

B. Metal Chelates

Various chelating ligands, (e.g., β-diketones, thiosemicarbazones, 8-hydro-xychinoline derivatives, dithiocarbamates, etc.) have been used in chromatographic separations. Dialkyl dithiocarbamates (*12*) can serve as an example of organic ligands forming complexes with a large number of metal ions [115,116]. The coordination occurs via the sulfur atoms, and the molecule of dithiocarbamate acts as a bidentate ligand (e.g., with bivalent cation M^{2+}) (*13*). The relative stability of complexes with various cations are in the order $Mn^{2+} \sim Zn^{2+} < Fe^{3+} < Cd^{2+} < Pb^{2+} < Ni^{2+} \sim Co^{2+} < Cu^{2+} < Ag^{2+} < Hg^{2+}$ [117]. This means that dithiocarbamates behaves as soft bases forming stable complexes with soft metal ions (see Table 3).

(*12*)   (*13*)

Dithiocarbamates can be analyzed by RPLC with added proper salt in mobile phase [118]; on the other hand, the dialkyldithiocarbamate ligands have been chemically bonded to the silica surface [119]. In such a case, various separation systems can be conceived:

$$P\text{—}M(H_2O)_n + A^- \rightleftharpoons P\text{—}M \cdot A(H_2O)_{n-1} + H_2O \qquad (26)$$

$$P\text{—}MA + L^- \rightleftharpoons P\text{—}M\text{—}L + A^- \qquad (27)$$

$$P\text{—}MA + M^{2+} \rightleftharpoons P\text{—}M^- + MA^+ \qquad (28)$$

The schemes given by Eqs. (26) and (27) respresent ligand exchange chromatography. The principle of this technique was first recognized by Helfferich (120) and used broadly by Davankov and co-workers (9) for the separation of optical isomers. Optical isomers of compounds containing one or more functional groups forming complexes with ions of transition metals (e.g., Cu^{2+}) are separated via interaction with immobilized complexes with asymmetric ligands [9,70]. As a first result, complete resolution of *D*, *L*-proline was reported (121,122). Ligand-exchange LC is discussed in detail in Chapter 4.

Generally, the technique of ligand-exchange chromatography involves separation mechanisms based on the complexation equilibria between competing ligands. The kinetics of complex formation/dissociation must be fast. Also, HSBA classification should be respected, since soft–soft and hard–hard interactions are supposed to be faster than the others. In addition, outer-sphere complexation can be utilized. Chow and Grushka [123] separated nucleotides on an immobilized $Co(en)_3^{3+}$ complex (*14*) prepared by reaction of ethylenediamine silica with $Co(en)_2Cl_2^+$. Nucleotides are known to form outer-sphere complexes with $Co(en)_3{}^{3+}$ as well as inner-sphere complexes with Mg^{2+}. The separation mechanism is based on the equilibria between the two forms of complexes with bonded $Co(en)_3{}^{3+}$ and Mg^{2+} added to the mobile phase.

$$\equiv Si - (CH_2)_3 {\overset{\displaystyle N}{\underset{\displaystyle N}{\rightleftharpoons}}} N - \overset{\displaystyle N}{\underset{\displaystyle N}{Co}} - N$$

$$(14)$$

Grushka and Chow [124] bonded the dithiocarbamate ligands to silica and used the column to separate nucleosides and nucleotides in the presence of Mg^{2+} ions. the retention of nucleosides was found to vary inversely with the Mg^{2+} ion concentration.

Cooke et al. [125] studied the RPLC of amino acids, dipeptides, and aromatic carboxylic acids under the presence of hydrophobic chelating agent (4-dodecyl-diethylenetriamine) and Zn^{2+} ions. The selectivity results from kinetically rapid formation of outer-sphere complexes. Inner-sphere complexation is generally slower process and can result in a loss of chromatographic efficiency.

As an example of the utilization of the covalent bonding in chromatography, the separation of sugars via esterification reaction with boronic acid can serve. Wulff and co-workers [126] developed sorbents with chiral cavities formed during polymerization process. The boronic acid chemically bonded in such cavities is able to form bidentate borates with only one enantiomeric form of analyte. The kinetic and steric requirements of this selective surface reaction result, however, in extremely low separation efficiency of the column as well as in excessive peak asymmetry.

IV. CONCLUSION

The formation of labile complexes as a part of a chromatographic separation mechanism allows very selective separations. Various types of EDA complexes can be used in GC and LC. The applicability of many types of complexes remains to be fully investigated, but both theory and practice predict that this technique will find many future uses.

The continuing development of bonded electron acceptors, electron donors, or complexing ligands represents a background for the next research. But in comparison with the classical RPLC or GC, the gain in chromatographic selectivity is often reduced by the low efficiency of the chromatographic columns. Thus detailed study of kinetics, solvent and ligand structure effects, the influence of temperature, and so on, will be required for the improvement of many promising separation systems.

REFERENCES

1. M. W. F. Nielen, R. W. Frei, and U. A. Th. Brinkmann, in *Selective Sample Handling and Detection in High-Performance Liquid Chromatography*, Part A (R. W. Frei and K. Zech, eds.), Elsevier, Amsterdam, 1988, Chap. 1.
2. W. B. Jensen, *The Lewis Acid–Base Concept: An Overview*, Wiley-Interscience, New York, 1980.
3. W. Guttmann, *The Donor-Acceptor Approach to Molecular Interactions*, Plenum Press, New York, 1978.
4. G. Klopman, *J. Am. Chem. Soc., 90*: 223 (1968).
5. E. Negishi, *Organometallics in Organic Synthesis*, Wiley, New York, 1980, Chap. I.
6. W. Holstein and H. Hemetsberger, *Chromatographia, 15*: 186 (1978).
7. J. Porath, *J. Chromatogr., 159*: 13 (1978).
8. V. A. Davankov, in *Advances in Chromatography*, Vol. 18 (J. C. Giddings, E. grushka, J. Cazes, and P. R. Brown, eds.), Marcel Dekker, New York, 1980, p. 139.
9. V. A. Davankov, V. A. Kurganov, and A. S. Bochkov, in *Advances in Chromatography*, Vol. 22 (J. C. Giddings, E. Grushka, J. Cazes, and P. R. Brown, eds.), Marcel Dekker, New York, 1983, p. 81.
10. V. Schurig and W. Burkle, *Angew. Chem., 90*: 132 (1978).
11. L. H. Klemm and D. Reed, *J. Chromatogr., 3*: 364 (1960).
12. R. W. Moshier and R. E. Sievers, *Gas Chromatogrpahy of Metal Chelates*, Pergamon Press, Oxford, 1965.
13. M. Godlewicz, *Nature, 164*: 1132 (1949).
14. R. Foster, *Organic Charge–Transfer Complexes*, Academic Press, London, 1969.
15. M. A. Slifkin, *Charge Transfer Interactions of Biomolecules*, Academic Press, London, 1971.
16. K. Tamaru and M. Ichikawa, *Catalysis by Electron Donor–Acceptor Complexes*, Kodansha, Tokyo, 1975.
17. K. Yoshihara, K. Futamura, and S. Nagakura, Chem. Lett., *1234* (1972).
18. H. A. Craenen, J. W. Verhoeven, and J. de Boer, *Recl. Trav. Chim. Pays-Bas, 91*: 405 (1972).
19. R. S. Mulliken, *J. Am. Chem. Soc., 74*: 811 (1952).
20. R. S. Mulliken, *J. Phys. Chem., 56*: 801 (1952).
21. G. R. Anderson, *J. Am. Chem. Soc., 92*: 3552 (1970).
22. H. Ratajczak, *J. Phys. Chem., 76*: 3000 (1972).
23. P. O. Lowdin, *Adv. Quantum Chem., 2*: 213 (1965).
24. P. R. Wells, *Linear Free Energy Relationships*, Academic Press, London, 1968.
25. E. Tomlinson, *J. Chromatogr., 113*: 1 (1975).
26. H. A. Benesi and J. H. Hildebrand, *J. Am. Chem. Soc., 70*: 2832 (1948).
27. R. O. C. Norman, *Proc. Chem. Soc.*, 151 (1958).
28. S. H. Langer, C. Zahn, and G. Pantazoplos, *J. Chromatogr., 3*:154 (1960).
29. A. R. Cooper, C. W. P. Crowne, and P. G. Farrel, *Trans. Faraday Soc., 62*: 2725 (1966).
30. A. R. Cooper, C. W. P. Crowne, and P. G. Farrel, *Trans. Faraday Soc., 63*: 447 (1967).

31. E. Gil-Av and J. Herling, *J. Phys. Chem.*, 66: 1208 (1962).
32. C. A. Wellington, *Adv. Anal. Chem. Instrum.*, 11: 237 (1973).
33. E. E. Martire and P. Riedl, *J. Phys. Chem.*, 72: 3478 (1968).
34. R. J. Laub and R. L. Pecsok, *J. Chromatogr.*, 113: 47 (1975).
35. C. P. W. Crowne, M. F. Harper, and P. G. Farrell, *J. Chromatogr. Sci.*, 14: 321 (1976).
36. C. Eon and B. L. Carger, *J. Chromatogr. Sci.*, 10: 140 (1972).
37. C. Eon, C. Pommier, and G. Guiochon, *J. Phys. Chem.*, 75: 2632 (1971).
38. L. H. Klemm, D. Reed, and C. D. Lind, *J. Org. Chem.*, 22: 739 (1957).
39. C. H. Lochmuller and E. C. Jensen, *J. Chromatogr.*, 216: 333 (1981).
40. L. Nondek, *J. Chromatogr.*, 373: 61 (1986).
41. J. T. Ayres and C. K. Mann, *Anal. Chem.*, 36: 2185 (1964).
42. G. Smets, V. Balogh, and Y. Castille, *J. Polym. Sci.*, Part C, 4: 1467 (1964).
43. J. Porath and B. Larsson, *J. Chromatogr.*, 155: 47 (1978).
44. J. -M. Egly and J. Porath, *J. Chromatogr.*, 168: 35 (1979).
45. P. Smidl, Ph.D. *Thesis*, Technical University, Prague, 1982.
46. P. Smidl and K. Pecka, *Chem. Listy*, 77: 468 (1983).
47. H. Stetter and J. Schroeder, *Angew. Chem., Int. Ed. Engl.*, 7: 48 (1968).
48. C. Arsenis and D. B. McCormick, *J. Biol. Chem.*, 241: 330 (1966).
49. S. Ray and R. W. Frei, *J. Chromatogr.*, 71: 451 (1972).
50. C. H. Lochmuller and C. W. Amoss, *J. Chromatrogr.*, 108: 85 (1975).
51. C. H. Lochmuller, R. R. Ryall, and C. W. Amoss, *J. Chromatrogr.*, 178: 298 (1979).
52. L. Nondek, and J. Malek, *J. Chromatogr.*, 155: 187 (1987).
53. H. Hemetsberger, H. Klaar, and H. Ricken, *Chromatographia*, 13: 277 (1980).
54. G. Eppert and I. Schinke, *J. Chromatogr.*, 60: 305 (1983).
55. S. Thompson and J. W. Reynolds, *Anal. Chem.*, 56: 2434 (1984).
56. S. A. Matlin, J. W. Lough, and D. G. Bryan, *J. High Resolut. Chromatogr. Chromatogr. Commun.*, 3: 33 (1980).
57. G. Felix and C. Bertrand, *J. High Resolut. Chromatogr. Chromatogr. Commun.*, 7: 160 (1984).
58. L. Nondek and R. Ponec, *J. Chromatogr.*, 294: 175 (1984).
59. K. J. Welch and N. E. Hoffman, *J. High Resolut. Chromatogr. Chromatogr. Commun.*, 9: 417 (1986).
60. W. Holstein, *Chromatographia*, 14: 468 (1981).
61. Ph. Jadaud, M. Caude, and R. Rosset, *J. Chromatogr.*, 393: 39 (1987).
62. G. Felix and C. Bertrand, *J. High Resolut. Chromatogr. Chromatogr. Commun.*, 8: 363 (1985).
63. G. Felix and C. Bertrand, *J. High Resolut. Chromatogr. Chromatogr. Commun.*, 10: 411 (1978).
64. C. H. Lee, T. C. Chang, S. L. Lee, and T. G. Den, *Fresenius Z. Anal. Chem.*, 328: 37 (1987).
65. D. Y. Pharr, P. C. Uden, and S. Siggia, *J. Chromatogr. Sci.*, 26: 432 (1988).
66. T. H. Mourey and S. Siggia, *Anal. Chem.*, 51: 763 (1979).
67. C. H. Lochmuller, A. S. Colborn, L. M. Hunnicut, and J. M. Harris, *Anal. Chem.*, 55: 1344 (1983).
68. M. Verzele and N. Van de Velde, *Chromatographia*, 20: 239 (1985).
69. S. L. Lee, C. H. Lee, and T. G. Den, *Fresenius Z. Anal. Chem.*, 328: 41 (1987).

70. G. Gubitz, in *Selective Sample Handling and Detection in High-Performance Liquid Chromatography*, Part A (R. W. Frei and K. Zech, eds.), Elsevier, Amsterdam, 1988, Chap. 3.
71. L. Salem, *J. Am. Chem. Soc.*, *90*: 543 (1968).
72. I. Fleming, *Frontier Orbitals and Organic Chemical Reactions*, Wiley, New York, 1979.
73. L. Nondek and M. Minarik, *J. Chromatogr.*, *324*: 261 (1985).
74. G. P. Blumer and M. Zander, *Fresenius Z. Anal. Chem.*, *288*: 277 (1977).
75. P. L. Grizzle and J. S. Thomson, *Anal. Chem.*, *54*: 1071 (1982).
76. R. E. Merrifield and W. D. Phillips, *J. Am. Chem. Soc.*, *80*: 2778 (1958).
77. R. Foster, I. Horman, and J. W. Morris, unpublished work (see ref. 12, p. 199).
78. L. Nondek, M. Minarik, and J. Malek, *J. Chromatogr.*, *178*: 427 (1979).
79. M. Charton, *J. Org. Chem.*, *31*: 2991, 2996 (1966).
80. H. M. Rosenberg, E. C. Eimutis, and D. Hale, *Can. J. Chem.*, *45*: 2859 (1967).
81. W. E. Hammers, A. G. M. Theeuwes, W. K. Brederode, and C. L. de Ligny, *J. Chromatogr.*, *234*: 321 (1982).
82. L. Nondek and M. Minarik, *4th Danube Symposium on Chromatography*, Bratislava, 1983, Abstract E27.
83. L. Nondek P. Dienstbier, and R. Rericha, *J. High Resolut. Chromatogr. Chromatogr. Commun.*, *11*: 217 (1988).
84. S. S. Yang and R. K. Gilpin, *J. Chromatogr.*, *408*: 93 (1987).
85. J. C. van Miltenburg and W. E. Hammers, *J. Chromatogr.*, *268*: 147 (1983).
86. Ph. Jadaud, M. Caude, and R. Rosset, *J. Chromatogr.*, *439*: 195 (1988).
87. N. M. D. Brown, R. Foster, and C. A. Fyfe, *J. Chem. Soc. B*, 406 (1967).
88. R. S. Drago, T. F. Bolles, and R. J. Niedzielski, *J. Am. Chem. Soc.*, *88*: 2717 (1966)
89. J. R. Bishop and L. E. Sutton, *J. Chem. Soc.*, 6100 (1964).
90. L. Nondek, unpublished results.
91. Y. Y. Filakov and A. Y. Borovikov, *Zh. Obshch. Khim.*,*48*: 248 (1978).
92. D. C. Hunt, P. J. Wild, and N. T. Crosby, *J. Chromatogr.*, *130*: 320 (1977).
93. H. Hemetsberger and H. Ricken, *Chromatographia*, *15*: 236 (1982).
94. L. J. Murphy, S. Siggia, and P. C. Uden, *J. Chromatogr.*, *366*: 161 (1986).
95. H. W. Quinn and J. H. Tsai, *Adv. Inorg. Chem. Radiochem.*, *12*: 217 (1969).
96. O. K. Guha and J. Janak, *J. Chromatogr.*, *68*: 325 (1972).
97. W. Szczepaniak, J. Nawrocki, and W. Wasiak, *Chromatographia*, *12*: 484 (1979)
98. F. R. Hepner, K. N. Trueblod, and H. J. Lucas, *J. Am. Chem. Soc.*, *74: 1333 (1952).*
99. B. B. Baker, *Inorg. Chem.*, *3*; 200 (1964).
100. R. J. Cvetanovic, F. J. Duncan, W. E. Falconer, and R. S. Irwin, *J. Am. Chem. Soc.*, *87*: 1827 (1965).
101. T. Fueno, O. Kajimoto, T. Okuyama, and J. Furukawa, *Bull. Chem. Soc. Jpn.*, *41*: 785 (1968).
102. H. Hosoya and S. Nagakura, *Bull. Chem. Soc. Jpn.*, *37*: 249 (1964) .
103. M. A. Muhs and F. T. Weiss, *J. Am. Chem. Soc.*,*84*: 4697 (1962) .
104. B. Smith and R. Ohlson, *Acta Chem. Scand.*, *16*: 1743 (1962) .
105. T. Fueno, O. Kajimoto, and J. Furukawa, *Bull. Chem. Soc. Jpn.* , *41*: 782 (1968) .
106. T. Fueno, T. Okuyama, T. Deguchi, and J. Furukawa, *J. Am. Chem. Soc.* , *87*: 170 (1965) .

107. W. Featherstone and A. J. S. Sorie, *J. Chem. Soc.*, 5235 (1964).
108. V. Schurig, R. C. Chang, A. Zlatkis, E. Gil-Av, and F. Mikes, *Chromatographia, 6*: 223 (1973).
109. V. Schurig, *Chromatographia, 13*: 263 (1980).
110. V. Schurig and R. Weber, *J. Chromatogr.*, *217*: 51 (1981).
111. H.Traitler and M. Rossier, *J. High Resolut. Chromatogr. Chromatogr. Commun., 5*: 189 (1982).
112. W. Szczepaniak, J. Nawrocki, and W. Wasiak, *Chromatographia, 12*: 559 (1979).
113. G. E. Baiulescu and V.A. Ilie, *Stationary Phases in Gas Chromatography*, Pergamon Press, Oxford, 1975.
114. M .G. M. de Ruyter and A.P. de Leenbeer, *Anal. Chem., 51*: 43 (1979).
115. A. Hulanicki, *Talanta, 14*: 1371 (1967).
116. J. A. Cras and J. Willemse, in *Comprehensive Coordination Chemistry: The Synthesis, Reactions, Properties and Applications of Coordination Compounds*, Vol. 2 (G. Wilkinson, R. D. Gillard, and J. A. McCleverts, eds.), Pergamon Press, Oxford, 1988.
117. G. Eckert, *Z. Anal. Chem., 155*: 23 (1956).
118. F. Vlacil and V. Hamplova, *Collect. Czech. Chem. Commun., 50*: 2221 (1985).
119. H. Irth, *Ligand Exchange Principles for Trace Enrichment and Detection in Liquid Chromatography*, thesis, Free University, Amsterdam, 1989.
120. F. Helfferich, *Nature, 189*: 1001 (1961).
121. V. A. Davankov and A. V. Semchkin, *J. Chromatogr., 141*: 313 (1977).
122. S. V. Rogozhin and V. A. Davankov, *Chem. Commun.*, 490 (1971).
123. F. K. Chow and E. Grushka, *J. Chromatogr., 185*: 361 (1979).
124. E. Grushka and F. K. Chow, *J. Chromatogr., 199*: 283 (1980).
125. N. H. C. Cooke, R. L. Viavattene, R. Eksteen, W. S. Wong, G. Davies, and B. L. Karger, *J. Chromatogr., 149*: 391 (1978).
126. G. Wulff, W. Wesper, R. Grobe-Einsler, and A. Sarhan, *Macromol. Chem., 178*: 2799 (1977).

2

Survey of Packings in Donor–Acceptor Complex Chromatography

Guy Felix
Bordeaux I University, Talence, France

I. INTRODUCTION

Donor–acceptor complex chromatography plays an important role in chromatography methods. The various types of packings used are numerous. For clarity, we report here only on those publications that describe the complete synthesis of packings and exclude those where the syntheses are incomplete or reported completely in other articles, except for a few of practical interest.

The chapter is divided into four parts, corresponding to those chromatography methods that use the principles of the donor–acceptor complex: gas chromatography, thin-layer chromatography, ligand-exchange chromatography, and organic donor–acceptor liquid chromatography.

II. GAS CHROMATOGRAPHY

A. Silver Chromatography

In 1955, Bradford et al. [1] found that a 30% solution of saturated silver nitrate in ethylene glycol mixed with kieselguhr gave excellent separations of traces of ethane in ethylene. Several studies using silver salts as stationary phases have been published.

Various procedures have been used for the preparation of packings. Most commonly, a saturated solution of silver nitrate is mixed with the support in the desired amounts (Table 1). (All tables can be found at the end of this chapter beginning with pg. 49.) Solvents are generally polyalcohols such as ethylene glycol. An alternative procedure has been developed that uses solutions of low concentration in silver nitrate (Table 2). Several papers have described the dissolution of the salt in nitrile solvents, which have the ability to complex silver ion (Table 3) in a more stable moiety.

Bendell et al. [30] have used a stationary phase prepared from 25% w/w solution of 40% silver tetrafluoroborate in β, β'-oxypropionitrile for the separation of octenes. Only one organic silver complex was described. Poly(p-phenylene, bis-methoxyphenyldithiophosphinato)silver was coated on Chromosorb W acid-washed dimethylchlorosilane (AW-DMCS) (4% w/w) and used for the separation of olefins [114].

B. Chromatography on Other Metals

A wide range of metals have been used as absorbents in gas chromatography, in both their inorganic and organic forms. Thus numerous separations were obtained with alkaline, alkaline earth, and transition metal salts as well as with rare earths and other elements. In some cases the columns were packed with pure salts, or more generally with salts coated on a support. In the latter case a solution of salt in an appropriate solvent was mixed with the support; then the solvent was removed.

1. Metals of Groups Ia and IIa; Rare Earths

Group Ia. The various packings made with alkaline metal salts are summarized in Tables 4 to 7. Terphenyl isomers were resolved on rubidium chloride packings with 25 wt % salt on Chromosorb P [45], and alcohols, alkanes, and aliphatic amines are retained on pure rubidium nitrate [49]. RbB(C6H5)4 (in acetone) was impregnated on silanized Chromosorb W (15%) and studied as stationary phases for the separation of different families, such as ketones and alcohols [66].

Group IIa. MgCl2/6H2O was used to study the retention times of aliphatic and aromatic hydrocarbons [76,77]. Several organic salts of alkaline earth metals were used: magnesium di-n-hexylphosphinate coated on silanized Porasil C (5 wt %) [75] or on Chromosorb W hexamethyldisilazane (HMDS) (10 wt %) [119,120] to separate olefins and aromatics. The retention times of alkanes, ketones, alcohols, and aromatics were studied on magnesium n-nonyl-β-diketone coated on Celite (20 wt %) [86,97]. Other alkaline earth metal salts are summarized in Tables 8 to 10.

By cation-exchange modification of zeolites Y (Na$^+$ by Mg$^+$, Ca$^+$, Sr$^+$, Ba$^+$) in various degrees of substitution, Andronikashvili et al. have made supports capable of separating C1-C4 hydrocarbons [72]. By dissolving in water a eutectic mixture composed of sodium, potassium, and lithium nitrates (18.2/54.5/27.3 wt %) and then mixing with the support, several packings were made. Various amounts of eutectic (2 to 28.6%) were used to study the separations of polyphenyl compounds [43,51]. Another eutectic mixture (LiCl/KCl 58.5:41.5 mol %) impregnated at 20% on Chromosorb G was used by Guiochon for the separation of thallium and nobium chlorides [59].

Rare earths.   Cadogan and Sawyer coated silica gel with 10% mixtures of LaCl3/NaCl containing 19.4, 35.5, and 81% LaCl3, respectively, for the separation of substituted benzenes [63]. On the other hand, rare earth chelates such as tris(2,2,6,6-tetramethyl 3,5-heptanedione) [67], tris(3-trifluoroacetyl-d-camphorate) [67], bis(diisobutyrylmethane [67], and tris(1,1,1,2,2,2,3,3-heptaflu or-7,7-dimethyl 4,6-octanedione) [67,68], generally dissolved in squalane, were coated on Gas Chrom Z (15 wt %) to separate ketones and alcohols. More recently, Golding used a solution of bis(3-trifluoroacetyl-(1R)-camphorato) europium or praseodymium in squalane deposited (15% w/w) on Chromosorb W HP to resolve racemic epoxides [74].

2. Transition Metals and Groups IIIb, IVb, Vb, and VIb

Inorganic metal salts.   The inorganic salts of transition metals are efficient adsorbants in gas chromatography. Some of their uses are summarized in Table 11.

Another type of column packing has been made by treating H-form Amberlyst 15 with an aqueous solution of nickel nitrate. The resin has an exchange capacity for nickel of 5.9 mEq/g and was used for the separation of carbon monoxide and carbon dioxide present in air [100]. Other inorganic metal salts of groups IIIb and Vb are also used as adsorbents. With aluminum sulfate, coated on silanized Porasil (10% w/w), free energies of adsorption for aromatic compounds with various functional groups were measured [57]. Thalium nitrate mixed with Chromosorb W (30% w/w) was used as a packing to study the separation of various hydrocarbons [85]. A saturated solution of thallium nitrate in diethylene glycol or polyethylene glycol 400 coated on G Gel (25% w/w) to separate terpenes, olefins, and aromatic hydrocarbons [92]. Antimony oxide coated on Chromosorb (25 wt %) resolved terphenyl isomers [45].

Several salts containing two metals were also used as adsorbent. Generally, one of the metals was a transition metal; the other was an alkaline or alkaline earth metal (Table 12).

Guran and Rogers used pure Cu2HgI4 and Ag2HgI4 as adsorbants to study the variations of capacity ratios of alkanes, alkenes, alkyldienes, cyclenes, cyclanes, and aromatics [85,91]. By contrast, Guiochon used a mixture of chloride salts to separate metallic chloride salts of groups IVa and Va (Table 13).

In the same way, eutectic mixtures of chloride salts coated on supports were also used: PbCl2/LiCl2 (20% w/w) to separate ZrCl4/HfCl4 [59], and PbCl2/BiCl3 and

CdCl$_2$/KCl (70% w/w) to achieve the separation of TiCl$_4$/SbCl$_3$ and SnCl$_4$/SbCl$_3$, respectively [82].

Organic metal salts.   Several copper salts, both organic and inorganic, were used as adsorbants, either in pure form [83,84] or coated on Chromosorb W [89] (Table 14). Clathrate-forming Werner complexes have also been used as stationary phases, especially dithiocyanatotetrakis(1-arylalkylamine) nickel (Table 15). The effect of replacing nickel by cobalt and iron has also been examined [83]. Finally, organic rhodium carbonyl complexes were used, generally dissolved in squalane for the separation of various compounds, especially olefins (Table 16).

Most of the transition metals in the form of organic salts have been investigated as adsorbants. Chromium phthalocyanine mixed with silicone oil DC 200 was dispersed in Chromosorb T (4 wt %) to study the separation of amines, alcohols, and ether [90], and poly(chromiumdiphenylphosphinate) was deposited on DMCS Spherosil XOB 030 carrier (1 to 3 wt %) to determinate retention times and standard adsorption entropies of various hydrocarbons [117,118].

Manganese stearate coated on Celite (19 wt %) was used to study the separation of alkanes, ketones, amines, alcohols, and alkylbenzenes [80]. Thermodynamic data for adduct formation between esters, ketones, alcohols, aldehydes, sulfides, ethers, and manganese trifluorocamphorate coated on Chromosorb P (15 wt %) have been measured [112]. Epoxides were separated on bis-3-perfluoroacyl-(1 R)-camphorates of manganese (0.05 to 0.15) in squalane [123]. Iron phthalocyanine was used in the same way and for the same studies as chromium phthalocyanine [90]. Numerous cobalt (Table 17), nickel (Table 18), copper (Table 19), and zinc (Table 20) complexes were coated on various supports.

Organic derivatives of transition metals of period V were also used in gas chromatography. Rhodium benzoate coated on Chromosorb P or AW-DMCS (4 % w/w) interacts rapidly and reversibly with olefins, ethers, ketones, and esters [109]. Palladium and cadmium salts of poly(p-phenylenebis(methoxyphenyl)dithiophosphinic acid) coated on Chromosorb W AW-DMCS (4 % w/w) were used for the separation of alkylamines [106] or olefins and sulfur compounds [114]. In addition, complexes of bis(2-thienyl-1-naphtyldithiophosphinato) palladium or platinum packed in the same way were used to separate olefins and sulfur compounds [114]. Solutions (10^{-1}, 5×10^{-2}, 10^{-2}) of cadmium stearate in Quadrol coated on Chromosorb W (20 wt %) were used to study the gas chromatographic behavior of aliphatic amines [99].

By cation-exchange modification of type Y zeolites, cadmium can be incorporated (5,19% of substitution) and used as an adsorbent for the separation of light hydrocarbons [71]. For the separation of C$_5$–C$_6$ hydrocarbons, a copolymer of mercurated styrene/divinylbenzene containing 0.8 mmol of Hg can be used [113].

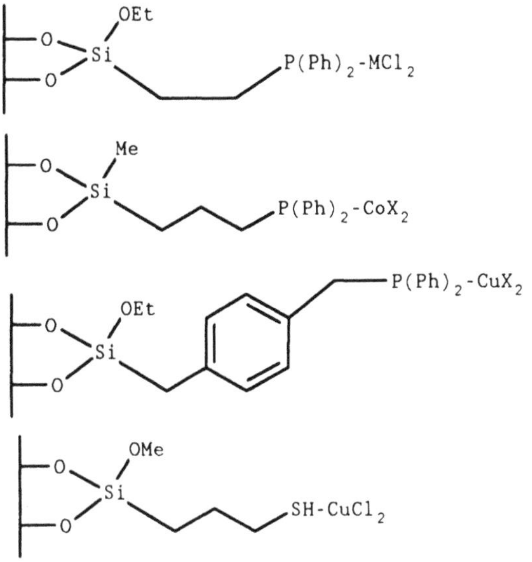

I   (Si with OEt) — $P(Ph)_2$-MCl_2 M = Cu, Ni   [125]

I   (Si with Me) — $P(Ph)_2$-CoX_2 X = Cl, Br   [126]

II   (Si with OEt) — $P(Ph)_2$-CuX_2 X = Cl, Br   [127]

V   (Si with OMe) — SH-$CuCl_2$ [128]

FIGURE 1 Structures of bonded stationary phases.

3. Other Elements

Homologous series of various compounds (alkanes, aromatic hydrocarbons, ketones, alcohols) were chromatographied using a phase made with 20% of bis(alumino-*n*-nonyl β-diketonate) on Celite [86,97] or 4% aluminum phthalocyanine on Chromosorb T [90]. Cvetanovic and colleagues [88,135] studied the interactions of olefins and related hydrocarbons with molecular iodine coated 40 and 20 wt % on various silanizated supports (Celite, Chromosorb, etc.).

4. Bonded Metals Chromatography

Transition metal salts of cobalt, nickel, and copper have been bonded on supports. The metal salts are complexed with a diphenylphosphine derivatives [125–127] or thiol groups [128] chemically bonded on silica. The structure of the grafts is shown in Figure 1. The metal content of each stationary phase and their different uses are summarized in Table 21.

C. Organic Acceptor Chromatography

Several organic electron acceptors have been used, coated on gas chromatography supports for the separation of electron-rich molecules. These compounds were generally aromatic derivatives substituted by electron attractors such as chloride [129,131,132,141-144,146] or nitro groups [130,136,140]. The first series of organic compounds used in donor–acceptor (D-A) gas chromatography were tetrachlorophthalates esters (Table 22), phthalates esters (Table 23), and phthalates derivatives (Table 24).

Esters such as catechol dibenzoate and catechol dibutyrate, mixed with ground unglazed tile (17.6 % w/w), have been used as stationary phases for the separation of alkyl benzenes [137]. Several organic aromatic compounds substituted by nitro groups have also been used. With 2,4,7-trinitrofluorenone coated on firebrick, Norman resolved the nitrotoluene isomers [130]. Alkylanilines [140] and alkylbenzenes [146] were also separated with 2,4,7-nitrofluorenone coated on silanized (HMDS) Celite (4 to 5% w/w) or silanized (DMCS) Chromosorb W, respectively. Trinitrobenzene was used, coated in 40 to 60 wt % on firebrick or Chromosorb W to study the separation of branched olefins [136], and 2,4-dinitrochlorobenzene coated on Celite 545 (15% w/w) was used for the separation of alkylbenzenes [135].

Langer and colleagues [134] have determinated, by gas–liquid chromatography, the thermodynamic constants of some aromatic hydrocarbons and halogenated aromatic compounds in solution, using benzyldiphenyl, phenanthrene, or 7,8-benzoquinoline coated on silanized firebrick. On an other hand, separations of alkylindoles, quinoleines, pyridines, and polyaromatic hydrocarbons have been realized on stationary phases made with polyoxyalkylene adipate (Reoplex 400) coated in 20 to 30% w/w on Celite [133]. Nitrile compounds such as thiodipropionitrile (38% on Chromosorb) [147] or oxydipropionitrile (30% on Sterchamol) [148] were also used in gas chromatography as electron acceptors.

III. THIN-LAYER CHROMATOGRAPHY

A. Silver Chromatography

In 1962, Morris used impregnated silica gel plates with silver nitrate for the separation of methyl ester isomers of higher fatty acids [149]. The impregnation was realized by spraying the plates with a saturated aqueous solution of silver nitrate and then drying for 1 h in an oven at 110°C. Several authors used this method because of its greater convenience (Table 25).

Generally, the silver salt was dissolved in a solvent or a mixture or solvents in the desired concentration and then incorporated in the silica gel slurry. After spreading, the plates were heated in an oven at 110°C. Some of these studies are summarized in Tables 26 and 27.

Another way to impregnate silica gel plate is to suspend it in a 12.5% solution of silver nitrate in water [165] or to plunge it very rapidly into a solution of 1:1 ethanol/20% w/v silver nitrate in water [178]. Andersen and Leovey [207] incorporated silver nitrate by dipping the plates for 5 s in 5% AgNO3 in 1:1 ethanol/acetonitrile, whereas Andreev used pure acetonitrile [258].

Separations were also carried out with developing solvents containing silver salt. For example, Enzell and colleagues have separate methyl esters of resin acids [150] or terpenes [226] on hexadecane-impregnated paper eluted with a solution of silver fluoroborate in methanol. In a similar way, chromatograms of lipids were developed with 70 to 90% aqueous methanol saturated with dodecane and silver nitrate [187]. The separation of fatty acid methyl esters on DMCS-treated silica gel chromatoplate was achieved by developing with water, methyl cyanide, or methanol (2.5/2.5/95) saturated with silver nitrate [168]. Several authors used aluminum oxide rather than silica gel (Table 28).

Several silver salts have been used, including silver perchlorate [232] or silver iodate [234] dissolved (55% w/v) in water/acetone (1:10) and mixed with silica gel (15% w/w) for the separation of terpenoides. Pyridine homologs were separated on silver oxide, generated directly with a 5% silver nitrate silica gel or aluminum oxide plates by adding a 5% NaOH solution [201].

B. Other Metals Chromatography

As for D-A gas chromatography, both mineral and organic metal salts have been used. Generally, transition metal salts have been coated on chromatoplates. In some cases alkaline earth metal salts such as $BaSO_4$ and $CaSO_4$ (30%) gave separations of aromatic amines [238] or calcium oxalate (30%) which seem more suitable for the separation of substituted anilines [263] and amino acids [247].

For coating transition metal salts, the same techniques as those used for silver thin-layer chromatography (TLC) have been used. Separation of hexoamines was realized on TLC plates prepared by spraying with a methanol cupric sulfate solution (0.05%) [240].

A solution obtained using equal volumes of palladium chloride ($2.0 \times 10^{-4} M$) and calcein (1.6×10^{-4}) in a phosphate buffer (0.04 M, pH 7.2) diluted with acetone has given a spray that has been useful in determining organosulfur compounds [244]. In the same way, zinc chloride and mercuric, cadmium, and copper acetates (1 wt % in methanol) loaded on silica gel plates were used for the determination of sulfur compounds in high-boiling petroleum distillates [250]. Some determinations of sulfides in lubricant-based oils were described using a $PdCl_2$-impregnated silica gel plates achieved by spraying on an aqueous solution of palladium chloride [257].

Prostaglandins were chromatographied on plates coated by dipping in a 10% aqueous ferric chloride solution [245]. The chromatographic behavior of various amino acids has been studied on paper strips drenched in zirconium phosphate [264] and zirconium oxichloride [248]. Good separations of fatty acid esters were obtained on silica gel sheets impregnated by submersion in an aetonitrile solution of cobalt nitrate [258]. Separations obtained with transition metal salts coated on chromatoplates are listed in Tables 29 to 32.

The separation of substituted anilines by zinc nitrate coated on silica gel plates has also been realized on molecular sieves and on aluminum oxide [238]. Zinc dust impregnated on silica gel plates in 10% w/w has been used to achieved separation of polynitroluenes [237].

Three other metal salts have been coated on silica gel: mercuric acetate (25%), aluminum sulfate (30%), and pyridinium tungstoarsenate (1%). These phases were used to study the separation of alkylphenyl sulfides [251], aromatic amines [239], and sulfonamides [252], and amino acids [253], respectively.

Marino and colleagues have prepared thin layers of cation exchangers of polystyrene/iminodiacetic acid type (Chelex 100) coordinatively bound to Co, Ni, Cu, and Zn. They were used for the analysis of very dilute solutions of diamines, amino acids, and carboxylic acids [255].

C. Organic Donor–Acceptor Chromatography

1. Acceptor Chromatography

The techniques used to coat organic acceptors on chromatoplates are the same as for metal salts. Libickova has impregnated silica gel plates by spraying them with a solution of the acceptor in a suitable solvent (Table 33) [272]. Several authors preferred to dip the plates in a solution of the acceptor (Table 34). However, impregnation by mixing a solution of acceptor with silica gel has generally been used. The various acceptors are summarized in Tables 35 (silica gel as support) and 36 (aluminum oxide as support).

Only one example of acceptor bonded to the support has been described. Riboflavin grafted on cellulose or on silica gel by derivation of an aminopropyl silica gel was used to study separations of pyrimidines and purines [276].

2. Donor Chromatography

To achieve separation of polynitroaromatic compounds, Frank-Neumann and Jössang used as eluant a solution of cyclohexane/chloroform (50:50) saturated in anthracene [266].

3. Chiral Chromatography

Wainer and colleagues have obtained good separations of racemic 2,2,2-trifluoro-1-(9-anthryl)ethanol on a chiral plate containing (R)-N-(3,5-dinitrobenzoyl)phenylglycine. The resolving agent is ionically bonded to γ-aminopropyl silanized silica gel coated on microscope slides [277].

Several authors have used a combination of reversed-phase plates (RP-18) with chiral eluants. For example, the mobile phase contained 8 mM of N,N-di-n-propyl-L-alanine, 4 mM of copper acetate, and 0.3 M sodium acetate dissolved in water/acetonitrile (70:30 [280] or 60:40 [282]) for the separation of dansyl amino acids. In another case, amino acids were separated on plates, first immersed (1 min) in a 0.25% copper acetate solution (methanol/water 1:9 v/v) and after drying, in a 0.8% methanol solution of the chiral selector: (2S, 4R,2'RS)-4-hydroxy-1-(hydroxydodecyl)proline [281].

IV. LIGAND-EXCHANGE CHROMATOGRAPHY

Three methods are generally used in ligand-exchange chromatography with a metal. The metal can be bonded or coated on the support, dissolved in the mobile phase, or both. The metals most often used are silver and copper.

A. Silver Chromatography

1. Silver Coated on Support

In liquid chromatography, silver salts are coated on supports by the same techniques as those used with TLC. One method consists of dissolving silver salt with a solvent and mixing the solution with the support. After evaporation of the solvent, the coated support is filled in an open column or packed in a stainless steel column and used. The various silver columns are listed in Tables 37 to 39.

Other silver salts have been used: for example, silver halides coated indirectly by impregnation of silica gel with saturated $AgNO_3$ solution in methanol and after washing with water, addition of potassium halide solutions to obtain AgBr or AgI supports, or HCl solution to obtain AgCl supports [303,304]. These phases contain about 1% silver salts and are used to separate aromatic amines [304] and xanthines [303]. Several stationary phases have been made by an in situ method which consists of passing a solution of silver salt through a silica- or a cation-exchange-bonded silica gel packed column (Table 40).

Resins have also been used as support. The procedure is the same as for silica gel. Generally, the hydrogen form of the resin is stirred with aqueous silver nitrate solution, and after decantation, washed with distilled water and equilibrated with the eluant. In some cases the in situ method is used [313,322]. Separation of fatty acid esters [320,321,327,324–331], unsaturated fatty alcohol acetates [322,323], sulfur-containing amino acids [313], and prostaglandins [324] have been realized on these silver-loaded resins.

2. Silver in Mobile Phase

Dutton and colleagues have extracted and separated geometric isomers and isologs of fatty acid esters by countercurrent distribution using 0.2 M of silver nitrate in 90% methanol and petroleum ether [335,337–339].

The partition chromatography technique was also used with glass fiber paper impregnated with alkane (C_{10}, C_{12}) and aqueous methanol containing silver nitrate (separation of triglycerides) [340] or silver fluoroborate (separation of sesquiterpenes) [336] as mobile phase. Separation of sesquiterpenes has also been realized with the same mobile phase on an open column filled with polyvinyl chloride impregnated with petroleum ether [334]. However, the most commonly used method is to have silver salt in the mobile phase of a reversed-phase liquid chromatography system (Table 41). Kissinger and Robins have separated prostaglandins in silver normal-phase chromatography using a silica gel column and silver nitrate in the mobile phase ($CH_2Cl_2/CH_3CN/H_2O/CH_3COOH$) [350].

B. Copper Chromatography

1. Copper Linked with Support

Copper has been bonded directly by reaction of the copper salt with an organic support. On copper/Sephadex G-25, free amino acids and peptides have been fractionated [352,365,385,387]. On copper/Chelex 100, free amino acids [353,354,356, 364,379,387,590], dansyl amino acids [362], dipeptides [590], peptides [387], nucleosides [355], oxipurines [359], and enkephalin [389] have been separated. Other resins loaded with copper ions, such as Wolfatil MC 50 (separation of nucleotide [369]) and Bio Rex 63 and 70 (separation of amino acids [379]), have been also used.

Rosset, and colleagues linked copper on silica gels by treatment with an ammonia solution of copper sulfate and have demonstrated the ability of the support to separate free amino acids [380,383,384,397], peptides [384,397], and sugars [397]. Zinc silicate strips dipped in copper nitrate solution were used to study the separation of amines [403].

As for Chelex 100, the iminodiacetic acid groups bonded on various supports can be used for the immobilization of copper salts. For example, Porath has separated proteins [363,368] on copper/iminodiacetate Sepharose 4B. Davankov, in his studies to resolve amino acids, used copper/iminodiacetate cross-linked polystyrene [370]. Unger demonstrated the good stability with regard of hydrolysis of a copper/iminodiacetate-bonded silica gel [395], and Krauss used it for the separation of nucleic acid constituents [396]. In the same way, Szczepaniak resolved some racemic amino acids on an ion exchanger containing iminodi(methanephosphonic) groups on their aminocopper form [394].

Unger have synthesized mono-, di-, and triamino silica gel to complex copper salts and separated amino acids [395]. Chow and Grushka used a coordinated copper ion bonded to a γ-aminopropyl silica gel for the separation of aromatic amines [277]. This method was extended to di- and triamino silica gel for aliphatic amines [400]. To study the separation of dialkyl sulfides, Ando and co-workers [398] prefer to coordinate copper ions with 8-quinolinol, 8-quinolinethiol, and N-butyl-1-napthylmethylamine bonded on silica gel. Uracil derivatives are resolved on a copper/8-hydroxyquinoline silica gel [401].

The best way to graft copper ion is to complex it with an amino acid bonded on the support. Several publications have described the synthesis of copper/amino acid bonded phases on cross-linked polystyrene [371], on chloromethylated cross-linked polystyrene [373], and on polyvinylpyridine [399]. Free amino acids are separated on these various stationary phases (Table 42).

L-Proline-bonded polyacrylamide coated on silica gel has given, after copper complexation, good separations of amino acids [391]. L-Glycine bonded on Sephadex has been used for group separation of peptides [402].

2. Copper in Mobile Phase

D,L-Tryptophan and D,L-Tyrosin are separated on L-proline bonded on silica gel with copper ions in the mobile phase [513,514].Several authors have preferred to introduce the chiral agents in the mobile phase. Generally, 2 mol of an amino acid derivative for 1 mol of copper salt was used. In most cases the solutes are resolved in reversed-phase chromatography on alkyl-bonded phases. Examples of the resolution of free amino acids are listed in Table 43.

Amino acid derivatives such as dansyl amino acids have also been resolved by reversed-phase chromatography with copper sulfate and L-proline [409,426,469], L-hydroxyproline [426], or L-histidine [469]. Copper-L-proline eluant separated α-substituted ornithine and lysine on a preparative scale in a reversed-phase mode [428]. The optical purity of 4-thiazolidinecarboxylic acid have been determinated with a copper complex of (2S, 4R, 2'RS)-N-(2'-hydroxydodecyl)-4-hydroxy-proline [420]. To avoid the loss of copper ion, some authors have introduced, in addition to the mobile phase, copper chloride [414] or copper acetate [423] (dansyl derivatives).

Instead of using amino acid in the eluant, several authors have bonded silica gel with amino acids, the copper salts being in the mobile phase. Davankov resolved amino acids on L-proline/polystyrene grafted on silica gel with copper sulfate in the eluants [417]. Karger used undecanoyl-L-valine or undecyl-L-proline bonded on silica gel with copper acetate in the mobile phase to separate racemic dansyl amino acids [415] and amino alcohols [427]. To obtain high selectivity in the resolution of free amino acids, it seems preferable to have stationary and mobile phases that contain copper ion (Table 44).

Dansyl amino acids were also resolved on L-valine [433] and L-proline [433,434] bonded on silica gel with copper sulfate present in the two phases. Another method, consisting of coated N-decyl-L-histidine [449] or other N-alkylamino acids [516] on a reversed-phase silica gel, has been proposed by Davankov. The resolution of free amino acids has been obtained in the presence of copper acetate in the eluants. Other amino acid separations have been realized on N-β-aminoethyl/γ-aminopropyl glass beds and copper chloride [431]. This system has also been used to separate amino sugars [431].

Separation of substituted aromatic amines have been obtained on silica gel bonded with dithiocarbamate or diketone systems and copper sulfate [432]. The final chromatographic system used with copper is to link copper on the support directly and to have copper ion in the mobile phase (Table 45).

C. Metals Chromatography

The metal salts, generally transition metals, are used in the same manner as for silver or copper chromatography. As with gas and thin-layer chromatography, some authors occassionally use pure metal salts.

1. Pure Metal Salts as Stationary Phase

2,4-Dinitrophenylhydrazone of ketones and aldehydes has been separated on pure $ZnCO_3$ [450], whereas the separation of some organic sulfides has been studied on hydrated $ZnCl_2$ [455]. Nitroalkyl [451] or aromatic derivatives [462,463] and *cis/trans* azo compounds [456] have been separated on pure tetra(-4-methyl-pyridino)nickel dithiocyanate. Nitroaromatic compounds have also been separated by a tetra(γ-picoline)dithiocyanate complex of Co, Fe, and Ni [453,454].

2. Metals Linked with Support

To study ligand-sorption and ligand-exchange equilibria, Helfferich [507,508] converted the H^+ form of a cation-exchanger resin (Amberlite IRC 50) to its Ni^+ form by passing a nickel ammonia solution. Later, several authors used the selectivity of cation-exchange resin-metal forms to separate organic compounds (Table 46).

3. Metal Salts Coated on Support

When covalent or ionic binding is not possible, the coating method is used. The various metal salts coated on support are listed in Table 47.

4. Metal Salts in Mobile Phase

As in silver or copper chromatography, several authors have dissolved metal salts in the mobile phase to resolve enantiomeric or isomeric mixtures by reversed-phase chromatography (Table 48). Metal ions are sometimes linked on the support and also added to the mobile phase. Zinc ions have been linked with various func-

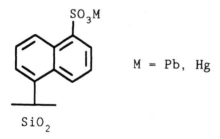

M = Pb, Hg

FIGURE 2   Napthylsulfonated silica gel.

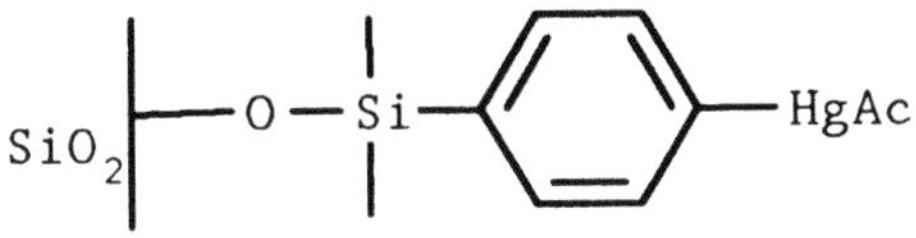

FIGURE 3   Organomercuric bonded phase.

tional resins to resolve di- and polyamines with 0.0004 M of zinc sulfate in the mobile phase [485]. Sulfa drugs have been separated on zinc or cadmium salts complexed on diamino or triamino silica gel using 0.02 M of metal sulfate in the mobile phase [487].

5. Metal Bonded on Support

Banner and colleagues [457] have separated olefins from alkanes on lead or mercuric sulfonate salts bonded on silica gel by the intermediary of naphthyl groups (Fig. 2). Chmielowiec [465] studied the separation of polyaromatic hydrocarbons or heterocycles on a support obtained by mercuration of a phenyl silica gel (Fig. 3).

V. ORGANIC DONOR–ACCEPTOR LIQUID CHROMATOGRAPHY

A.   Acceptor Stationary Phases

The simplest manner is the use of pure powdered acceptors as stationary phases. Buu-Hoï and Jacquignon used tetrahalophtalic anhydride (Cl, Br, I) as the corresponding imides in a simple open column to separate polyaromatic hydrocarbons (PAH) and azaaromatics [521]. Generally, acceptors are coated or bonded on the supports.

1. Acceptors Coated on Supports

The earliest work in this field was reported in 1949 by Godlewicz [520]. He separated a lubricating oil on trinitrobenzene or picric acid coated on silica gel. Klemm et al. used picric acid or 2,4,7-trinitrofluorenone impregnated in 4% w/w on silica gel to resolve α, β-substituted naphthalenes [522]. Karger and co-workers [527] have investigated the influence of temperature, the water content of solvent, and the degree of coating of 2,4,7-trinitrofluorenone. Burger and Tomlinson [540] have studied the interactions of methyl-substituted benzenes in HPLC on silica gel impregnated with tetracyanoethylene. Köhler separated PAH on silica gel coated with polyvinylpyrrolidone [555].

Felix and co-workers used silica gel impregnated with 20% w/w of caffeine to separate acenaphthenesilyl compounds on a preparative scale [545]. In preparative separations, strong solvents are needed to elute most solutes and also the coated

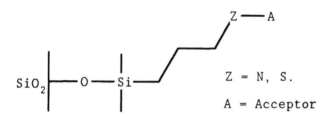

FIGURE 4 Structure of acceptor stationary phases synthetized by derivatization of amino-, chloro-, thiol-bonded silica gel.

acceptor. To overcome this problem, Sondheimer and Pollak [523] used 7-(2,3-dihydroxypropyl)theophylline as coating agent to separate various cinnamic acids (cyclohexane/chloroform 3:1) with elution of less than 10 mg of acceptor per liter. On a preparative scale, Jewell [529] used 2,4,7-trinitrofluorenone impregnated on alumina to extract mono- and diaromatics of petroleum products.

2. Acceptors Bonded on Supports

Several acceptors have been coupling on Sephadex G-25 gel. For example, acriflavin was used for separation of nucleosides, nucleotides, and cyclic nucleotides [532], and 2,4-dinitrothiophenol resolved aromatic amino acids [541].

Polymers were also used as supports. Ayres and Mann [524] described the synthesis of a new support obtained by nitration of cross-linked divinylbenzene/polystyrene copolymers. Newman and Junjappa have synthesized a copolymer by reaction of terephtalyl chloride and d, l-di-β-hydroxyethyl-2,4,5,7-tetra-nitrofluorenylidenaminooxysuccinate and used it to purify methylbenz[a]anthracene [525].

The most commonly employed support is silica gel. Two methods are generally used: (a) the derivatization of an amino- (Table 49), chloro-, or thiol-bonded silica gel with the acceptor molecule, and (b) the total synthesis of the acceptor-organosilyl derivative, then grafting of the silica gel with the synthesized silyl compound (Table 50). In the first case, the general structure of the bonded silica gel is as shown in Figure 4.

Langer et al. [531] have obtained, by reaction of a chloro-bonded silica gel with several pyrimidine bases (uracil, thymine, cystosine), new stationary phases to separate some nucleic acids. By derivatization of the same support with pentafluoroaniline, Felix and co-workers [559] synthesized a weak acceptor stationary phase to achieve the separation of PAH. Apfel et al. [557] used a 8-bromooctylsilica to graft theophylline. With this support they have separated aromatic carboxylic acids. Welch and Hoffman preferred to derive a mercaptopropyl silica gel with

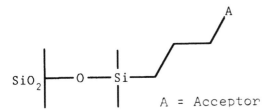

A = Acceptor

FIGURE 5   General structure of silica gel bonded with the acceptor molecule (except for acceptor notice, indicated by * in Table 50).

picryl chloride, 2,4-dinitrochlorobenzene, or 3,5-dinitrobenzoylchloride [556] to obtain supports for PAH separations.

In the second case, the structure of the bonded silica gel is as shown in Figure 5. Ray and Frei [526] have prepared a support, by a method generally less used, consisting of the reaction of silica gel with p-nitrophenylisocyanate. On this support, polychlorobiphenyls and polynuclear azaaromatics have been separated.

B. Donor Stationary Phases

In comparison with acceptor bonded on silica gel, there are only a few donors of silica gel type. Mourey and Siggia have demonstrated that for the separation of nitrobenzenes, the partition mechanism on an aryl ether phase is altered by a donor–acceptor contribution [561]. Lochmüller and co-workers [562] have bonded a pyrene nucleus on silica gel and obtained good separations of nitropolynuclear aromatics.

C. Chiral Donor–Acceptor Stationary Phases

The resolution of enantiomers by donor–acceptor chromatography is a special application of some acceptors described in Sections V.A and V.B. The chiral acceptor (or donor) coated or chemically bonded to the support interacts with the racemic donor (or acceptor), forming molecular complexes of different stabilities, giving differences in capacity ratios of each enantiomer. 2,4,5,7-Tetranitrofluorenylideneaminoxypropionic acid (TAPA) (Fig. 6) is the first resolutive agent coated on silica or alumina used to resolve racemic mixture (Table 51).

To resolve helicene, Gil-Av and co-workers used riboflavin (8% w/w) coated on silica gel [567]. Actually, donor–acceptor resolutive agents were bonded on the supports in the same way as those described for the acceptors (Section V.A.2).

From 3-mercaptopropyl silica gel, Pirkle and co-workers [568,569] have grafted (R)-(-)2,2,2-trifluoro-1-[9-(10-α-bromomethyl)anthryl]ethanol. They used the chiral support to resolve 3,5-dinitrobenzoyl (DNB) sulfoxide enantiomers. Salvadori and co-workers have separated binaphthol derivatives [584] or arylcarbinol [594] on bonded *Cinchona* alkaloids.

By reaction of (-)*trans*-1,2-cyclohexanediamine on 3-glycidoxypropyl silica gel, Sinibaldi and colleagues obtained a new phase to separate 2,2'-dihydroxy-1,1,binaphthyl enantiomers [579]. Arm and colleagues have grafted 1-(α-naphthyl)ethylamine to resolve DNB–amine derivatives or DNB–amino acid methyl esters [587].

The most common method used is to derive aminopropyl silica gel. Generally, the chiral agent is linked covalently (Tables 52 and 53), but Pirkle and Schreiner resolved bi-β-naphthols on a DNB–phenylglycine ionically bonded aminopropyl silica gel [571].

Iwaki and colleagues [590] synthesized a poly-α-naphthylethylamine by reaction of polyaminosilica gel with α-naphthylethylisocyanate. Resolution of *p*-bromophenylcarbamyl amino acids was in reversed-phase.

The problem with the derivation method is the substitution of all reactive groups. To overcome this difficulty, several authors prefer to realize the total synthesis of the graft (Table 54). Matlin and co-workers [592] have bonded silica gel with *N*-2,3-dinitrophenyl-L-phenylalanine to resolve 7,10-dicyanohexahelicene.

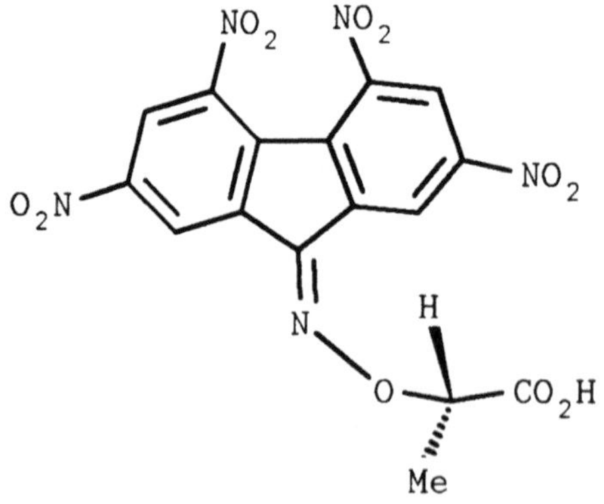

FIGURE 6 Structure of TAPA.

TABLE 1 Stationary Phases Made with Saturated Solution of Silver Nitrate

Solvent	Salt/support concentration	Compounds separated	Ref.
Ethylene glycol	30%	Ethane/ethene	1
	33%	Methylcyclohexenes	5
	33%	Methylcyclopentenes	7
	40%	Hydrocarbons C_4-C_8	6
	60%	*cis/trans*-olefins	10
	25%	Deuterated ethylenes	29
	40%	Olefins	20
	25%	Methylcyclobutanes	21
	17%	Methylpentenes	11
	5, 17%	*cis/trans*-Olefins	23
	1-4 M	Deuterated olefins	27
	83%	*cis/trans*-Olefins	24
	33%	Methylenecyclohexane	42
Diethylene glycol	40%	Alicyclic olefins	2
		syn/anti-Norbornene	13
	35%	*endo/exo*-Olefins	16
Triethylene glycol	3-10%	Olefins	3
Tetraethylene glycol	28.5%	Olefins	4
Glycerol	40%	Hydrocarbons C_4-C_8	6
Propylene glycol		Alkanes/olefins	9
Carbitol		Hydrocarbons C_4	14
Water	1-80%	Pentenes, hexenes	18

TABLE 2 Stationary Phases Made with Various Concentrations of Silver nitrate

Solvent	Salt/solvent concentration	Salt/support concentration	Compounds separated	Ref.
Ethylene glycol	27, 42%	20%	Olefins	12
	30%	33%	Cyclohexenes	17
	30%	33%	p-Menthenes	17
	1.77 N	25%	Methylcyclo C_4–C_6	19
	0.2, 0.4 N	40%	Alkynes C_6–C_8	34
	0.68, 0.76 N	40%	Alkynes C_6–C_8	34
	0.25, 1.2 M	40%	Olefins	35
	3 M	35%	Olefins	41
Tetraethylene glycol	30%	35%	Alkylcycloalkenes	15
Triethylene glycol	30%	33%	Methylcycloheptenes	28
	4 M	40%	Nitriles	31
	0.08 g/L		Hydrocarbons	33
		4.6 mEq/g	Olefins	37
Water		3.2 mEq/g	Olefins	37
		5.0 mEq/g	Olefins	37
	20 g/L	0.16 mEq/g	Olefins	38
		0.3–1.0%	Aromatics	40
Propanol	0.08 g/L		Hydrocarbons	33
N-Methyl acetamide	10%	20%	Hexenes	36

TABLE 3 Stationary Phases made with Silver Nitrate Solution in Cyanides

Solvent	Salt/solvent Concentration	Salt/support Concentration	Compounds separated	Ref.
Benzylcyanide	40%	40%	Isoprene	8
	30%	33%	Cyclohexenes	17
		10%	cis/trans-Pentene	39
		0.564 μmol	NH$_3$	22
p-Xylylcyanide	33%	36%	Hydrocarbons C$_1$, C$_2$	26
Acetonitrile		17, 50%	Hydrocarbons C$_8$	32
		33%	cis/trans-Methylpentene	32
		50, 75%	Hydrocarbons C$_6$, C$_7$	32

TABLE 4 Stationary Phases Made with Lithium Metal Salts

Li salts	Salt/support concentration	Compounds separated	Ref.
Fluoride	25%	Terphenyl isomers	45
Chloride	20%	Polyphenyls	46, 51
	20%	Polyaromatics	47
	25%	Terphenyl isomers	45
	50%	Polyphenyls	51
	Pure	Alcohols, alkanes, amines	49
	0.44–2 M	Alkylpyridines	73
Bromide	10%	Olefins, alkanes	57
	10%	Substituted benzenes	57
Iodide	25%	Terphenyl isomers	45
Sulfate	25%	Terphenyl isomers	45
Nitrate	Pure	Alcohols, alkanes, amines	49
Carbonate	25%	Terphenyl isomers	45
Bis(2-ethylhexyl) phosphate	20%	Polychlorobenzenes	65

TABLE 5   Stationary Phases Made with Sodium Metal Salts

Na salts	Salt/support concentration	Compounds separated	Refs.
Chloride	10%	Substituted benzenes	54, 57, 62, 63
	10%	Olefins, alkanes	55, 57, 78
	Pure	Ketones, alcohols, esters	48
Bromide	10%	Olefins, alkanes	55
	Pure	Ketones, alcohols, esters	48
Iodide	25%	Terphenylisomers	45
	Pure	Ketones, alcohols, esters	48
Sulfate	5, 10, 20, 30%	Hydrocarbons-C_6	52
	10%	Substituted benzenes	53, 54, 57
	10%	Olefins, alkanes	55, 57, 78
	25%	Terphenyl isomers	45
Carbonate	25%	Terphenyl isomers	45
	28%	O-terphenyls	51
	Pure	O-terphenyls	51
Phosphate	5, 10, 20, 30%	Hydrocarbons-C_6	52
	10%	Substituted benzenes	57
	10%	Olefins, alkanes	57, 78
	25%	Terphenyl isomers	45
Bis(2-ethylhexyl) phosphate	20%	Polychlorobenzenes	65
Metaborate	25%	Terphenyl isomers	45
Tetraphenylborate	15%	Alcohols, esters, ketones	66
	15%	Alkanes, haloalkanes	66
Silicate	25%	Terphenyl isomers	45

TABLE 6 Stationary Phases Made with Potassium Metal Salts

K salts	Salt/support concentration	Compounds separated	Ref.
Fluoride	25%	Terphenyl isomers	45
Chloride	25%	Terphenyl isomers	45
	Pure	Alcohols, ketones, esters	48
Bromide	25%	Terphenyl isomers	45
	Pure	Alcohols, ketones, esters	48
Iodide	Pure	Alcohols, ketones, esters	48
Sulfate	25%	Terphenyl isomers	45
Carbonate	25%	Terphenyl isomers	45
	28%	O-terphenyls	51
Phosphate	25%	Terphenyl isomers	45
Bis(2-ethylhexyl) phosphate	20%	Polychlorobenzenes	65
Tetraborate	25%	Terphenyl isomers	45
Tetraphenylborate	15%	Alcohols, esters, ketones	66
	15%	Alkanes, haloalkanes	66

TABLE 7 Stationary Phases Made with Cesium Metal Salts

Cs salts	Salt/support concentration	Compounds separated	Ref.
Fluoride	10%	Fluoroketones	79
Chloride	10%	Polyphenyls	56
	15%	Polyphenyls	50
	20%	Alkylbenzenes	44
	20%	Polyphenyls	46, 51
	20%	Olefins, alkanes	58
	25%	Terphenyls isomers	45
	50%	Polyphenyls	51
	Pure	Alcohols, ketones, esters	49
	Pure	Olefins, alkanes	58
	Pure	O-terphenyls	51
Tetraphenylborate	15%	Alcohols, esters, ketones	66
	15%	Alkanes, haloalkanes	66

TABLE 8   Stationary Phases Made with Calcium Metal Salts

Ca salts	Salt/support concentration	Compounds separated	Ref.
Fluoride	10%	Fluoroketones	79
Chloride	10%	Olefins, substituted benzenes	71
	20%	Polyphenyls	46
	25%	Terphenyl isomers	45
	35%	Polyphenyls	46
	50%	Polyphenyls	46
Di-*n*-hexylphosphinate	5.16%	Olefins	75
Stearate	20%	Aliphatic amines	99

TABLE 9   Stationary Phases Made with Strontium Metal Salts

Sr salts	Salt/support concentration	Compounds separated	Ref.
Chloride	Pure	Alcohols, aliphatic acids	60
	Pure	Substituted benzenes	60
Bromide	Pure	Alcohols, aliphatic acids	60
	Pure	Substituted benzenes	60
Iodide	Pure	Alcohols, aliphatic acids	60
	Pure	Substituted benzenes	60
Nitrate	25%	Terphenyl isomers	45
Di-*n*-hexylphosphinate	5.16%	Olefins	75

TABLE 10 Stationary Phases Made with Barium Metal Salts

Ba salts	Salt/support concentration	Compounds separated	Refs.
Chloride	5%	Alkylbenzenes	70
	10%	Bromothiophenes, alkylbenzenes	69, 70
	20%	Alkylbenzenes	70
	25%	Terphenyl isomers	45
	Pure	Alcohols, aliphatic acids	60
	Pure	Substituted benzenes	60
	Pure	Bromothiophenes, alkylbenzenes	69
Bromide	Pure	Alcohols, aliphatic acids	60
	Pure	Substituted benzenes	60
Iodide	Pure	Alcohols, aliphatic acids	60
	Pure	Substituted benzenes	60
Sulfate	Pure	Alkylbenzenes, cyclic olefins	61
Nitrate	25%	Terphenyl isomers	45
Di-*n*-hexylphosphinate	5.16%	Olefins	75

TABLE 11 Stationary Phases Made with Inorganic Transition Metal Salts

Metal salts	Salt/support concentration	Compounds separated	Refs.
VCl_2	Pure	Olefins, aromatics	105
$Cr(SO_4)_3$	10%	Olefins, substituted aromatics	57
$MnCl_2$	10%	Olefins, haloaromatics	70, 116
$MnCl_2$	10-70%	Alkylpyridines	72
$MnCl_2$	Pure	Olefins, aromatics	105, 115
$CoCl_2$	5%	Aromatics, bromothiophenes	69
$CoCl_2$	10%	Hydrocarbons C_4, thiophenes	68
$CoCl_2$	10%	Olefins, haloaromatics	70, 116
$CoCl_2$	10%	Aromatics, bromothiophenes	69
$CoCl_2$	20%	Aromatics, bromothiophenes	69
$CoCl_2$	25%	Aromatics, bromothiophenes	69
$CoCl_2$	10-70%	Alkylpyridines	72
$CoCl_2$	Pure	Olefins, aromatics	105, 115
$CoCl_2$	Pure	Hydrocarbons C_4, thiophenes	68
$CoSO_4$	10%	Olefins, substituted aromatics	57
$NiCl_2$	5%	Aromatics, bromothiophenes	69
$NiCl_2$	10%	Hydrocarbons C_4, thiophenes	68
$NiCl_2$	10%	Aromatics, bromothiophenes	69
$NiCl_2$	20%	Aromatics, bromothiophenes	69
$NiCl_2$	25%	Aromatics, bromothiophenes	69
$NiCl_2$	10-70%	Alkylpyridines	72
$NiCl_2$	Pure	Hydrocarbons C_4, thiophenes	68
$NiBr_2$	10-70%	Alkylpyridines	72
$NiSO_4$	10%	Olefins, substituted aromatics	57
$NiClO_4$	26%	Olefins, substituted aromatics	57
$Cu(NH_3)_4(NO_3)_2$	Pure	Alkanes, substituted benzenes	70
$CuSO_4, H_2O$	Pure	Alkanes, substituted benzenes	70, 77
$CuSO_4, 5H_2O$	Pure	Alkanes, substituted benzenes	70, 77
$ZnCl_2$	10%	Olefins, haloaromatics	70, 116
$ZnCl_2$	60%	$NbCl_5, TaCl_5, ZrCl_4, HfCl_4$	59
$Zn_3(PO_4)_2$	25%	Terphenyl isomers	45
$CdSO_4$	28%	Terphenyles	51
$CdSO_4$	Pure	Alkanes, substituted benzenes	70, 77
$CdSO_4$	Pure	Alkanes, olefins	58
$CdSO_4, 8H_2O$	Pure	Alkanes, substituted benzenes	70, 77
$HgClO_4/HClO_4$	50%	Hydrocarbon type analysis	81, 101
$HgClO_4$	$10^{-5}\ g^{-1}$	Hydrocarbon type analysis	88
	$1.5 \times 10^{-5}\ g^{-1}$	Hydrocarbon type analysis	88
$HgSO_4/H_2SO_4$	70%	Hydrocarbon type analysis	101

TABLE 12 Stationary Phases Made with Two Inorganic Metal Salts

Metal salts	Salt/support concentration	Compounds separated	Refs.
Na$_2$MoO$_4$	10%	Olefins, aromatics	57, 78
NaWO$_4$	25%	Terphenyl isomers	45
LiFeCl$_4$	40-70%	NbCl$_5$, TaCl$_5$, SbCl$_3$	95
NaFeCl$_4$	40-70%	NbCl$_5$, TaCl$_5$, SbCl$_3$	95
KFeCl$_4$	40-70%	NbCl$_5$, TaCl$_5$, SbCl$_3$	95
TlFeCl$_4$	40-70%	NbCl$_5$, TaCl$_5$, SbCl$_3$	95
LiAlCl$_4$	40-70%	NbCl$_5$, TaCl$_5$, SbCl$_3$	95
NaAlCl$_4$	40-70%	NbCl$_5$, TaCl$_5$, SbCl$_3$	95
KAlCl$_4$	40-70%	NbCl$_5$, TaCl$_5$, SbCl$_3$	95
TlAlCl$_4$	40-70%	NbCl$_5$, TaCl$_5$, SbCl$_3$	95

TABLE 13 Stationary Phases Made with a Mixture of Metallic Chloride Salts

Metal salts	Salt/support concentration	Compounds separated	Refs.
AgCl/ZnCl$_2$	60%	ZrCl$_4$	59
PbCl$_2$/AgCl	60%	NbCl$_5$–NfCl$_4$	59
ZnCl$_2$/SnCl$_2$	60%	NbCl$_5$–ZrCl$_4$	59
ZnCl$_2$/SnCl$_2$	Pure	ZrCl$_4$	94
PbCl$_2$/TlCl$_2$	60%	NbCl$_5$, TaCl$_5$, ZrCl$_4$, HfCl$_4$	59

TABLE 14 Stationary Phases Made with Copper Complexes

Metal salts[a]	Salt/support concentration	Compounds separated	Refs.
Cu(Phen)$_2$Cl$_2$	Pure	Alkanes, aromatics, alcohols	83, 84
Cu(Py)$_4$(NO$_3$)$_2$	10%	Olefins, ketones, esters	89
Cu(Py)$_4$(NO$_3$)$_2$	Pure	Alkanes, substituted benzenes	70, 77
CuBiPy(NO$_3$)$_2$	10%	Olefins, Ketones, esters	89
CuBiPy(NO$_3$)$_2$	Pure	Alkanes, aromatics, alcohols	83, 84
Cu(Phen)$_2$(NO$_3$)$_2$	Pure	Alkanes, aromatics, alcohols	83, 84
Cu(Quin)$_2$(NO$_3$)$_2$	10%	Olefins, ketones, esters	89
Cu(Py)$_4$SO$_4$	Pure	Alkanes, substituted benzenes	70, 77
Cu(Phen)$_2$SO$_4$	Pure	Alkanes, aromatics, alcohols	83, 84

[a]Py, pyridine; Phen, 1,10-phenanthroline; Quin, quinoleine; BiPy, 2,2′-Bipyridine.

TABLE 15 Stationary Phases Made with Nickel Complexes

Ni complexes[a]	Salt/support concentration	Compounds separated	Ref.
Ni(NCS)₂(MPy)₄	30%	Aromatics, alkanes, olefins	95
Ni(NCS)₂(MPy)₄	30%	Substituted aromatics	83
Ni(NCS)₂(EPy)₄	30%	Substituted aromatics	83
Ni(NCS)₂(PEA)	30%	Aromatics, alkanes, olefins	95

[a]MPy, 4-methylpyridine; EPy, 4-ethylpyridine; PEA, 1-phenylethylamine.

TABLE 16 Stationary Phases Made with Rhodium Complexes

Rh complexes[a]	Concentration on support	Compounds separated	Refs.
(CO)₂ACAC	15	Olefins	103, 104
(CO)₂TFAC	0.06 m	Olefins, epoxides	122
(CO)₂TFAC	15	Olefins, esters	103, 104, 110

[a]ACAC, acetylacetonate; TFAC, 3-trifluoroacetylcamphorate.

TABLE 17 Stationary Phases Made with Cobalt Complexes

Co complexes[a]	Salt/support concentration	Compounds separated	Refs.
Stearate	19%	Alkanes, alkylbenzenes, ketones, alcohols, amines	80
Phthalocyanine	Pure	Alkanes, alkylbenzenes, phenols, alcohols, terpenes	102
Phthalocyanine	4%	Alcohols, amines, ethers	90
Poly(R-DTP)	Pure	Alkylamines	106
Poly(R-DTP)	4%	Amines, alkylphosphits	107, 111
Bis(N-DTP)	4%	Amines, alkylphosphits	114
TFAC	15%	Esters, ketones, alcohols, aldehydes, sulfides, heterocycles	112

[a]R-DTP, *p*-phenylenebis(methoxyphenyl)dithiophosphinic acid; N-DTP, 2-thienyl-1-naphtyl-dithiophosphinic acid; TFAC, trifluoroacetylcamphorate.

TABLE 18 Stationary Phases Made with Nickel Complexes

Ni complexes[a]	Salt/support concentration	Compounds separated	Refs.
Stearate	20%	Alkanes, alkylbenzenes, ketones, alcohols, amines	80, 99
(NdC)$_2$	20%	Alkanes, ketones, alcohols, aromatics	86, 97
Phthalocyanine	Pure	Alkanes, alkylbenzenes, phenols, alcohols, terpenes	102
Phthalocyanine	4%	Alcohols, amines, ethers	90
DMG	4.5, 8.2, 26.8%	Hydrocarbons, alcohols, ethers, esters, aldehydes	98
CHD	12.7, 27.7%	Hydrocarbons, alcohols, ethers, esters, aldehydes	98
SAL	3.8, 12.6, 25.7%	Hydrocarbons, alcohols, ethers, esters, aldehydes	98
Poly(R-DTP)	Pure	Alkylamines	106
Poly(R-DTP)	4%	Amines, alkylphosphits	107, 111
Bis(N-DTP)	4%	Amines, alkylphosphits	114
CHB	0.5%	Sulfur compounds	108
TFAC	15%	Aldehydes, ketones, alcohols, esters, sulfides, heterocycles	112
TFAC-1R	0.06 m	Olefins, epoxides	122

[a]NdC, n-nonyl β.-diketone; DMG, bis(dimethylglyoximato); CHD, bis(1,2-cyclohexanedione-dioximato); SAL, bis(salicylaldimino); CHB, cyclohexylbutyrate; R-DTP, p-phenylenebis(methoxy-phenyl)dithiophosphinic acid; N-DTP, 2-thienyl-1-naphtyldithiophosphinic acid; TFAC, trifluoro-acetylcamphorate; TFAC-IR, trifluoroacetyl-(1R)-camphorate.

TABLE 19 Stationary Phases Made with Copper Complexes

Cu complexes[a]	Salt/support concentration	Compounds separated	Refs.
Stearate	19%	Alkanes, alkylbenzenes, ketones, alcohols, amines	80
Phthalocyanine	Pure	Alkanes, alkylbenzenes, phenols, alcohols, terpenes	102
Phthalocyanine	4%	Alcohols, amines, ethers	90
(NdC)$_2$	20%	Alkanes, ketones, alcohols	97
Poly(DHP)	10%	Esters, olefins	119, 120
		Chlorocarbons	121

[a]NdC, n-nonyl β-diketone; DHP, di-n-hexylphosphinate.

TABLE 20 Stationary Phases Made with Zinc Complexes

Zn complexes[a]	Salt/support concentration	Compounds separated	Refs.
Sterate	20%	Alkanes, alkylbenzenes, ketones, alcohols, amines	80, 99
Phthalocyanine	Pure	Alkanes, alkylbenzenes, phenols, alcohols, terpenes	102
Phthalocyanine	4%	Alcohols, amines, ethers	90
(NdC)$_2$	20%	Alkanes, ketones, alcohols, aromatics	86, 97
Poly(R-DTP)	Pure	Alkylamines	106
nPrS	0.05%	Sulfur compounds	108
CHB	0.01%	Sulfur compounds	108

[a]NdC, n-nonyl β-diketone; CHB, cyclohexylbutyrate; R-DTP, p-phenylenebis(methoxyphenyl) dithiophosphinic acid; nPrS, n-propylsulfur.

TABLE 21 Stationary Phases Made with Metal Complexes Chemically Bonded on Silica Gel

Metal salts	Percent of metal on silica	Compounds separated	Ref.
CuCl$_2$	0.27	Alkanes, olefins, dienes, ketones, cycloalkanes, chloroalkanes	125
NiCl$_2$	0.35	Alkanes, olefins, dienes, cycloalkanes, cycloalkenes	125
CoCl$_2$	0.82	Olefins, dienes, cycloalkanes, cycloalkenes	126
CoBr$_2$	0.88	Olefins, dienes, cycloalkanes, cycloalkenes	126
CuCl$_2$	0.7	Olefins, dienes, chloroalkanes, chloroalkenes	127
CuBr$_2$	0.42	Olefins, dienes, chloroalkanes, chloroalkenes	127
CuCl$_2$	2.24	Olefins, dienes, aromatics	128

TABLE 22 Stationary Phases Made with Tetrachlorophthalates Esters (TCPs) Coated on Support

TCPs	TCP/support concentration	Compounds separated	Refs.
Me, Pr	20%	*m-/p*-Xylenes	129
	20%	Alkylbenzenes	132
Di-Me/di-*n*-Pr (90/10)	25%	Alkylbenzenes	132
Di-*n*-Pr	7%	*m-/p*-Xylenes, alkylbenzenes	129, 132
	10%	Alkylbenzenes	132
	15%	Alkyl, halogenobenzenes	138
	0–20%	Alkylbenzenes	142
Di *n*-Bu	15%	Alkyl, halogenobenzenes	138
	0.12–0.89 *M*	2-Ethylthiophene	143
	40%	Heterocycles	144
Di-*n*-amyl	15%	Alkyl, halogenobenzenes	138
Di-*n*-octyl	15%	Alkyl, halogenobenzenes	138
n-Nonyl	5–10%	Substituted anilines	138
	5–10%	Hydrocarbons, heterocycles	138

TABLE 23 Stationary Phases Made with Phthalate Esters (PhEs) Coated on Support

Phthalate esters	PhE/support concentration	Compounds separated	Ref.
Di-*n*-butyl	40%	Alkylbenzenes	131
Di-*n*-Bu bis-glycolate	17.6%	Alkylbenzenes	137
Di-*n*-decyl	15%	Alkylbenzenes	131
Di-*n*-cyclohexyl	17.6%	Alkylbenzenes	137
Di-*n*-phenyl	17.6%	Alkylbenzenes	137
Di-*n*-(*β*-phenylethyl)	17.6%	Alkylbenzenes	137
Di-*n*-benzyl	17.6%	Alkylbenzenes	137

TABLE 24 Stationary Phases Made with Phthalate Derivatives (PhDs) Coated on Support

Phthalate derivatives	PhD/support concentration	Compounds separated	Ref.
Di-*n*-hexylisophthalate	17.6%	Alkylbenzenes	137
Di-*n*-hexylterephthalate	17.6%	Alkylbenzenes	137
Phthalic acid	0.5, 1, 5%	Hydrocarbons, ethers, ketones, alcohols, esters	139
Isophthalic acid	0.5, 1, 5%	Hydrocarbons, ethers, ketones, alcohols, esters	139
Terephthalic acid	0.5, 1, 5%	Hydrocarbons, ethers, ketones, alcohols, esters	139
Phthalic anhydride	0.5, 1, 5%	Hydrocarbons, ethers, ketones, alcohols, esters	139
Dipotassium phthalate	0.5, 1, 5%	Hydrocarbons, ethers, ketones, alcohols, esters	139
Tetra-Bu pyromellitate	10%	Alkylbenzenes	145

TABLE 25 Silica Gel Plates Impregnated by Spraying with a Solution of Silver Nitrate

Solvent	Salt/solvent concentration	Compounds separated	Ref.
Water	10%	Cholesterol esters	210
Water	25%	Cholesterol esters	210
Water	25%	Cholesteol/cholestanol	216
Water	Saturated	Higher fatty acid esters	149
Water/MeOH	5%	Sterols/stanols	213
Water/MeOH	5%	Tetracyclic triterpenes	229
MeOH	1%	Sulfur compounds	250
H$_2$O/EtOH 50:50	5%	Vegetable/mineral oils	173
H$_2$O/EtOH 50:50	10%	Fatty acid esters	164
H$_2$O/EtOH 10:90	Saturated	Sterols/steroids	208

TABLE 26 Silica Gel Plates Impregnated with a Water Solution of Silver Nitrate

Salt/water concentration	Salt/support concentration	Compounds separated	Refs.
2–30%	1–15%	Synthetic steroids	221
	2.5%	Olefins	231
	0–3%	Terpenes C_{10}, C_{15}, C_{20}, alcohols	228
2%	4%	Prostaglandins	204, 205
2.4%	5%	Isono oil	151
2.4%	5%	Oxygenated fatty acids	159
	5%	Oleic/linoleic acids	166
7.5%	15%	Equine estrogens	222
3.6%	8%	Phosphatidylserines	176
4%	8.5%	Fatty acid mixtures	174
5%	10%	Isomeric octadecenoates	169
	10%	Furanoid fatty acids	164
	10%	Methyl octadecadienoates	171, 177
	10%	Triglycerides, phospholipids	184, 193
	12.5%	Phosphatidyl fatty acids	170
12%	14.5%	Cyclopropene fatty acids	167
	15%	Oleic/linoleic acids, glycerides	155, 188
	15%	Alkylbenzenes	233
	17%	Oleic/linoleic acids	157
12%	19%	Triglycerides	183
5%	20%	Sterol acetates	219
9%	20%	Fatty acid esters	152
10%		Lipids, steroidal alkaloids	186, 209
10%	20%	Hydroxyprogesterones	218
12%	20%	Cholesterol mixtures	211
12.5%	20%	Fatty acid mixtures	158, 161
12.5%	20%	Glycerides, triglycerides	182, 194
		Glycerides, triglycerides	179
12.5%	20%	Glyceryl monoethers	191
	20%	*a*-Eleostearic acid	160
	20%	Oxygenated olefinic acids	172
	20%	Methyl octadecadienoates	175
	20%	Sterol esters	223
	20%	Anacardic acid	235
	20–30%	Prostaglandins	203
12%	21%	Cholesterols	225
12%	22%	Fatty acid mixtures	154
10%	23%	Lecithins	189

TABLE 26 (Continued)

Salt/water concentration	Salt/support concentration	Compounds separated	Refs.
11.5%	23%	Aldehydes and ketones DNPH[a]	197
12.5%	24%	Lecithins	192
19%	24%	Phosphatidyl glycerol	190
12.5%	25%	Sterols, monoterpenes	217, 230
	25%	Aldehydes and ketones DNPH[a]	198
24%	25%	Alkylphenylsulfides	251
12.5%	26%	Allylic–propenylic isomers	236
15%	27%	Conjugated estrogens	224
18%	30%	Lecithins, lipids	180, 181, 185
18%	30%	dyes, sterol acetates	200, 215
	30%	Fatty acid mixtures	156
	30%	Prostaglandins	202
52%	50%	Fatty acid esters	162
68%	72%	Fatty acid esters	162

[a]DNPH, 2,4-dinitrophenylhydrazone.

TABLE 27   Silica Gel Plates Impregnated with Silver Nitrate Dissolved in Various Mixtures of Solvents

Solvent	Salt/solvent concentration	Salt/support concentration	Compounds separated	Ref.
H_2O/86% EtOH	10%		Arachidonic metabolites	206
H_2O/60% EtOH	54%	13%	Sesquiterpenes	227
Acetone/10% H_2O	29%	13%	Sesquiterpenes, diterpenes	232
Acetone/10% H_2O	29%	13%	Terpenoides	234
NH_3	5%	10%	Isomeric octadecenoates	169
H_2O/5% NH_4OH	1%	1.5%	Ethynyl steroids	212
H_2O/30% NH_4OH	12.5%	66%	Fatty acid mixtures	161

TABLE 28   Aluminum Oxide Plates Impregnated with Silver Nitrate Dissolved in Various Mixtures of Solvents

Solvent	Concentration in solvent	Concentration on support	Compounds[a] separated	Ref.
H_2O/MeOH 1:2	17%	28.5%	Fatty acid methyl esters	153
Water	13%	20%	Aldehydes and ketones DNPH	195
Water	25%	30%	Aldehydes DNPH	196
		25%	Aldehydes and ketones DNPH	198
		44%	Aldehydes and ketones DNPH	199
Water	50%	22%	Sterols	220

[a]DNPH, 2,4-dinitrophenylhydrazone.

TABLE 29   Silica Gel Plates Coated with Manganese Salts

Mn salts	Salt/support concentration	Compounds separated	Ref.
Sulfate	1%	Amino sugars	262
Sulfate	5%	Dyes	256
Formiate	14.3%	Aromatic amines	243
Acetate	5%	Dyes	256
Acetate	20%	Aromatic amines	243
Na_2EDTA	33%	Aromatic amines	243

TABLE 30   Silica Gel Plates Coated with Copper Salts

Cu salts	Salt/support concentration	Compounds separated	Ref.
Chloride	0.1%	Sulfamides	261
Sulfate/polyamide	25%	Amino acids	259
Sulfate/$MoO_4(NH_4)_2$	25%	Dyes	260
Sulfate	33%	Aromatic amines	239
Sulfate	0.1%	Sulfamides	261
Acetate	0.1%	Sulfamides	261

TABLE 31   Silica Gel Plates Coated with Zinc Salts

Zn salts	Salt/support concentration	Compounds separated	Refs.
Chloride	6–30%	Chlorinated anilines	241
Chloride	30%	Chlorinated anilines	243
Nitrate	6%	Substituted anilines	238
Nitrate	30%	Chlorinated anilines	241
Sulfate		Amino sugars	262
Sulfate	5%	Dyes	256
Sulfate	30%	Aromatic amines	239
Carbonate	60%	Substituted phenols, diols	246, 247
Acetate	5%	Dyes	256
Acetate	30%	Chlorinated anilines	241

TABLE 32 Silica Gel Plates Coated with Cadmium Salts

Cd salts	Salt/support concentration	Compounds separated	Refs.
Nitrate	6%	Substituted anilines	238
Nitrate		Amino sugars	262
Sulfate	30%	Aromatic amines	239, 242, 243
Sulfate	5%	Dyes	256
Sulfate		Amino sugars	262
Phosphate	30%	Aromatic amines	242
Acetate	5%	Dyes	256
Acetate	10%	Sulfamides	254
Acetate	25%	Alkylphenyl sulfides	251
Acetate	30%	Aromatic amines	242
Acetate		Amino sugars	262

TABLE 33 Silica Gel Plates Coated with Organic Acceptor by Spraying

Acceptor	Solvent	Accep./solvent concentration	Compounds[a] separated
Tetramethyluric acid	Water	0.05 M	PAHs
Caffeine	Water	0.05 M	PAHs
Tetramethyluric acid	Water	Saturated	PAHs
Caffeine	Water	Saturated	PAHs
Pyromellitic dianhydride	Ethyl acetate	Saturated	PAHs
Chloranil	Chloroform	Saturated	PAHs
Tetracyanoethylene	Acetone	0.05 M	PAHs

[a]PAHs, polyaromatic hydrocarbons.

TABLE 34   Silica Gel Plates Coated with Organic Acceptor by Dipping

Acceptor	Solvent	Accep./solvent concentration	Compounds separated	Refs.
m-dinitrobenzene	Acetone	3%	Anilines	270
Trinitrobenzene	Benzene	2%	PAHs	268
Trinitrobenzene	Benzene	3%	PAHs	266
Trinitrobenzene	Acetone	3%	Anilines	270, 271
Picryl chloride	Acetone	3%	Anilines	271
2,3,7-Trinitrofluorenone	Benzene	2%	PAHs	268

TABLE 35   Silica Gel Plates Coated with Organic Acceptor

Acceptor[a]	Compounds separated	Refs.
m-Dinitrobenzene	Aromatic amines	274
Trinitobenzene	Purines, pyrimidines	276
Picric acid	PAHs	265, 269
Styphnic acid	PAHs	265
2,4,7-Trinitrofluorenone	PAHs	265, 268
Caffeine	PAHs	265, 267, 273
p-benzoquinone	PAHs, purines, pyrimidines	275, 276
Bromanil	PAHs, purines, pyrimidines	275, 276
Chloranil	PAHs, purines, pyrimidines	275, 276, 279
DDQ	PAHs	275
Tetracyanoethylene	Purines, pyrimidines, PAHs	276, 279
Riboflavine	Purines, pyrimidines	276
Nucleic acid bases	PAHs	278, 279
Amino acids	PAHs	279

[a]DDQ, 2,3-dichloro-5,6-dicyanobenzoquinone.

TABLE 36 Aluminum Oxide Plates Coated with Organic Acceptor

Acceptor	Compounds separated	Refs.
Styphnic acid	PAHs	265
2,4,7-Trinitrofluorenone	PAHs	265, 268
Caffeine	PAHs	273

TABLE 37 Silica Gel Columns Impregnated with a Water Solution of Silver Nitrate

Salt/water concentration	Salt/support concentration	Compounds separated	Refs.
6, 30%	0.3–0.4%	Aminoaromatic compounds, drugs	308*
	2–30%	Fatty acid esters	328*
	5%	Fatty acid esters	310*
20%	6.5%	cis/trans-5-Cyclodecen-1-ol	283
	7%	Fatty acid esters	310*
14%	10%	Fatty acid esters	284
	13%	Cyclopropane fatty acid ester	289
50%	17%	Fatty acid esters	293
40%	20%	Terpenes, pheromones	288, 299
50%	25%	Fatty acid esters	295, 297
50%	27%	Isano acids esters	287
50%	33%	Fatty acid esters	285
40%	44%	Triflycerides	296
20%	50%	Sterol esters	286
30%	50%	Tritium-labeled fatty acids	290
30%	50%	cis/trans-Monoethenoid acids	291
50%	50%	Fatty acid esters	294
	60%	Fulmar oil esters	292

*, packed in a stainless steel column.

TABLE 38 Silica Gel Columns Impregnated with a Acetonitrile Solution of Silver Nitrate

Salt/CH$_3$CN concentration	Salt/support concentration	Compounds separated	Ref.
1%	4%	Sterols	309*
6%	5, 10, 20%	Urushiol diacetate	312*
4%	20%	Z/E olefin isomers	302*
	20%	Thiophene compounds	318*
20%	25%	Fatty acid esters	298
	40%	Z/E olefin isomers	316

*, packed in a stainless steel column.

TABLE 39 Silica Gel Columns Impregnated with a Methanol Solution of Silver Nitrate

Salt/MeOH concentration	Salt/support concentration	Compounds separated	Ref.
Saturated	0.7%	Polynuclear azaheterocyclics	300*
0.6%	4%	Fatty acid esters	305*
	5%	Methylcycloalkanes	301*
2%	10%	Triglycerides	311*
	20%	Pheromones	306*

*, packed in a stainless steel column.

TABLE 40 Columns Coated In Situ With Silver Salts

Support	Salt	Solvent	Concentration on support	Compounds separated	Refs.
Silica	AgNO$_3$	MeOH	0.5%	Gibberellin esters	307
Silica	AgNO$_3$	MeOH	0.4%	Sesquiterpene alcohols	314
Silica	AgNO$_3$	CH$_3$CN	2.5%	Z/E alkene-ol acetates	315
Silica	AgNO$_3$	CH$_3$CN	2.5%	Hinesol/β-eudesmol	317
Silica	AgClO$_4$	Toluene	6.0%	Sesquiterpenes	314
SA	Ag$^+$			Fatty acid esters, lipids	325, 512
SCX	AgNO$_3$	Water	1 M	Prostaglandins	326
CAT	AgNO$_3$	Water	1 M	Prostaglandins	332, 333

[a]SA, CAT, strongly acid; SCX, strongly basic.

TABLE 41 Silver Salts in the Mobile Phase with a Reversed-Phase Liquid Chromatography System

Support	Solvent	Salt	Salt/support concentration	Compounds separated	Ref.
Porapak Q	Propanol-1/water (66/33)	AgNO$_3$	0.08 g/mL	n-Decene	341
C$_{18}$	Propanol-2/water (55/45)	AgNO$_3$	7, 15 g/L	Fatty acid esters	342
C$_{18}$	Methanol/water (95/5)	AgNO$_3$	0.2 g/L	Vitamin D	343
C$_{18}$	Methanol/water (20/80)	AgClO$_4$	0.5 M	Prostaglandins	344
C$_{18}$	Methanol/water (75/25)	AgClO$_4$	0.01 N	Olefins, azaaromatics	345
C$_{18}$	Methanol/water (90/10)	AgNO$_3$	0.8%	Fatty acid esters	346
C$_{18}$	Methanol/water (80/20)	AgNO$_3$	0.06 M	Retinyl esters	347
C$_{18}$	Methanol/water (70/30)	AgNO$_3$	0.01 M	Olefins	351
C$_{18}$	Methanol/CH$_3$Cl (85/15)	AgNO$_3$	7.5 mM	Menaquinones	348
C$_8$	Methanol	AgNO$_3$	10 mM	Menaquinones	348
C$_{18}$	Methanol	AgNO$_3$	50 mM	Pheromones	349

TABLE 42 Copper/Amino Acid Complexes Bonded on the Support: Resolution of Racemic Free Amino Acids

Support[a]	Amino acids	Refs.
SDVB	L-Proline	361, 370, 374
	L-Proline	376, 377
	L-Hydroxyproline	375, 376, 377
	L-Azetidine carboxylic acid	376, 377
	L-Allohydroxyproline	377
	N-Benzyl-L-proline	388
SDVB 0.8%	L-Proline	357, 358
	L-Hydroxyproline	392, 393
	N-Carboxymethyl-L-valine diethyl ester	360
	N-Carboxymethyl-L-aspartic acid triethyl ester	360
SDVB 1.8%	N-Carboxymethyl-L-valine diethyl ester	360
	N-Cargboxymethyl-L-aspartic acid triethyl ester	360
PCMS	L-Leucine	366
	L-Hydroxyproline	381, 392
	(S)-1-Phenyl-ethylamine	382
	N,N,-Bis[(S)-1-phenylethyl]ethylenediamine	382
	N-Benzylidene-(R)-1,2-propanediamine	382
	N-Benzyl-2-(R)-benzylideniminopropylamine	382
AMBA	L-Proline	367, 378
PAA	L-Proline	386
	L-Phenylalanine	393

[a]SDVB-0.8, styrene/0.8% divinylbenzene copolymer; SDVB-1.8, styrene/1.8% divinylbenzene compolymer; PS, polystyrene; PCSM, polychloromethylstyrene; AMBA, acrylamide/methylenebisacrylamide copolymer; PAA, polyacrylamide.

TABLE 43   Reversed-Phase Chromatography of Racemic Free Amino Acids Using Copper/Amino Acids in Eluants

Cu salts	Chrial agents	Refs.
Acetate	L-Proline	407, 408
	N,N-Di-n-propyl-L-alanine	422, 470
	L-Phenylalanine	424
	N-Methyl-L-phenylalanine	424
	N,N-Dimethyl-L-phenylalanine	424
	N,N-Dimethyl-L-valine	519
Chloride	L-Aspartylethylamide	410
	L-Aspartyl-n-butylamide	410
	L-Aspartyl-n-hexylamide	410
	L-Aspartyl-n-octylamide	410
	L-Glutamylcyclohexylamide	410
	N,N-(Dihydroxyethyl)glycine	412
	Nitrolotriacetic acid	412
	Ethylenediaminetetraacetic acid	412
	Diethylenetriaminepentaacetic acid	412
	N-Hydroxyethyleethylenediaminetriacetic acid	412
Perchlorate	N,N,N,' N,'-tetramethyl-(R)-propanediamine-1,2	411
	L-Aspartyl-L-phenylalanine methyl ester	427
Sulfate	L-Phenylalanine	408
	L-Proline	404[a], 413
	L-Histidine	415
	L-Histidine methyl ester	419
	1-Alkylsulfonates	425
Ammoniac	L-Aspartame	405
	L-Aspartylcyclohexylamide	406
	N,N-Dialkyl-α-amino acids	416

[a]Resin support.

TABLE 44   Resolution of Racemic Free Amino Acids on Copper Systems

Supports[a]	Ligands	Cu salts	Refs.
SDVB-0.3	(R)-N,N,-dibenzyl-1,2-propanediamine	Acetate	435
Silica gel	L-Proline	Acetate	437
Cl-silica gel	L-Proline	Acetate	515, 517
	L-Hydroxyproline	Acetate	518
Silica gel	L-Hydroxyproline	Acetate	437
	PCSM-L-Proline	Acetate	441
	PCSM-L-Hydroxyproline	Acetate	441
SP-Bentonite	L-Lysine	Acetate	446
Montmorillonite	L-Lysine	Acetate	446
Silica gel	(–)-trans-1,2-Cyclohexanediamine	Acetate	447
Glass	N-β-Aminoethyl-γ-aminopropyl	Chloride	431
Silica gel	PCSM-L-methionine-d,l-sulfoxide	Chloride	442
	Iminodiacetic acid	Perchlorate	440
G-Gel-60	L-Hydroxyproline	Sulfate	436
Silica gel	L-Proline	Sulfate	438
	L-Hydroxyproline	Sulfate	438
	L-Azetidine carboxylic acid	Sulfate	438
	L-Pipecolic acid	Sulfate	438
	L-Phenylalanine	Sulfate	438
	(1R,2S)-2-Amino-1,2-diphenylethanol	Sulfate	448
	L-Histidine	Ammoniac	439
	Various L-amino acids	Ammoniac	445

[a]SDVB-0.3, polystyrene/0.3% divinylbenzene copolymer; Cl-silica gel, 3-chloropropylsilyl-bonded silica gel; G-Gel-60, poly(2,3-epoxipropylmethacrylate); PCSM, polychloromethylstyrene.

TABLE 45   Copper Ion Directly Linked on the Support

Supports[a]	Cu salts	separated	Ref.
Dowex 50-X8 (1)	Chloride	D,L-Aminodiols	429
Bio-Rex 63 (1)	Ammoniac	Amino acids	430
Bio-Rex 70 (3)	Ammoniac	Amino acids	430
	Sulfate	Aromatics, thioaromatics	466
	Sulfate	Amino sugars, amio acids	484
Aminex Q-150S (1)	Sulfate	Amino sugars, amino acids	484
Chelex 100 (2)	Ammoniac	Amino acids	430
Silica gel	Ammoniac	Sugars	443
	Ammoniac	Amino acids	444

[a]1, sulfonic resin; 2, iminodiacetate resin; 3, carboxylic resin.

TABLE 46   Metal Linked on the Support

Metal salts[a]	Supports[b]	Compounds separated	Refs.
Ti(SO$_4$)$_2$	Amberlite CG-120	Hydroxyaromatic acids	480
Fe^3 + OPTA	Dowex 1-X2	N-Benzoyl-DL-amino acids	475
FeCl$_3$	Amberlite CG-120 (1)	Nitrosonaphthol isomers	476
	Amberlite CG-120 (1)	Hydroxyaromatic acids	480
	Chelex 100 (3)	Substituted phenols	493, 498
	8-Quinolinol silica	Phenols	499
Fe(NO$_3$)$_3$	Dowex-50W (1)	Hydroxyaromatic acids	504
Fe^{3+}	Amberlite CG-50 (4)	o,m,p-Hydroxybenzoic acids	479
Ru(Phen)$_3$$^{2+}$	Montmorillonite	Co, Rh(acac)$_3$	495, 500
Ru(Phen)$_3$$^{2+}$	Montmorillonite	Cr(acac)$_3$, aromatics	501, 511
[Co(en)$_3$]$^{3+}$	Bio-Rex 70 (2)	Amino acids	507
Co^{3+}	Silice diamino silica (5)	Dinucleotides	490, 503
		Co^{3+}, Cr^{3+} carboxylic salts	494
NiCl$_2$	Chelex 100 (3)	Amino and carboxylic acids	482, 483
Ni(NH$_3$)$_n$$^{2+}$	Dowex-50W (1)	Amines	471, 472
	Bio-Rex 63 (1)	Amines	471
	Bio-Rex 70 (2)	Amines	471, 472
		Amphetamines, amino sugars	478, 484
		Amino acids	484
	Zirconium phosphate	Amines	471, 472
	Amberlite XE 219 (2)	Amines	472
Ni^{2+}	Dowex A-1 (3)	Polyamines	473
	Chelex 100 (3)	Polyamines	473, 481
	Dowex-50W (1)	Aziridines, ethanolamines	481
	Amberlite CG-50 (4)	Aziridines, ethanolamines	481
	L-Histidine resin	Amino acids	418
Ni(Phen)$_3$$^{2+}$	Montmorillonite	Ru(acac)$_3$	509, 510
Zn(Ac)$_2$		γ-Aminobutyric acid	477
Cd^{2+}	Bio-Rex 70 (2)	Amphetamines	478
Hg(Ac)$_2$	Amberlyst XN-500	Aza, oxo, thioaromatics	474
Hg^{2+}	Amberlite CG-50	o,m,p-Hydroxybenzoic acids	479
Al(NO$_3$)$_3$	Dowex-50W (1)	Hydroxyaromatic acids	504
Ce(NO$_3$)$_3$	Dowex-50W (1)	Hydroxyaromatic acids	504

[a]OPTA, N-(2-hydroxyethyl)propylenediamine triacetate; en, ethylenediamine; Phen, 1,10-phenan-throline.

[b]1, Sulfonic resin; 2, carboxylic resin; 3, iminodiacetate resin; 4, polymethacrylate resin; 5, 3-(2-aminoethylamino)propylsilyl silica gel.

TABLE 47 Metal Salts Coated on the Support

Metal salts	Supports salts	Percent of salts on support	Compounds separated	Refs.
FeCl$_3$	Silica	10	Prostaglandins	458
Co(NO$_3$)$_2$	Cellulose		Alkyl amines	460
PdCl$_2$	Silica	5	Sulfur compounds	467, 468, 501
CdCl$_3$	Silica	0.3–0.6	Chloroaniline isomers	461
CdI$_2$	Silica	10	Aromatic amines	459
HgCl$_2$	Cellulose		Alkyl amines	460
Hg(Ac)$_2$	Silica	20	Alkyl sulfides	452
SbCl$_3$	Cellulose		Alkyl amines	460

TABLE 48 Reversed-Phase Chromatography of racemic or isomeric Mixtures Using Metal Salts in Eluants

Metal salts	Concentration	Compounds separated	Ref.
Ni(Ac)$_2$	0.015 M	Aromatic amines	486
Ni(Ac)$_2$	0.015 M	Aminophenols	508
L-Pro-C$_8$-amide-Ni^{2+}	4 mM	D,L-Dansyl amino acids	492
L-Pro-C$_{10}$-amide-Ni^{2+}	4 mM	D,L-Dansyl amino acids	492
C$_2$H$_4$PtCl$_3$	0.01 N	Olefins, dienes	464
C$_2$H$_4$PtCl$_2$R	0.01 N	Olefins, dienes	464
C$_2$H$_4$PtCl$_2$R$_1$	0.01 N	Olefins, dienes	464
C$_3$-C$_8$-dien-Zn^{2+}	0.8 mM	D,LDansyl amino acids	489
C$_3$-C$_8$-R-dien-Zn^{2+}	0.65 mM	D,LDansyl amino acids	491
C$_3$-C$_8$-R-dien-Cd^{2+}	0.65 mM	D,LDansyl amino acids	491

[a]R, NH$_2$—CH(CH$_3$)—C$_6$H$_5$; R$_1$, O$_2$C—CH(NH$_3$)$_2$—C$_6$H$_5$; C$_3$-C$_8$-dien, L-2-ethyl-4-octyl-diethyltriamine; C$_3$-C$_8$-R-dien, L-2-alkyl-4-octyl-diethyltriamine (alkyl = ethyl, isopropyl, isobutyl); L-Pro-C$_8$-amide, L-Propyl-n-octylamide, L-Pro-C$_{10}$-amide, L-Propyl-n-dodecylamide.

TABLE 49 Acceptor Stationary Phases Synthesized by Derivatization of an Amino-Bonded Silica Gel

Acceptors	Compounds[a] separated	Refs.
2,4,5,7-Tetranitrofluorenimine	PAHs	528, 543
2,4-Dinitroaniline	PAHs	530, 542
		543, 549
Picramine	PAHs	533, 537
		543, 549
Polynitrofluorenone	Methylbenzanthracenes	535
2,4,7-Trinitrofluorenimine	PAHs	542
Bis(3-nitrophenyl)sulfone	PAHs	542
2,4-Dinitrobenzene sulfonamide	Aza aromatics	546
Theophylline-7-Acetic amide	Various benzoic acids	548
2,4,6-Trinitrobenzenesulfonamide	PAHs	554

[a]PAHs, polyaromatic hydrocarbons.

TABLE 50 Acceptors Bonded on Silica Gel

Acceptors	Compounds[a] separated	Ref.
N-Uracil	Adenine derivatives	534
O-(2,4,5,7-Tetranitrofluorene)oxime	PAHs	536
N-Pyrrolidone	PAHs, shale oil	538
N-(3,5-Dinitrobenzamide)	PAHs	544
Picrylether	PAHs	546
n-Octylpicramine*	PAHs	549
n-Octyl-N-(2,4-dinitroaniline)*	PAHs	549
N-Pentafluorobenzamide	PAHs	550
7-Theophylline	PAHs	551
p-Acetylbenzene	Substituted anilines	553
1-Theobromine	PAHs	552
Ethylpentafluorobenzene*	PAHs	558
Tetrachlorophtalimide	PAHs, coal	560

[a]PAHs, polyaromatic hydrocarbons.

TABLE 51 TAPA Coated on Support

Supports	TAPA concentration	Racemic mixtures	Ref.
Silica gel	8.9%	Naphthyl-2-butyl ethers	563
	10	Tetrahydroxy[2,2]paracyclophanes	564
	10–25%	Helicenes	565
		9-sec-butylphenanthrene	596
Alumina		d,l-cis/trans-Naphthylstilbene	566

TABLE 52 Chiral Acceptors Bonded on Amino Silica Gel

Chiral acceptors	Compounds separated	Refs.
N-3,5-Dinitrobenzoylphenylglycine	9-Anthrylfluoroalcohols	569, 572
	Gossypol	580
TAPA	cis/trans Di-OH, di-H PAHs[a]	571
	Gossypol	580
	Helicenes	595
(S)-2-(4-Chlorophenyl)isovaleric acid	DNB amines, DNB acids[b]	574
(S)-N-(3,5-Dinitrobenzoyl)leucine	α-Naphthamides	577
N-3,5-Dinitrobenzoylphenylalanine	Gossypol	580
N-2,4-Dinitrophenylalanine	Azahelicene	593

[a]PAHs, polyaromatic hydrocarbons.
[b]DNB, dinitrobenzamido.

TABLE 53 Chiral Donors Bonded on Amino Silica Gel

Chiral donors	DNB-enantiomers separated	Ref.
(S)-1-(α-Naphthyl)ethylamine	Amines, amino acids	573
	Amino acids	586
Triazine(S)-1-(α-naphthyl)ethylamine	Amines, amino acids	575
1-(α-Naphthyl)ethylamine, L-valine	Amines, alcohols, acids	589
(S)-1-(α-Naphthyl)ethylamineurea	Amino acids[a]	591

[a]Phenylthiohydantoin form.

TABLE 54 Chiral Donors Bonded on Silica Gel

Chiral donors	DNB-enantiomers separated	Refs.
1-Phenylethylamine	α-Naphthylethylamines	576
6,7-Dimethyl-1-naphthylalkylamine	Amino acids	578, 581
6,7-Dimethyl-1-naphthylethylamineurea	α-Alkylamines	582
6,7-Dimethyl-1-naphthylhydantoin	Amino acids	583
Benzyloxycarbonyl-D-phenylglycine	Amino acid propylesters	585
Benzyloxycarbonyl-L-phenylalanine	Amino acid propylesters	585
Benzyloxycarbonyl, L-isoleucine	Amino acid propylesters	585
tert-Butyloxycarbonyl-D-phenylglycine	Amino acid propylesters	585
β-Naphthyl-L-valine	Amino acids, thiols	588
β-Naphthyl-L-valine	Amines	588

REFERENCES

1. B. W. Bradford, D. E. Harvey, and D. F. Chalkley, *J. Inst. Pet., 41*: 80 (1955).
2. A. C. Cope, C. L. Bumgardner, and E. E. Schweizer, *J. Am. Chem. Soc., 79*: 4729 (1957).
3. H. M. Tenney, *Anal. Chem., 30*: 2 (1958).
4. A. C. Cope and E. M. Acton, *J. Am. Chem. Soc., 80*: 355 (1958).
5. E. Gil-Av, J. Herling, and J. Shabtai, *J. Chromatogr., 1*: 508 (1958).
6. M. E. Bednas and D. S. Russell, *Can. J. Chem., 36*: 1272 (1958).
7. E. Gil-Av and J. Herling, J. Shabtai, *J. Chromatogr., 2*: 406 (1959).
8. F. Armitage, *J. Chromatogr., 2*: 655 (1959).
9. R. H. Luebbe and J. E. Willard, *J. Am. Chem. Soc., 81*: 761 (1959).
10. B. S. Rabinovitch and K. W. Michel, *J. Am. Chem. Soc., 81*: 5065 (1959).
11. B. Smith and R. Ohlson, *Acta Chem. Scand., 13*: 1253 (1959).
12. G. R. Primavesi, *Nature, 184*: 250 (1959).
13. R. R. Sauers, *Chem. Ind.,* 176 (1960).
14. J. N. Butler and G. B. Kistiakowsky, *J. Am. Chem. Soc., 82*: 759 (1960).
15. A. C. Cope, P. T. Moore, and W. R. Moore, *J. Am. Chem. Soc., 82*: 1744 (1960).
16. A. C. Cope, D. Ambros, E. Ciganek, C. F. Howell, and Z. Jucura, *J. Am. Chem. Soc., 82*: 1750 (1960).
17. J. Herling, J. Shabtai, and E. Gil-Av, *J. Chromatogr., 8*: 349 (1962).
18. J. J. Duffield and L. B. Rogers, *Anal. Chem., 34*: 1193 (1962).
19. E. Gil-Av and J. Herling, *J. Phys. Chem., 66*: 1209 (1962).
20. M. A. Muhs and F. T. Weiss, *J. Am. Chem. Soc., 84*: 4697 (1962).
21. J. Shabtai, J. Herling, and E. Gil-Av, *J. Chromatogr., 11*: 32 (1963).
22. L. A. du Plessis, *J. Gas Chromatogr., 1*: 6 (1963).
23. R. J. Cvetanovic, F. J. Duncan, and W. E. Falconer, *Can. J. Chem., 41*: 2095 (1963).
24. E. K. C. Lee and F. S. Rowland, *Anal. Chem., 36*: 2181 (1964).
25. J. Shabtai, *Isr. J. Chem., 1*: 300 (1963).
26. A. Zlatkis, G. S. Chao, and H. R. Kaufman, *Anal. Chem., 36*: 2354 (1964).
27. R. J. Cvetanovic, F. J. Duncan, W. E. Falconer, and R. S. Irwin, *J. Am. Chem. Soc., 87*: 1827 (1965).
28. J. Shabtai, *J. Chromatogr., 18*: 302 (1965).
29. H. J. Ache and A. P. Wolf, *J. Am. Chem. Soc., 88*: 888 (1966).
30. E. Bendel, B. Fell, W. Gartzen, and G. Kruse, *J. Chromatogr., 31*: 531 (1967).
31. H. Schnecko, *Anal. Chem., 40*: 1391 (1968).
32. A. Zlatkis and I. M. R. de Andrade, *Chromatographia, 2*: 298 (1969).
33. J. Janak, Z. Jagaric, and M. Dressler, *J. Chromatogr., 53*: 525 (1970).
34. R. Queignec and B. Wojtkowiak, *Bull. Soc. Chim. Fr.,* 3829 (1970).
35. C. L. de Ligny, T. van't Verlaat, and E. Karthaus, *J. Chromatogr., 76*: 115 (1972).
36. M. Kraitr, R. Komers, and F. Cuta, *J. Chromatogr., 86*: 1 (1973).
37. R. F. Hirsch, H. C. Stober, M. Kowblansky, F. N. Hubner, and A. W. O'Connell, *Anal. Chem., 45*: 2100 (1973).
38. M. Kraitr, R. Komers, and F. Cuta, *Anal. Chem., 46*: 974 (1974).
39. P. Magidman, R. A. Barford, D. H. Saunders, and H. L. Rothbart, *Anal. Chem., 48*: 44 (1976).

40. F. B. Wampler, *Anal. Chem.*, *48*: 1644 (1976).
41. S. P. Wasik and R. L. Brown, *Anal. Chem.*, *48*: 2218 (1976).
42. E. Gil-Av, J. Herling, and J. Shabtai, *Chem. Ind.*, 1483 (1957).
43. W. W. Hanneman, C. F. Spencer, and J. F. Johnson, *Anal. Chem.*, *32*: 1386 (1960).
44. B. Versino, F. Geiss, and G. Barbero, *Z. Anal. Chem.*, *201*: 20 (1964).
45. J. A. Fabre and L. R. Kallenbach, *Anal. Chem.*, *36*: 63 (1964).
46. P. W. Solomon, *Anal. Chem.*, *36*: 476 (1964).
47. O. T. Chortyk, W. S. Schlotzhauer, and R. L. Stredman, *J. Gas Chromatogr.*, *3*: 394 (1965).
48. B. T. Guran and L. B. Rogers, *Anal. Chem.*, *39*: 632 (1967).
49. R. L. Grob, G. W. Weinert, and J. W. Brelich, *J. Chromatogr.*, *30*: 305 (1967).
50. F. Onuska, J. Janak, K. Tesarik, and A. V. Kiselev, *J. Chromatogr.*, *34*: 81 (1968).
51. F. Geiss, B. Versino, and H. Schlitt, *Chromatographia*, *1*: 9 (1968).
52. G. L. Hargrove and D. T. Sawyer, *Anal. Chem.*, *40*: 409 (1968).
53. D. J. Brookman and D. T. Sawyer, *Anal. Chem.*, *40*: 1368 (1968).
54. D. T. Sawyer and D. J. Brookman, *Anal. Chem.*, *40*: 1847 (1968).
55. D. J. Brookman and D. T. Sawyer, *Anal. Chem.*, *40*: 2013 (1968).
56. F. Onuska and J. Janak, *Chem. Zvesti*, 22: 929 (1968).
57. A. F. Isbell and D. T. Sawyer, *Anal. Chem.*, *41*: 1381 (1969).
58. B. Versino and F. Geiss, *Chromatographia*, 2: 254 (1969).
59. C. Pommier, C. Eon, H. Fould, and G. Guiochon, *Bull. Soc. Chim. Fr.*, 1401 (1969).
60. R. L. Grob, R. J. Gondek, and T. A. Scales, *J. Chromatogr.*, *53*: 477 (1970).
61. L. D. Belyakova, G. A. Soloyan, and A. V. Kiselev, *Chromatographia*, *3*: 254 (1970).
62. D. F. Cadogan and D. T. Sawyer, *Anal. Chem.*, *42*: 190 (1970).
63. D. F. Cadogan and D. T. Sawyer, *Anal. Chem.*, *43*: 941 (1971).
64. J. P. Okamura and D. T. Sawyer, *Anal. Chem.*, *43*: 1730 (1971).
65. V. A. Ilie, M. Boroanca, and G. E. Baiulescu, *Chim. Anal.*, *1*: 33 (1971).
66. G. E. Baiulescu and V. A. Ilie, *Anal. Chem.*, *44*: 1490 (1972).
67. B. Feibush, M. F. Richardson, and R. E. Sievers, C. S. Springer, Jr., *J. Am. Chem. Soc.*, *94*: 6717 (1972).
68. J. J. Brooks and R. E. Sievers, *J. Chromatogr. Sci.*, *11*: 303 (1973).
69. L. D. Belyakova, A. M. Kalpakian, and A. V. Kiselev, *Chromatographia*, 7: 14 (1974).
70. L. D. Belyakova and A. M. Kalpakian, *J. Chromatogr.*, *91*: 699 (1974).
71. N. H. C. Cooke, E. F. Barry, and B. S. Solomon, *J. Chromatogr.*, *109*: 57 (1975).
72. T. Andronikashvili, C. V. Tsitsishvili, and L. Laperashvili, *Chromatographia*, 8: 223 (1975).
73. M. Leddet, D. Cuzin, and F. Coussemant, *Bull. Soc. Chim. Fr.*, *9–10*: 866 (1977).
74. B. T. Golding, P. J. Sellard, and A. K. Wong, *J. Chem. Soc., Chem. Commun.*, 570 (1977).
75. W. Nawrocki, W. Szczepaniack, and W. Wasiak, *J. Chromatogr.*, *188*: 323 (1980).
76. A. G. Altenau and L. B. Rogers, *Anal. Chem.*, *35*: 915 (1963).
77. A. G. Altenau and L. B. Rogers, *Anal. Chem.*, *36*: 1726 91964).
78. D. J. Brookman and D. T. Sawyer, *Anal. Chem.*, *40*: 106 (1968).
79. R. P. Hirschman, H. L. Simon, L. R. Anderson, and W. B. Fox, *J. Chromatogr.*, *50*: 118 (1970).
80. D. W. Barber, C. S. G. Phillips, G. F. Tusa, and A. Verdin, *J. Chem. Soc.*, 18 (1959).

81. D. M. Coulson, *Anal. Chem., 31*: 906 (1959).

82. R. S. Juvet and F. M. Wachi, *Anal. Chem., 32*: 290 (1960).

83. A. G. Altenau and L. B. Rogers, *Anal. Chem., 37*: 1432 (1965).

84. A. C. Bhattacharyya and A. Bhattacharjee, *Anal. Chem., 41*: 2055 (1969).

85. B. T. Guran and L. B. Rogers, *J. Gas Chromatogr., 3*: 269 (1965).

86. G. P. Cartoni, A. Liberti, and R. Palombari, *J. Chromatogr., 20*: 278 (1965).

87. D. J. McEwen, *Anal. Chem., 38*: 1047 (1966).

88. R. J. Cvetanovic, F. J. Duncan, W. E. Falconer, and W. A. Sunder, *J. Am. Chem. Soc., 88*: 1602 (1966).

89. A. G. Altenau and C. Mettitt, *J. Gas Chromatogr., 5*: 30 (1967).

90. R. L. Pecsok and E. M. Vary, *Anal. Chem., 39*: 289 (1967).

91. B. T. Guran and L. B. Rogers, *J. Gas Chromatogr., 5*: 574 (1967).

92. W. E. Falconer and R. J. Falconer, *J. Chromatogr., 27*: 20 (1967).

93. D. V. Banthorpe, C. G. Atford, and B. R. Hollebone, *J. Gas Chromatogr., 6*: 61 (1968).

94. C. Pommier, C. Eon, H. Fould, and G. Guiochon, *C. R. Acad. Sci., 268C*: 1553 (1969).

95. R. S. Juvet, Jr., V. R. Shaw, and M. A. Khan, *J. Am. Chem. Soc., 91*: 3788 (1969).

96. A. C. Bhattacharyya and A. Bhattacharjee, *J. Chromatogr., 41*: 446 (1969).

97. V. A. Ilie, *Rev. Chim., 20*: 43 (1969).

98. R. T. Pflaum and L. E. Cook, *J. Chromatogr., 50*: 120 (1970).

99. R. C. Castells and J. A. Catoggio, *Anal. Chem., 42*: 1268 (1970).

100. K. Ohzeki and T. Kambara, *J. Chromatogr., 55*: 319 (1971).

101. N. L. Soulages and A. M. Brieva, *J. Chromatogr. Sci., 9*: 492 (1971).

102. C. Vidal-Madjar and G. Guiochon, *J. Chromatogr. Sci., 9*: 664 (1971).

103. E. Gil-Av and V. Schurig, *Chem. Commun.,* 650 (1971).

104. E. Gil-Av and V. Schurig, *Anal. Chem., 43*: 2030 (1971).

105. R. L. Grob and E. J. McGonigle, *J. Chromatogr., 59*: 13 (1971).

106. W. Kuchen, J. Delventhal, and H. Keck, *Angew. Chem. Int. Ed. Engl., 11*: 435 (1972).

107. J. Delventhal, H. Keck, and W. Kuchen, *Angew. Chem. Int. Ed. Engl., 11*: 830 (1972).

108. W. Bruenning and I. M. R. De Antrade Bruening, *An. Asoc. Bras. Quim., 28*: 61 (1972).

109. V. Schurig, J. L. Bear, and A. Zlatkis, *Chromatographia, 5*: 301 (1972).

110. V. Schurig, R. C. Chang, A. Zlatkis, E. Gil-Av, and F. Mikes, *Chromatographia, 6*: 223 (1973).

111. J. Delventhal, H. Keck, and W. Kuchen, *J. Chromatogr., 77*: 422 (1973).

112. V. Schurig, R. C. Chang, and A. Zlatkis, *J. Chromatogr., 99*: 147 (1974).

113. W. Szczepaniak and W. Nawrocki, *Chem. Anal. (Warsaw), 19*: 375 (1974).

114. J. Delventhal, H. Keck, and W. Kuchen, *J. Chromatogr., 95*: 238 (1974).

115. E. J. McGonigle and R. L. Grob, *J. Chromatogr., 101*: 39 (1974).

116. E. F. Barry and N. H. C. Cooke, *J. Chromatyogr., 104*: 161 (1975).

117. W. Szczepaniak and W. Nawrocki, *J. Chromatogr., 138*: 337 (1977).

118. W. Szczepaniak and W. Nawrocki, *J. Chromatogr., 168*: 89 (1979).

119. W. Szczepaniak and W. Nawrocki, *Chem. Anal. (Warsaw), 23*: 795 (1978).

120. W. Szczepaniak and W. Nawrocki, *J. Chromatogr., 168*: 97 (1979).

121. W. Nawrocki, W. Szczepaniak, and W. Wasiak, *J. Chromatogr., 178*: 91 (1979).

122. V. Schuring, *Chromatographia, 13*: 263 (1980).

123. V. Schurig and R. Weber, *J. Chromatogr., 217*: 51 (1981).

124. A. Yamagishi and R. Oshnishi, *Angew. Chem. Int. Ed. Engl., 22*: 162 (1983).

125. W. Wasiak, *Chem. Anal. (Warsaw)*, 29: 211 (1984).
126. W. Wasiak and W. Szczepaniack, *Chromatographia*, 18: 205 (1984).
127. W. Wasiak and W. Szczepaniak, *J. Chromatogr.*, 364: 259 (1986).
128. W. Wasiak, *Chromatographia*, 23: 423 (1987).
129. S. H. Langer, C. Zahn, and F. Pantazoplos, *Chem. Ind.*, 1145 (1958).
130. R. O. C. Norman, *Proc. Chem. Soc.*, 151 (1958).
131. F. A. Fabrizio, R. W. King, C. C. Cerato, and J. W. Loveland, *Anal. Chem.*, 31: 2060 (1959).
132. S. H. Langer, C. Zahn, and G. Pantazoplos, *J. Chromatogr.*, 3: 154 (1960).
133. S. H. Langer and J. H. Purnell, *J. Phys. Chem.*, 67: 263 (1963).
134. K. Malinowska, *Chem. Anal. (Warsaw)*, 9: 353 (1964).
135. R. J. Cvetanovic, F. J. Duncan, and W. E. Falconer, *Can. J. Chem.*, 42: 2410 (1964).
136. N. Petsev and Chr. Dimitrov, *J. Chromatogr.*, 23: 382 (1966).
137. A. R. Cooper, C. W. P. Crowne, and P. G. Farrell, *J. Chromatogr.*, 29: 1 (1967).
138. J. E. Heveran and L. B. Rogers, *J. Chromatogr.*, 25: 213 (1966).
139. A. R. Cooper, C. W. P. Crowne, and P. G. Farrell, *Trans. Faraday Soc.*, 62: 2725 (1966).
140. S. H. Langer, B. M. Johnson, and J. R. Conder, *J. Phys. Chem.*, 72: 4020 (1968).
141. D. F. Cadogan and J. J. Purnell, *J. Chem. Soc. A*, 2133 (1968).
142. C. Eon, C. Pommier, G. Guiochon, *C. R. Acad. Sci.*, 270: 1436 (1970).
143. C. Eon, C. Pommier, and G. Guiochon, *J. Phys. Chem.*, 75: 2632 (1971).
144. J. P. Sheridan, D. E. Martire, and Y. B. Tewari, *J. Am. Chem. Soc.*, 94: 3294 (1972).
145. J. P. Sheridan, M. A. Capeless, and D. E. Martire, *J. Am. Chem. Soc.*, 94: 3298 (1972).
146. J. P. Sheridan, D. E. Martire, and F. P. Banda, *J. Am. Chem. Soc.*, 95: 4788 (1973).
147. R. J. Laub and R. L. Pecsok, *Anal. Chem.*, 46: 1214 (1974).
148. H. Kelker, *Angew. Chem.*, 6: 218 (1959).
149. L. J. Morris, *Chem. Ind.*, 1238 (1962).
150. P. Daniels and C. Enzell, *Acta Chem. Scand.*, 16: 1530 (1962).
151. L. J. Morris, *J. Chem. Soc.*, 5779 (1963).
152. O. S. Privett, M. L. Blank, and O. Romanus, *J. Lipid Res.*, 4: 260 (1963).
153. D. F. Zinkel and J. W. Rowe, *J. Chromatogr.*, 13: 74 (1964).
154. L. D. Bergelson, E. V. Byatlovitskaya, and V. V. Voronkova, *J. Chromatogr.*, 15: 191 (1964).
155. F. D. Gunstone, F. B. Padley, and I. Qureshi, *Chem. Ind.*, 483 (1964).
156. P. F. V. Ward, T. W. Scott, and R. M. C. Dawson, *Biochem. J.*, 92: 60 (1964).
157. F. D. Gunstone, R. J. Hamilton, and I. Qureshi, *J. Chem. Soc.*, 319 (1965).
158. E. Dunn and P. Robson, *J. Chromatogr.*, 17: 501 (1965).
159. J. L. Morris and D. M. Wharry, *J. Chromatogr.*, 20: 27 (1965).
160. K. L. Mikolajczak and M. O. Bagby, *J. Am. Oil Chem. Soc.*, 42: 43 (1965).
161. R. Wood and F. Snyder, *J. Am. Oil Chem. Soc.*, 43: 53 (1966).
162. N. Ruseva-Atanasova and J. Janak, *J. Chromatogr.*, 21: 207 (1966).
163. M. M. Paulose, *J. Chromatogr.*, 21: 141 (1966).
164. L. J. Morris, M. O. Marshall, and W. Kelly, *Tetrahedron Lett.*, 4249 (1966).
165. A. M. Lees and E. D. Korn, *Biochim. Biophys. Acta*, 116: 403 (1966).
166. P. F. Wilde and R. M. C. Dawson, *Biochem. J.*, 98: 469 (1966).
167. P. K. Raju and R. Reiser, *Lipids*, 1: 10 (1966).

168. W. O. Ord and P. C. Bamford, *Chem. Ind.*, 277 (1967).
169. L. J. Morris, D. M. Wharry, and E. W. Hammond, *J. Chromatogr.*, *31*: 69 (1967).
170. S. M. Hopkins, G. Shehan, and R. L. Lyman, *Biochim. Biophys. Acta*, *164*: 272 (1968).
171. W. W. Christie, *J. Chromatogr.*, *34*: 405 (1968).
172. R. Kleiman, G. F. Spencer, L. W. Tjarks, and F. R. Earle, *Lipids*, *6*: 617 (1971).
173. C. Subhash, C. Srinivasulu, and S. N. Mahapatra, *J. Chromatogr.*, *106*: 475 (1975).
174. P. A. Dudley and R. A. Anderson, *Lipids*, *68*: 113 (1975).
175. M. S. F. Lie Ken Jie and C. H. Lam, *J. Chromatogr.*, *124*: 147 (1976).
176. N. Salem, Jr., L. G. Abood, and W. Hoss, *Anal. Biochem.*, *76*: 407 (1976).
177. M. S. F. Lie Ken Jie and C. H. Lam, *J. Chromatogr.*, *129*: 181 (1976).
178. M. S. J. Dallas and F. B. Padley, *Lebensm. Wiss. Technol.*, *10*: 328 (1977).
179. C. B. Barret, M. S. J. Dallas, and F. B. Padley, *Chem. Ind.*, 1050 (1962).
180. H. P. Kaufmann, H. Wessels, and C. Bondopadhyaya, *Fette Siefen Anstrichm.*, *65*: 543 (1963).
181. B. de Vries and G. Jurriens, *Fette Seifen Anstrichm.*, *65*: 725 (1963).
182. C. B. Barret, M. S. J. Dallas, and F. B. Padley, *J. Am. Oil Chem. Soc.*, *40*: 580 (1963).
183. H. P. Kaufmann and H. Wessels, *Fette Seifen Anstrichm.*, *66*: 81 (1964).
184. E. Vioque, M. P. Maza, and M. Calderon, *Grasas Aceites*, *15*: 173 (1964).
185. F. G. Den Boer, *Z. Anal. Chem.*, *205*: 308 (1964).
186. J. M. Cubero and H. K. Mangold, *Microchem. J.*, *9*: 227 (1965).
187. A. G. Vereshchagin, *J. Chromatogr.*, *17*: 382 (1965).
188. F. D. Gunstone and F. B. Pradley, *J. Am. Oil Chem. Soc.*, *42*: 957 (1965).
189. G. A. E. Arvidson, *J. Lipid Res.*, *6*: 574 (1965).
190. F. Haverkate and L. L. M. van Deenen, *Biochem. Biophys. Acta*, *106*: 78 (1965).
191. R. Wood and F. Snyder, *Lipids*, *1*: 62 (1966).
192. L. M. G. van Golde, R. F. A. Zwaal, and L. L. M. van Deenen, *Biochemistry*, *98*: 255 (1966).
193. O. Renkonen and I. Rikkinen, *Acta Chem. Scand.*, *21*: 2282 (1967).
194. D. T. Burns, R. J. Stretton, G. F. Shepherd, and M. S. J. Dallas, *J. Chromatogr.*, *44*: 399 (1969).
195. G. Urbach, *J. Chromatogr.*, *12*: 196 (1963).
196. K. de Jong, K. Mostert, and D. Sloot, *Recl. Trav. Chim.*, *82*: 837 (1963).
197. H. T. Badings and J. G. Wassink, *Neth. Milk Diary J.*, *17*: 132 (1963).
198. R. Denti and M. P. Luboz, *J. Chromatogr.*, *18*: 325 (1965).
199. D. Sloot, *J. Chromatogr.*, *24*: 451 (1966).
200. J. W. Copius-Peereboom and H. W. Beekes, *J. Chromatogr.*, *20*: 43 (1965).
201. S. Tabak and M. R. M. Verzola, *J. Chromatogr.*, *51*: 334 (1970).
202. B. Samuelsson, *J. Biol. Chem.*, *238*: 3229 (1963).
203. K. Green and B. Samuelsson, *J. Lipid Res.*, *5*, 117 (1964).
204. E. W. Horton and I. H. M. Main, *Br. J. Pharmacol. Chemother.*, *30*: 582 (1967).
205. H. A. Davis, E. W. Horton, K. B. Jones, and J. P. Quilliam, *Br. J. Pharmacol.*, *42*: 569 (1971).
206. J. E. Greenwald, M. S. Alexander, M. Vanrollins, L. K. Wong, and J. R. Bianchine, *Prostaglandins*, *21*: 33 (1981).
207. N. H. Andersen and E. M. K. Leovey, *Prostaglandins*, *6*: 361 (1974).
208. J. Avignan, D. S. Goodman, and D. Steinberg, *J. Lipid Res.*, *4*: 100 (1963).

209. K. Schreiber, O. Aurich, and G. Osske, *J. Chromatogr., 12*: 63 (1963).
210. J. L. Morris, *J. Lipid Res., 4*: 357 (1963).
211. J. R. Claude and J. L. beaumont, *Ann. Biol. Clin., 22*: 815 (1964).
212. A. Ercoli, R. Vitali, and R. Gardi, *Steroids, 3*: 479 (1964).
213. R. Ikan and M. Cudzinovski, *J. Chromatogr., 18*: 422 (1965).
214. T. M. Lees, M. J. Lynch, and F. R. Mosher, *J. Chromatogr., 18*: 595 (1965).
215. J. W. Copius-Peereboom and H. W. Beekes, *J. Chromatogr., 17*: 99 (1965).
216. A. S. Truswell and W. D. Mitchell, *J. Lipid Res., 6*: 438 (1965).
217. N. W. Ditullio, C. S. Jacobs, Jr., and W. L. Holmes, *J. Chromatogr., 20*: 354 (1965).
218. B. P. Lisboa, *Steroids, 8*: 319 (1966).
219. H. W. Wroman and C. F. Cohen, *J. Lipid Res., 8*: 150 (1967).
220. R. Kammereck, W. H. Lee, A. Paliokas, and G. J. Schroepfer, *J. Lipid Res., 8*: 282 (1967).
221. P. J. Stevens, *J. Chromatogr., 36*: 253 (1968).
222. L. E. Crocker and B. A. Lodge, *J. Chromatogr., 62*: 158 (1971).
223. E. Haahti, T. Nikkari, and K. Juva, *Acta Chem. Scand., 17*: 538 (1967).
224. L. E. Crocker and B. A. Lodge, *J. Chromatogr., 69*: 419 (1972).
225. J. R. Claude, *J. Chromatogr., 17*: 596 (1965).
226. H. S. Baretto and C. Enzell, *Acta Chem. Scand., 15*: 1313 (1961).
227. A. S. Gupta and S. Dev, *J. Chromatogr., 12*: 189 (1963).
228. E. Stahl and H. Vollman, *Talanta, 12*: 525 (1965).
229. R. Ikan, *J. Chromatogr., 17*: 591 (1965).
230. M. V. Schantz, S. Juvonen, and R. Hemming, *J. Chromatogr., 20*: 618 (1965).
231. I. Jardine and F. J. McQuillin, *J. Chem. Soc., 458* (1966).
232. R. S. Prasad, A. S. Gupta, and S. Dev, *J. Chromatogr., 92*: 450 (1974).
233. R. P. Enganhouse, E. C. Ruth, and I. R. Kaplan, *Anal. Chem., 55*: 2120 (1983).
234. J. C. Kohli and K. K. Badaisha, *J. Chromatogr., 320*: 455 (1985).
235. J. H. P. Tyman and N. Jacobs, *J. Chromatogr., 54*: 83 (1971).
236. G. M. Nano and A. Martelli, *J. Chromatogr., 21*: 349 (1966).
237. K. Yasuda, *J. Chromatogr., 13*: 78 (1964).
238. K. Shimomura and H. F. Walton, *Sep. Sci., 3*: 493 (1968).
239. K. Yasuda, *J. Chromatogr., 60*: 144 (1971).
240. M. D. Martz and A. F. Krivis, *Anal. Chem., 43*: 790 (1971).
241. K. Yasuda, *J. Chromatogr., 72*: 413 (1972).
242. K. Yasuda, *J. Chromatogr., 72*: 142 (1972).
243. K. Yasuda, *J. Chromatogr., 87*: 565 (1973).
244. R. W. Frei, B. L. MacLellan, and J. D. MacNeil, *Anal. Chim. Acta, 66*: 139 (1973).
245. J. A. F. Wickramasinghe and S. R. Shaw, *Prostaglandins, 4*: 903 (1973).
246. S. P. Srivastava and V. K. Dua, *Z. Anal. Chem., 275*: 29 (1975).
247. S. P. Srivastava and V. K. Dua, *Z. Anal. Chem., 286*: 247 (1977).
248. N. J. Singh, Rajeev, and S. N. Tandon, *Indian J. Chem., 15B*: 581 (1977).
249. S. P. Srivastava and V. K. Dua, *Indian J. Chem., 15A*: 761 (1977).
250. T. Kaimal and A. Matsunaga, *Anal. Chem., 50*: 268 (1978).
251. V. Horak, M. De Valle Guzman, and G. Weeks, *Anal. Chem., 51*: 2248 (1979).
252. S. P. Srivastava, L. S. Chauhan, and A. K. Mital, *Anal. Lett., 12*: 235 (1979).
253. S. P. Srivastava, V. K. Dua, and K. Gupta, *Chromatographia, 12*: 605 (1979).

254. S. P. Srivastava, V. K. Dua, R. N. Mehrotra, and R. C. Saxena, *J. Chromatogr., 176*: 145 (1979).

255. M. L. Antonelli, A. Marino, A. Messina, and B. M. Petronio, *Chromatographia, 13*: 167 (1980).

256. S. P. Srivastava, L. S. Chauhan, and V. K. Dua, *J. Liq. Chromatogra., 3*: 1929 (1980).

257. A. Matsunaga and S. Kusayanagi, *J. Jpn. Pet. Inst., 24*: 298 (1981).

258. L. V. Andreev, *J. High Resolut. Chromatogr. Chromatogr. Commun., 6*: 575 (1983).

259. S. P. Srivastava, R. Bhushan, and R. S. Chauhan, *J. Liq. Chromatogr., 7*: 1359 (1984).

260. S. P. Srivastava, R. Bhushan, and R. S. Chauhan, *J. Liq. Chromatogr., 8*: 1255 (1985).

261. S. P. Srivastava, Reena, *Anal. Lett., 18*: 239 (1985).

262. Reena, *Anal. Lett., 18*: 753 (1985).

263. S. P. Srivastava and V. K. Dua, *Z. Anal. Chem., 276*: 382 (1975).

264. P. Catelli, *J. Chromatogr., 9*: 534 (1962).

265. A. Berg and J. Lam, *J. Chromatogr., 16*: 157 (1964).

266. M. Franck-Neumann and P. Jössang, *J. Chromatogr., 14*: 283 (1964).

267. J. Lam and A. Berg, *J. Chromatogr., 20*: 168 (1965).

268. R. G. Harvey and M. Halonen, *J. Chromatogr., 25*: 294 (1966).

269. H. Kessler and E. Müller, *J. Chromatogr., 24*: 469 (1966).

270. A. K. Dwivedy, D. B. Parihar, S. P. Sharma, and K. K. Verma, *J. Chromatogr., 29*: 120 (1967).

271. D. B. Parihar, S. P. Sharma, and K. K. Verma, *J. Chromatogr., 31*: 120 (1967).

272. V. Libickova, M. Stuchlik, and L. Krasnec, *J. Chromatogr., 45*: 278 (1969).

273. J. Lam, *Planta Med., 24*: 107 (1973).

274. J. P. Sharma and S. Ahuja, *Z. Anal. Chem., 267*: 368 (1973).

275. G. H. Schenk, G. L. Sullivan, and P. A. Fryer, *J. Chromatogr., 89*: 49 (1974).

276. M. A. Slifkin, W. A. Amarasiri, C. Schandorff, and R. Bell, *J. Chromatogr., 235*: 389 (1982).

277. I. W. Wainer and C. A. Brunner, T. D. Doyle, *J. Chromatogr., 264*: 154 (1983).

278. M. A. Slifkin and S. H. Liu, *J. Chromatohgr., 269*: 103 (1983).

279. M. A. Slifkin and S. H. Liu, *J. Chromatogr., 303*: 190 (1984).

280. N. Grinberg and S. Weinstein, *J. Chromatogr., 303*: 251 (1984).

281. K. Günther, J. Martens, and M. Schickedanz, *Angew. Chem. Int. Ed. Engl., 23*: 506 (1984).

282. S. Weinstein, *Tetrahedron Let., 25*: 985 (1984).

283. H. L. Goering, W. D. Closson, and A. C. Closson, *J. Am. Chem. Soc., 83*: 3507 (1961).

284. H. Wagner, J. D. Goetschel, and P. Lesch, *Helv. Chim. Acta, 46*: 2986 (1963).

285. B. de Vries, *J. Am. Oil Chem. Soc., 40*: 184 (1963).

286. E. Haahti, T. Nikkari, and K. Juva, *Acta Chem. Scand., 17*: 538 (1963).

287. F. D. Gunstone and A. J. Sealy, *J. Chem. Soc.*, 5772 (1963).

288. T. Norin and L. Westfelt, *Acta Chem. Scand., 17*: 1828 (1963).

289. S. Pohl, J. H. Law, and R. Ryhage, *Biochim. Biophys. Acta, 70*: 583 (1963).

290. D. S. Sgoutas and F. A. Kummerow, *J. Chromatogr., 16*: 448 (1964).

291. M. R. Subbaram and C. G. Youngs, *J. Am. Oil Chem. Soc., 41*: 150 (1964).

292. F. D. Gunstone and A. J. Sealy, *J. Chem. Soc.*, 4407 (1964).

293. D. Willner, *Chem. Ind.*, 1839 (1965).

294. R. L. Anderson and E. J. Hollenbach, *J. Lipid Res., 6*: 577 (1965).

295. N. Nicholaides and T. Ray, *J. Am. Oil Chem. Soc., 42*: 702 (1965).
296. A. Dolev and H. S. Olcott, *J. Am. Oil Chem. Soc., 42*: 624 (1965).
297. R. G. Powell and C. R. Smith, Jr., *Biochemistry, 5*: 625 (1966).
298. R. A. Stein and V. Slawson, *Anal. Chem., 40*: 2017 (1968).
299. Y. Tamaki, H. Noguchi, T. Yushima, and C. Hirano, *Appl. Entomol. Zool., 6*: 139 (1971).
300. R. Vivilecchia, M. Thiebaud, and R. W. Frei, *J. Chromatogr. Sci., 10*: 411 (1972).
301. F. Mikes, V. Schurig, and E. Gil-Av, *J. Chromatogr., 83*: 91 (1973).
302. R. R. Heath, J. H. Tumlinson, R. E. Doolittle, and A. T. Proveaux, *J. Chromatogr. Sci., 13*: 380 (1975).
303. R. Aigner, H. Spitzy, and R. W. Frei, *J. Chromatogr. Sci., 14*: 381 (1976).
304. R. Aigner, H. Spitzy, and R. W. Frei, *Anal. Chem., 48*: 2 (1976).
305. S. Lam and E. Grushka, *J. Chromatogr. Sci., 15*: 234 (1977).
306. R. R. Heath, J. H. Tumlinson, and R. E. Doolittle, *J. Chromatogr. Sci., 15*: 10 (1977).
307. E. Heftmann, G. A. Saunders, and W. F. Haddon, *J. Chromatogr., 156*: 71 (1978).
308. C. R. Vogt, J. S. Baxter, and T. R. Ryan, *J. Chromatogr., 150*: 93 (1978).
309. H. Colin, G. Guiochon, and A. Siouffi, *Anal. Chem., 51*: 1661 (1979).
310. M. Ozcimder and W. E. Hammers, *J. Chromatogr., 187*: 307 (1980).
311. E. C. Smith, A. D. Jones, and E. W. Hammond, *J. Chromatogr., 188*: 205 (1980).
312. Y. Yamauchi, R. Oshima, and J. Kumanotari, *J. Chromatogr., 198*: 49 (1980).
313. D. C. Shaw and C. E. West, *J. Chromatogr., 200*: 185 (1980).
314. S. Hara, A. Ohsawa, J. Endo, Y. Sashida, and H. Itokawa, *Anal. Chem., 52*: 428 (1980).
315. R. R. Health and P. E. Sonnet, *J. Liq. Chromatogr., 3*: 1129 (1980).
316. R. P. Evershed, E. D. Morgan, and L. D. Thompson, *J. Chromatogr., 237*: 350 (1982).
317. M. Morita, S. Mihashi, H. Itokawa, and S. Hara, *Anal. Chem., 55*: 412 (1983).
318. W. F. Joyce and P. C. Uden, *Anal. Chem., 55*: 540 (1983).
319. H. D. Friedel and R. Matusch, *J. Chromatogr., 407*: 343 (1987).
320. C. F. Wurster, Jr., J. H. Copenhaver, Jr., and P. R. Shafer, *J. Am. Oil Chem. Soc., 40*: 513 (1963).
321. E. A. Emken, C. R. Scholfield, and H. J. Dutton, *J. Am. Oil Chem. Soc., 41*: 388 (1964).
322. N. W. H. Houx, S. Voerman, and W. M. F. Jongen, *J. Chromatogr., 96*: 25 91974).
323. J. D. Warthen, Jr., *J. Chromatogr. Sci., 14*: 513 (1976).
324. M. V. Merritt and G. E. Bronson, *Anal. Chem., 48*: 1851 (1976).
325. N. W. H. Houx and S. Voerman, *J. Chromatogr., 129*: 456 (1976).
326. M. V. Merritt and G. E. Bronson, *Anal. Biochem., 80*: 392 (1977).
327. C. R. Scholfield and T. L. Mounts, *J. Am. Oil Chem. Soc., 54*: 319 (1977).
328. C. Battaglia and D. Frohlich, *Chromatographia, 13*: 428 (1980).
329. R. O. Adolf, H. Rakoff, and E. A. Emken, *J. Am. Oil Chem. Soc., 57*: 273 (1980).
330. R. O. Adolf and E. A. Emken, *J. Am. Oil Chem. Soc., 57*: 276 (1980).
331. R. O. Adolf and E. A. Emken, *J. Am. Oil Chem. Soc., 58*: 99 (1981).
332. W. S. Powell, *Anal. Biochem., 115*: 267 (1981).
333. W. S. Powell, *Methods Enzymol., 86*: 530 (1982).
334. J. Runeberg, *Acta Chem. Scand., 14*: 1288 (1960).
335. H. B. Dutton, C. R. Scholfield, and E. P. Jones, *Chem. Ind.*, 1874 (1961).
336. B. Wickberg, *J. Org. Chem., 27*: 4652 (1962).

337. C. R. Scholfield, E. P. Jones, R. O. Butterfield, and H. J. Dutton, *Anal. Chem., 35*: 386 (1963).
338. C. R. Scholfield, E. P. Jones, R. O. Butterfield, and J. J. Dutton, *Anal. Chem., 35*: 1588 (1963).
339. E. P. Jones, C. R. Scholfield, V. L. Davison, and H. J. Dutton, *J. Am. Oil Chem. Soc., 42*: 727 (1965).
340. A. G. Vershchagin and G. V. Novitskaya, *J. Am. Oil Chem. Soc., 42*: 970 (1965).
341. J. Janak, Z. Jagaric, and M. Dressler, *J. Chromatogr., 53*: 525 (1970).
342. G. Schomburg and K. Zegarski, *J. Chromatogr., 114*: 174 (1975).
343. R. J. Tscherne and G. Capitano, *J. Chromatogr., 136*: 337 (1977).
344. D. J. Weber, *J. Pharm. Sci., 66*: 744 (1977).
345. B. Vonach and G. Schomburg, *J. Chromatogr., 149*: 417 (1978).
346. H. W. S. Chan and G. Levett, *Chem. Ind.*, 578 (1978).
347. M. G. M. de Ruyter and A. P. de Leenheer, *Anal. Chem., 51*: 43 (1979).
348. D. O. Mack, *J. Liq. Chromatogr., 3*: 1005 (1980).
349. P. L. Phelan and J. R. Miller, *J. Chromatogr. Sci., 19*: 13 (1981).
350. L. D. Kissinger and R. H. Robins, *J. Chromatogr., 321*: 353 (1985).
351. L. A. d'Avila, H. Colin, and G. Guiochon, *J. Liq. Chromatogr., 10*: 71 (1987).
352. S. Fazakerley and D. R. Best, *Anal. Biochem., 12*: 290 (1965).
353. A. Siegel and E. T. Degens, *Science, 151*: 1098 (1966).
354. N. R. M. Buist and D. O'Brien, *J. Chromatogr., 29*: 398 (1967).
355. G. Goldstein, *Anal. Biochem., 20*: 447 (1967).
356. J. Boisseau and P. Jouan, *J. Chromatogr., 54*: 231 (1971).
357. V. A. Davankov and S. V. Rogozhin, *J. Chromatogr., 60*: 280 (1971).
358. S. V. Rogozhin and V. A. Davankov, *J. Chem. Soc. Chem. Commun.*, 490 (1971).
359. J. C. Wolford, J. A. Dean, and G. Goldstein, *J. Chromatogr., 62*: 48 (1971).
360. R. V. Snyder, R. J. Angelici, and R. B. Meck, *J. Am. Chem. Soc., 94*: 2660 (1972).
361. V. A. Davankov, S. V. Rogozhin, and A. V. Semechkin, *J. Chromatogr., 82*: 359 (1973).
362. J. F. Bellinger and N. R. M. Buist, *J. Chromatogr., 87*: 513 (1973).
363. J. Porath, J. Carlsson, I. Olsson, and G. Belfrage, *Nature, 258*: 598 (1975).
364. B. Hemmasi and E. Bayer, *J. Chromatogr., 109*: 43 (1975).
365. H. E. Gallo-Torres, O. N. Miller, and J. Ludorf, *Anal. Biochem., 64*: 260 (1975).
366. E. Tsuchida, H. Nishikawa, and E. Terada, *Eur. Polym. J., 12*: 611 (1976).
367. B. Lefebvre, R. Audebert, and C. Quiveron, *Isr. J. Chem., 15*: 69 (1977).
368. B. Lönnerdal, J. Carlsson, and J. Porath, *FEBS Lett., 75*: 89 (1977).
369. G. Krauss and H. Reinbothe, *Anal. Biochem., 78*: 1 (1977).
370. A. V. Semechkin, S. V. Rogozhin, and V. A. Davankov, *J. Chromatogr., 131*: 65 (1977).
371. M. A. Petit and J. Jozefonvicz, *J. Appl. Polym. Sci., 21*: 2589 (1977).
372. F. K. Chow, and E. Grushka, *Anal. Chem., 49*: 1756 (1977).
373. I. A. Yamskov, B. B. Berezin, and V. A. Davankov, *Makromol. Chem., 179*: 2121 (1978).
374. J. Jozefonvicz, M. A. Petit, and A. Szubarga, *J. Chromatogr., 147*: 177 (1978).
375. V. A. Davankov and Yu. A. Zolotarev, *J. Chromatogr., 155*: 285 (1978).
376. V. A. Davankov and Yu. A. Zolotarev, *J. Chromatogr., 155*: 295 (1978).

377. V. A. Davankov and Yu. A. Zolotarev, *J. Chromatogr., 155*: 303 (1978).
378. B. Lefebre, R. Audebert, and C. Quiveron, *J. Liq. Chromatogr., 1*: 761 (1978).
379. M. Doury-Berthod, C. Pointrenaud, and B. Tremillon, *J. Chromatogr., 179*: 37 (1979).
380. E. Schimdt, A. Foucault, M. Caude, and R. Rosset, *Analusis, 7*: 366 (1979).
381. N. F. Myasoedov, O. B. Kuznetsova, O. V. Petrenik, V. A. Davankov, and Yu. A. Zolotarev, *J. Labelled Compd. Radiopharmacol., 17*: 439 (1980).
382. A. A. Kurganov, L. Y. Zhuchkova, and V. A. Davankov, *Makromol. Chem., 180*: 2101 (1979).
383. M. Caude and A. Foucault, *Anal. Chem., 51*: 459 (1979).
384. F. Guyon, A. Foucault, and M. Caude, *J. Chromatogr., 186*: 677 (1979).
385. E. Rothenbühler, R. Waibel, and J. Solms, *Anal. Biochem., 97*: 367 (1979).
386. D. Muller, J. Jozefonvicz, and M. A. Petit, *J. Inorg. Nucl. Chem., 42*: 1083 (1980).
387. B. Sampson and G. B. Barlow, *J. Chromatogr., 183*: 9 (1980).
388. J. Jozefonvicz, D. Muller, and M. A. Petit, *J. Chem. Soc. Dalton, Trans.,* 76 (1980).
389. R. Maurer, *J. Biochem. Biophys. Methods, 2*: 183 (1980).
390. M. L. Antonelli, R. Bucci, and V. Carunchio, *J. Liq. Chromatogr., 3*: 885 (1980).
391. J. Boué, R. Audebert, and C. Quiveron, *J. Chromatogr., 204*: 185 (1981).
392. Y. A. Zolotarev, N. F. Myasoedov, V. I. Penkina, O. R. Petrenik, and V. A. Davankov, *J. Chromatogr., 207*: 63 (1981).
393. Y. A. Zolotarev, N. F. Myasoedov, V. I. Penkina, I. N. Dostovalov, O. R. Petrenik, and V. A. Davankov, *J. Chromatogr., 207*: 231 (1981).
394. W. Szczepaniak and W. Ciszewska, *Chromatographia, 15*: 38 (1982).
395. M. Gimpel and K. K. Unger, *Chromatographia, 16*: 117 (1982).
396. G. J. Krauss, *Z. Anal. Chem., 317*: 676 (1984).
397. F. Guyon, M. Caude, and R. Rosset, *Analusis, 12*: 321 (1984).
398. H. Takayanagi, O. Hatano, K. Fujimura, and T. Ando, *Anal. Chem., 57*: 1840 (1985).
399. D. Charmot, R. Audebert, and C. Quiveron, *Polym. Lett., 24*: 59 (1986).
400. H. Takayanagi, H. Tokuda, H. Uehira, K. Fujimura, and T. Ando, *J. Chromatogr., 356*: 15 (1986).
401. G. J. Krauss, *J. High Resolut. Chromatogr. Chromatogr. Commun., 9,* 419 (1986).
402. B. Monjon and J. Solms, *Anal. Biochem., 160*: 88 (1987).
403. D. K. Singh and A. Dabari, *Chromatographia, 23*: 93 (1987).
404. P. E. Hare and E. Gil-Av, *Science, 204*: 1226 (1979).
405. C. Gilon, R. Leshem, Y. Tapuhi, and E. Grushka, *J. Am. Chem. Soc., 101*: 7612 (1979).
406. C. Gilon, R. Leshem, and E. Grushka, *Anal. Chem., 52*: 1206 (1980).
407. E. Gil-Av, A. Tishbee, and P. E. Hare, *J. Am. Chem. Soc., 102*: 5115 (1980).
408. E. Oelrich, A. Preusch, and E. Wilhelm, *J. High Resolut. Chromatogr. Chromatogr. Commun., 3*: 269 (1980).
409. S. Lam and F. Chow, *J. Liq. Chromatogr., 3*: 1579 (1980).
410. C. Gilon, R. Leshem, and E. Grushka, *J. Chromatogr., 203*: 365 (1981).
411. A. A. Kurganov and V. A. Kavankov, *J. Chromatogr., 218*: 559 (1981).
412. C. C. T. Chinnick, *Analyst, 106*: 1204 (1981).
413. S. Lam, *J. Chromatogr., 234*: 485 (1982).
414. E. Grushka, S. Levin, and C. Gilon, *J. Chromatogr., 235*: 401 (1982).
415. S. Lam and A. Karmen, *J. Chromatogr., 239*: 451 (1982).
416. S. Weinstein, *Angew. Chem. Int. Ed. Engl., 21*: 218 (1982).

417. A. A. Kurganov, A. B. Tevlin, and V. A. Davankov, *J. Chromatogr., 261*: 223 (1983).
418. B. Feibush, M. J. Cohen, and B. L. Karger, *J. Chromatogr., 282*: 3 (1983).
419. S. Lam and A. Karmen, *J. Chromatogr., 289*: 339 (1984).
420. E. Busker and J. Martens, *Z. Anal. Chem., 319*: 907 (1984).
421. L. R. Gelber, B. L. Karger, J. L. Neumeyer, and B. Feibush, *J. Am. Chem. Soc., 106*: 7729 (1984).
422. M. Engel and S. A. Macko, *Anal. Chem., 56*: 2598 (1984).
423. R. Marchelli, A. Dossena, G. Casnati, F. Dallavalle, and S. Weinstein, *Angew. Chem. Int. Ed. Engl., 24*: 336 (1985).
424. R. Wernicke, *J. Chromatogr. Sci., 23*: 39 (1985).
425. S. Levin and E. Grushka, *Anal. Chem., 57*: 1830 (1985).
426. S. H. Lee, J. W. Ryu, and K. S. Park, *Bull. Korean Chem. Soc., 7*: 45 91986).
427. J. M. Broge and D. L. Leussing, *Anal. Chem., 58*: 2237 (1986).
428. J. Wagner, C. Gaget, B. Heintzelmann, and E. Wolf, *Anal. Biochem., 164*: 102 (1987).
429. J. Gaal and J. Inczedy, *J. Chromatogr., 102*: 375 (1974).
430. M. Doury-Berthod and C. Pointrenaud, B. Tremillon, *J. Chromatogr., 131*: 73 (1977).
431. R. G. Masters and D. E. Leyden, *Anal. Chim. Acta, 98*: 9 (1978).
432. F. K. Chow and E. Grushka, *Anal. Chem., 50*: 1346 (1978).
433. H. Engelhardt and S. Kromidas, *Naturwissenschaften, 67*: 353 (1980).
434. W. Linder, *Naturwissenschaften, 67*: 354 (1980).
435. V. A. Davankov and A. A. Kurganov, *Chromatographia, 13*: 339 (1980).
436. I. A. Yamskov, B. B. Berezin, V. A. Davankov, Yu. A. Zolotarev, I. N. Dostavolov, and N. F. Myasoedov, *J. Chromatogr., 217*: 539 (1981).
437. P. Roumeliotis, K. K. Unger, A. A. Kurganov, and V. A. Davankov, *Angew. Chem. Int. Ed. Engl., 21*: 930 (1982).
438. G. Gubitz, W. Jellenz, and W. Santi, *J. Liq. Chromatogr., 4*: 701 (1981).
439. N. Watanabe, *J. Chromatogr., 260*: 75 (1983).
440. M. Gimpel and K. K. Unger, *Chromatographia, 17*: 200 (1983).
441. A. A. Kurganov, A. B. Tevlin, and V. A. Davankov, *J. Chromatogr., 261*: 223 (1983).
442. B. B. Berezin, I. A. Yamskov, and V. A. Davankov, *J. Chromatogr., 261*: 301 (1983).
443. J. L. Leonard, F. Guyon, and P. Fabiani, *Chromatographia, 18*: 600 (1984).
444. A. Foucault and R. Rosset, *J. Chromatogr., 317*: 41 (1984).
445. H. Engelhardt, T. Konig, and S. Kromidas, *Chromatographia, 21*: 205 (1986).
446. F. Tsvetkov and U. Mingelgrin, *Clays Clay Miner., 23*: 391 (1987).
447. C. Corradini, F. Federici, M. Sinibaldi, and A. Messina, *Chromatographia, 23*: 118 (1987).
448. Y. Yuki, K. Saigo, H. Kimoto, K. Tachibana, and M. Hasegawa, *J. Chromatogr., 400*: 65 (1987).
449. V. A. Davankov, A. S. Bochkov, and Yu. P. Belov, *J. Chromatogr., 218*: 547 (1981).
450. H. van Duin, *Neth. Milk Dairy J., 12*: 74 (1958).
451. W. Kemula and D. Sybilska, *nature, 185*: 237 (1960).
452. W. L. Orr, *Anal. Chem., 39*: 1163 (1967).
453. W. Kemula, D. Sybilska, and K. Duszczyk, *Microchem. J., 11*: 296 (1966).
454. W. Kemula and D. Sybilska, *Anal. Chim. Acta, 38*: 97 (1967).
455. W. L. Orr, *Anal. Chem., 39*: 1163 (1967).
456. W. Kemula, Z. Borkoowska, and D. Sybilska, *Monatsh. Chem., 103*: 860 (1972).

457. D. C. Locke, J. J. Schermund, and B. Banner, *Anal. Chem., 44*: 90 (1972).
458. R. L. Spraggins, *J. Org. Chem., 38*: 3661 (1973).
459. D. Kunzru and R. W. Frei, *J. Chromatogr. Sci., 12*: 191 (1974).
460. R. A. A. Muzzarelli, A. F. Martelli, and O. Tubertini, *Analyst, 94*: 616 (1969).
461. C. R. Vogt, T. R. Ryan, and J. S. Baxter, *J. Chromatogr., 136*: 221 (1977).
462. M. Pawloowska, D. Sybilska, and J. Lipkowski, *J. chromatogr., 176*: 1 (1979).
463. M. Pawloowska, D. Sybilska, and J. Lipkowski, *J. Chromatogr., 176*: 43 (1979).
464. J. Kohler and G. Schomburg, *Chromatographia, 14*: 559 (1981).
465. J. Chielowiec, *J. Chromatogr. Sci., 19*: 296 (1981).
466. J. W. Vogh and J. E. Dooley, *Anal. Chem., 47*: 816 (1975).
467. M. Nishioka, R. M. Campbell, M. L. Lee, and R. N. Castle, *Fuel, 65*: 270 (1986).
468. J. T. Anderson, *Anal. Chem., 59*: 2207 (1987).
469. S. Lam, F. Chow, and A. Karmen, *J. Chromatogr., 199*: 295 (1980).
470. S. Weinstein, M. H. Engel, and P. E. Hare, *Anal. Biochem., 121*: 370 (1982).
471. J. J. Latterell and H. F. Wlaton, *Anal. Chim. Acta, 32*: 101 (1965).
472. A. G. Hill, R. Sedgley, and H. F. Walton, *Anal. Chim. Acta, 33*: 84 (1965).
473. K. Shimomura, L. Dickson, and H. F. Walton, *Anal. Chim. Acta, 37*: 102 (1967).
474. L. R. Snyder, *Anal. Chem., 41*: 314 (1969).
475. H. F. Humbel, D. Vonderschmitt, and K. Bernauer, *Helv. Chim. Acta, 53*: 1983 (1970).
476. K. Fujimura, M. Matsubaraz, and W. Funasaka, *J. Chromatogr., 59*: 383 (1971).
477. F. W. Wagner and R. L. Liliedahl, *J. Chromatogr., 71*: 567 (1972).
478. C. M. de Hernandez and H. F. Walton, *Anal. Chem., 44*: 890 (1972).
479. W. Funasaka, T. Hanai, K. Fujimura, and T. Ando, *J. Chromatogr., 78*: 424 (1973).
480. K. Fujimura, T. Koyama, T. Tanigawa, and W. Funasaka, *J. Chromatogr., 85*: 101 (1973).
481. K. Shimomura, T. J. Hsu, and H. F. Walton, *Anal. Chem., 45*: 501 (1973).
482. B. Hemmasi, *J. Chromatogr., 104*: 367 (1975).
483. R. Bedetti, V. Carunchio, and A. Marino, *J. Chromatogr., 95*: 127 (1974).
484. J. D. Navratil, E. Murgia, and H. F. Walton, *Anal. Chem., 47*: 122 (1975).
485. J. D. Navratil and H. F. Walton, *Anal. Chem., 47*: 2443 (1975).
486. L. A. Sternson and W. J. De Witte, *J. Chromatogr., 137*: 305 (1977).
487. N. H. C. Cooke, R. L. Viavattenne, R. Eksteen, W. S. Wrong, G. Davies, and B. L. Karger, *J. Chromatogr., 149*: 391 (1978).
488. N. Spassky, M. Reix, J. P. Guette, M. Guette, M. O. Sepulchre, and J. M. Blanchard, *C. R. Acad. Sci. Paris, 287*: 589 (1978).
489. W. Lindner, J. N. Le Page, G. Davies, D. E. Seitz, and B. L. Karger, *J. Chromatogr., 185*: 323 (1979).
490. K. F. Chow and E. Grushka, *J. Chromatogr., 185*: 361 (1979).
491. J. N. Le Page, W. Lindner, G. Davies, D. E. Seitz, and B. L. Karger, *Anal. Chem., 51*: 433 (1979).
492. Y. Tapuki, N. Miller, and B. L. Karger, *J. Chromatogr., 205*: 325 (1981).
493. B. M. Petronio, A. Lagana, and M. V. Russo, *Talanta, 28*: 215 (1981).
494. V. Carunchio, A. Messina, M. Sinibaldi, and C. Corradini, *J. Liq. Chromatogr., 5*: 819 (1982).
495. A. Yamagishi and R. Ohnishi, *J. Chromatogr., 245*: 213 (1982).
496. A. Yamagishi and R. ohnishi, *Inorg. Chem., 21*: 4233 (1982).

497. A. Yamagishi, R. Ohnishi, and M. Soma, *Chem. Lett., 85* (1982).
498. B. M. Petronio, E. De Caris, and L. Iannuzzi, *Talanta, 29*: 691 (1982).
499. K. D. Gundermann, H. P. Ansteeg, and A. Glitsch, *Proc. Int. Conf. Coal Sci.,* 631 (1983).
500. A. Yamagishi and R. Ohnishi, *Angew. Chem. Int. Ed. Engl., 22*: 162 (1983).
501. A. Yamagishi, *J. Chromatogr., 262*: 41 (1983).
502. G. J. Shahwan and J. R. Jezorek, *J. Chromatogr., 256*: 39 (1983).
503. M. Sinibaldi, V. Carunchio, A. Messina, and C. Corradini, *Ann. Chim., 74*: 175 (1984).
504. J. Mastowska and W. Pietek, *Chromatographia, 20*: 46 (1985).
505. F. Helfferich, *J. Am. Chem. Soc., 84*: 3237 (1962).
506. F. Helfferich, *J. Am. Chem. Soc., 84*: 3242 (1962).
507. J. Gaal and J. Inczedy, *Talanta, 23*: 78 (1976).
508. L. A. Sternson and W. J. Dewitte, *J. Chromatogr., 138*: 229 (1977).
509. A. Yamagishi, *J. Chem. Soc., Chem. Commun.,* 1168 (1981).
510. A. Yamagishi, *Inorg. Chem., 21*: 3393 (1982).
511. A. Yamagishi, *J. Chromatogr., 319*: 299 (1985).
512. W. W. Christie, *J. High Resolut. Chromatogr. Chromatogr. Commun., 10*: 148 (1987).
513. G. Gubitz, W. Jellenz, G. Löfler, and W. Santi, *J. High Resolut. Chromatogr. Chromatogr. Commun., 2*: 145 (1979).
514. A. Foucault, M. Caude, and L. Oliveros, *J. Chromatogr., 185*: 345 (1979).
515. K. Sugden, C. Hunter, and G. Lloyd-Jones, *J. Chromatogr., 192*: 228 (1980).
516. V. A. Davankov, A. S. Bochkov, A. A. Kurganov, P. Roumeliotis, and K. K. Unger, *Chromatographia, 13*: 677 (1980).
517. P. Roumeliotis, K. K. Unger, A. A. Kurganov, and V. A. Davankov, *J. Chromatogr., 255*: 51 (1983).
518. P. Roumeliotis, A. A. Kurganov, and V. A. Davankov, *J. Chromatogr., 266*: 439 (1983).
519. I. Benecke, *J. Chromatogr., 291*: 155 (1984).
520. M. Goldewicz, *Nature, 164*: 1132 (1949).
521. N. P. Buu-Hoï and P. Jacquignon, *Experiencia, 13*: 375 (1957).
522. L. H. Klemm, D. Reed, and C. D. Lind, *J. Org. Chem., 22*: 739 (1957).
523. E. Sondheimer and I. E. Pollak, *J. Chromatogr., 8*: 413 (1962).
524. J. T. Ayres and C. K. Mann, *Anal. Chem., 38*: 859 (1966).
525. M. S. Newman and H. Junjappa, *J. Org. Chem., 36*: 2606 (1971).
526. S. Ray and R. W. Frey, *J. Chromatogr., 71*: 451 (1972).
527. B. L. Karger, M. Martin, J. Loheac, and G. Guiochon, *Anal. Chem., 45*: 496 (1973).
528. C. H. Lochmüller and C. W. Amoss, *J. Chromatogr., 108*: 85 (1975).
529. D. W. Jewell, *Anal. Chem., 47*: 2048 (1975).
530. L. Nondek and J. Malek, *J. Chromatogr., 155*: 187 (1978).
531. J. Langer and Z. Kornetka, *Pol. J. Chem., 6*: 1303 (1978).
532. J. M. Egly, *FEBS Lett., 93*: 369 (1978).
533. S. A. Matlin, J. S. Tinker, A. Tito-Lioret, W. Lough, L. Chan, and D. G. Bryan, *Proc. Anal. Div. Chem. Soc., 16*: 354 (1979).
534. K. Szyfter and I. Langer, *J. Chromatogr., 175*: 189 (1979).
535. C. H. Lochmüller, R. R. Ryall, and C. W. Amoss, *J. Chromatogr., 178*: 298 (1979).
536. H. Hemstsberger, H. Klar, and H. Ricken, *Chromatographia, 13*: 277 (1980).

537. S. A. Matlin, W. Lough, and D. G. Bryan, *J. High Resolut. Chromatogr. Chromatogr. Commun., 3*: 33 (1980).
538. T. H. Mourey, S. Siggia, P. C. Uden, and R. J. Crosley, *Anal. Chem., 52*: 885 (1980).
539. W. Holstein, *Chromatographia, 14*: 468 (1981).
540. J. J. Burger and E. Tomlinson, *Anal. Proc., 126 (1982)*.
541. K. Ishihara, T. Iida, N. Muramoto, and I. Shinohara, *J. Chromatogr., 250*: 119 (1982).
542. W. E. Hammers, A. G. M. Theeuwes, W. K. Brederode, and C. L. de Ligny, *J. Chromatogr., 234*: 321 (1982).
543. G. Eppert and I. Schinke, *J. Chromatogr., 260*: 305 (1983).
544. G. Felix and C. Bertrand, *J. High Resolut. Chromatogr. Chromatogr. Commun., 7*: 160 (1984).
545. G. Felix and C. Bertrand, and A. Fevrier, *J. Liq. Chromatogr., 7*: 2383 (1984).
546. G. Felix and C. Bertrand, *J. High Resolut. Chromatogr. Chromatogr. Commun., 7*: 714 (1984).
547. L. Nondek and V. Chvalovsky, *J. Chromatogr., 312*: 303 (1984).
548. A. J. Repta, L. A. Sterson, K. A. Mareish, and N. M. Meltzer, *Int. J. Pharmacol., 18*: 277 (1984).
549. J. S. Thomson and J. W. Reynolds, *Anal. Chem., 56*: 2434 (1984).
550. G. Felix and C. Bertrand, *J. High Resolut. Chromatogr. Chromatogr. Commun., 8*: 362 (1985).
551. G. Felix and C. Bertrand, and F. Van Gastel, *Chromatographia, 20*: 155 (1985).
552. G. Felix and C. Bertrand, *J. Chromatogr., 319*: 432 91985).
553. D. Y. Pharr, P. C. Uden, and S. Siggia, *J. Chromatogr. Sci., 23*: 391 (1985).
554. L. Nondek, *J. High Resolut. Chromatogr. Chromatogr. Commun., 8*: 302 (1985).
555. J. Köhler, *Chromatographia, 21*: 573 (1986).
556. K. J. Welch and N. E. Hoffman, *J. High Resolut. Chromatogr. Chromatogr. Commun., 9*: 417 (1986).
557. M. Apfel, L. A. Sterson, A. J. Repta, and D. Mills, *J. Pharmacol. Biomed. Anal., 5*: 469 (1987).
558. G. Felix and C. Bertrand, *J. High Resolut. Chromatogr. Chromatogr. Commun., 10*: 411 (1987).
559. G. Felix, A. Thienpont, and C. Bertrand, *Chromatographia, 23*: 684 (1987).
560. P. Jadaud, M. Caude, and R. Rosset, *J. Chromatogr., 393*: 39 (1987).
561. T. H. Mourey and S. Siggia, *Anal. Chem., 51*: 763 (1979).
562. C. H. Lochmüller, M. L. Hunnicutt, and R. W. Beaver, *J. Chromatogr. Sci., 21*: 444 (1983).
563. L. H. Klemm and D. Reed, *J. Chromatogr., 3*: 364 (1960).
564. W. Rebafka and H. A. Staab, *Angew. Chem. Int. Ed. Engl., 12*: 776 (1973).
565. F. Mikes, G. Boshart, and E. Gil-Av, *J. Chem. Soc. Chem. Commun.,* 99 (1976).
566. B. Feringa and H. Wynberg, *J. Am. Chem. Soc., 99*: 602 (1977).
567. Y. H. Kim, A. Tishbee, and E. Gil-Av, *J. Am. Chem. Soc., 102*: 5915 (1980).
568. W. H. Pirkle and D. W. House, *J. Org. Chem., 44*: 1957 (1979).
569. W. H. Pirkle, D. W. House, and J. M. Finn, *J. Chromatogr., 192*: 143 (1980).
570. W. H. Pirkle and J. L. Schreiner, *J. Org. Chem., 46*: 4988 (1981).
571. Y. M. Kim, A. Tishbee, and E. Gil-Av, *J. Chem. Soc. Chem. Commun.,* 75 (1981).
572. W. H. Pirkle and J. M. Finn, *J. Org. Chem., 47*: 4037 (1982).

573. M. Oi, M. Nagase, and T. Doi, *J. Chromatogr.*, *257*: 111 (1983).

574. N. Oi, M. Nagase, Y. Inda, and T. Doi, *J. Chromatogr.*, *265*: 111 (1983).

575. N. Oi, M. Nagase, and Y. Sawada, *J. Chromatogr.*, *292*: 427 (1984).

576. R. Däppen, V. R. Meyer, and H. Arm, *J. Chromatogr.*, *295*: 367 (1984).

577. W. H. Pirkle and C. J. Welch, *J. Org. Chem.*, *49*: 138 (1984).

578. W. H. Pirkle, M. H. Hyun, and B. Bank, *J. Chromatogr.*, *316*: 585 (1984).

579. M. Sinibaldi, V. Carunchio, C. Corradini, and A. M. Girelli, *Chromatographia*, *18*: 459 (1984).

580. S. A. Matlin and R. Zhou, *J. High Resolut. Chromatogr. Chromatogr. Commun.*, *7*: 629 (1984).

581. W. H. Pirkle and M. H. Hyun, *J. Chromatogr.*, *322*: 287 (1985).

582. W. H. Pirkle and M. H. Hyun, *J. Chromatogr.*, *322*: 295 (1985).

583. W. H. Pirkle and M. H. Hyun, *J. Chromatogr.*, *322*: 309 (1985).

584. C. Rosini, P. Altemura, D. Pini, C. Bertucci, G. Zullino, and P. Salvadori, *J. Chromatogr.*, *348*: 79 (1985).

585. H. Berndt and G. Krüger, *J. Chromatogr.*, *348*: 275 (1985).

586. M. J. B. Lloyd, *J. Chromatogr.*, *351*: 219 (1986).

587. R. Däppen, V. R. Meyer, and H. Arm, *J. Chromatogr.*, *361*: 93 (1986).

588. W. H. Pirkle and Th. C. Pochapsky, *J. Am. Chem. Soc.*, *108*: 352 (1986).

589. N. Oi and H. Kitahara, *J. Liq. Chromatogr.*, *9*: 511 (1986).

590. K. K. Iwaki, S. Yoshida, N. Nimura, T. Kinoshita, K. Takeda, and H. Ogura, *J. Chromatogr.*, *404*: 117 (1987).

591. K. K. Iwaki, S. Yoshida, N. Nimura, T. Kinoshita, K. Takeda, and H. Ogura, *Chromatographia*, *23*: 727 (1987).

592. S. A. Matlin, A. Tito-Lloret, W. J. Lough, D. G. Bryan, T. Browne, and S. Mehani, *J. High Resolut. Chromatogr. Chromatogr. Commun.*, *4*: 81 (1981).

593. C. H. Lochmüller and R. R. Ryall, *J. Chromatogr.*, *150*: 511 (1978).

594. C. Rosini, C. Bertucci, D. Pini, P. Altemura, and P. Salvadori, *Tetrahedron Lett.*, 3361 (1985).

595. F. Mikes, G. Boshart, and E. Gil-Av, *J. Chromatogr.*, *122*: 205 (1978).

596. L. H. Klemm, K. B. Desai, and J. R. Spooner, *J. Chromatogr.*, *14*: 302 (1964).

3

Charge-Transfer Chromatography: Application to the Determination of Polycyclic Aromatic Compounds, Aromatic Amines and Azaarenes, and Biological Compounds

D. Cagniant
University of Metz, Metz, France

I. INTRODUCTION

The theoretical aspects of charge-transfer complexation (CTC) applied to chromatography were presented in Chapter 1. The main practical developments of chromatographic separations and identifications based on the formation of donor–acceptor complexes are the subject of Chapter 3.

Important classes of organic compounds can act as electron donors or electron acceptors: polycyclic aromatic hydrocarbons, polycyclic heterocyclic compounds (principally sulfur and nitrogen heterocycles), and biologically active molecules. In some cases of large biological compounds, a part of the molecule can behave as an electron donor and another part as an electron acceptor. These compounds are formed from both natural and industrial sources. Concerning polycyclic aromatic compounds (PACs)—hydrocarbons or heterocycles—the major sources lie in the world's industrial development. Until the beginning of the century there existed a natural balance between the production and natural degradation of PACs, which kept the background concentration low and fixed. But with increasing industrial development throughout the world, this natural balance has been disturbed and the amount of PACs is constantly rising.

Evaluation of the various classes of PACs is of interest in many fields of industrial chemistry, principally in the petroleum and coal industries. Furthermore, they are ubiquitous environmental pollutants, and many could have carcinogenic and/or mutagenic activities. Consequently, many studies have been undertaken to characterize PACs, often at very low levels, in complex environmental samples (airborne particulate matter, industrial effluents, foods, tobacco smoke, etc.). Owing to the complexity of the samples submitted to analysis, many of the procedures were carried out in a first step on model compounds. Their application to punctual industrial problems involving petroleum or coal products is considered in Chapter 6.

Another interesting point concerns biologically active molecules. During the last 10 years, many studies have been carried out dealing with the problems encountered in the analytical field of amino acids, proteins, nucleosides, nucleotides, and oligonucleotides as well as pharmaceutical drugs. At the same time, great improvements have been made in analytical chemistry, particularly in chromatographic technology (i.e., development of high-performance liquid chromatography since the 1970s, synthesis of new phases, perfecting of new detectors). Therefore, a great many analytical methods have been developed in the past 30 years and their number continues to increase. Our purpose is to show how charge-transfer chromatography contributes to the analysis of PACs and a number of biological compounds.

II. POLYCYCLIC AROMATIC AND HYDROAROMATIC HYDROCARBONS

A. Charge-Transfer Thin-Layer Chromatography

The separation and identification of polycyclic aromatic hydrocarbons (PAHs) from complex mixtures by means of thin-layer chromatography (TLC) is well known. As early as 1963, the R_f values of a number of hydrocarbons (and of some heterocyclic compounds) were reported by Kucharczyk et al. [1] with silica gel and alumina as adsorbents. However, the R_f values depend on experimental conditions, and the separation of PAHs is poor. The same observation results from the work of Harvey and Halonen [2] on hydroaromatic compounds. In all these experiments the spots are detected using spray reagents: tetracyanoethylene (TCNE) [1] and 2,4,7-trinitrofluorenone (TNF) [2]. The wide variation of colors exhibited by complexes of PAHs with TNF suggested a similar variation in the free energy of the charge-transfer (CT) complex formation. The idea that migration of aromatic hydrocarbons over impregnated silica gel adsorbent could be selectively retarded to a degree depending on the strength of the CT complex formed is the starting point of development of TLC by CTC [2]. The results reported in Table 1 confirm the efficiency of the technique, which has been described by various authors [2–6] between 1964 and 1966 and reviewed in 1969 by Foster [7]. In general, the methods were applied to synthetic mixtures of various aromatic hydrocarbons, ranging from alkyl benzenes and biphenyl to coronene.

Possible applications to the detection of PAHs in complex mixtures as encoun-

TABLE 1 Comparison of R_f values of PAHs and Some Hydroaromatic Derivatives on Silica Gel Plates With and Without Impregnation with TNF[a]

	$R_f \times 100$	
Hydrocarbon PAH	Silica gel	5% TNF
Phenanthrene[b]	48	14
Dihydro-	49	18
Tetrahydro-	53	27
Hexahydro-	54	44
Octahydro-	55	54
Chrysene[b]	35	1
Dihydro-	36	16
Tetrahydro-	36	29
Hexahydro-	38	34
Anthracene[b]	45	15
Dihydro-	43	37
Tetrahydro-	50	43
Hexahydro-	53	51

Source: Ref. 2.
[a]Solvent benzene/heptane 1:4.
[b]The reduced isomers are those with hydrogen: (1) at positions 9,10, 1,2,3,4, 1,2,3,4,9,10, and 1,2,3,4,4a,9,10,10a, respectively, for phenanthrene; (2) at positions 5,6, 5,6,11,12, and 4b,5,6,10b,11,12, respectively, for chrysene; (3) at positions 9,10, 1,4,9,10, and 1,4,5,8,9,10, respectively for anthracene.

tered, for example, in air pollution studies and in pyrolysis products—problems connected with environmental protection against carcinogenic PAHs—was predicted [3,4] and some experiments were described on coal tar [6].

1. Experimental Procedures Used in TLC by Charge-Transfer Complexation

Three methods have been used:

1. *Postcoating method, Method A.* A prepared plate is impregnated with a complex forming agent. One method of impregnation is by immersion [2,4]. Regular activated plates are dipped in a solution of the acceptor of known concentration for a given period and then dried. This is the method of choice since it is the most rapid and applicable to commercial plates and films. An alternative method making use of an ascending solution of the acceptor [3] is said to be less convenient, owing to the lack of control of acceptor concentration [2]. A modification of the latter procedure was described by Kessler and Müller [5]. Using commercial plates partly impregnated with a saturated solu-

tion of picric acid in benzene by the way of ascending migration at a level of only 8 cm, then dried, the authors performed TLC as usual. Finally, plates can be impregnated by spraying on a solution of the acceptor [8].

2. *Precoating method, Method B* [2,4,6,9]. A solution of the complex forming agent is added to the adsorbent during preparation of the plate, followed in some cases by activation.

3. *Addition of the donor (or acceptor) compound in the solvent of development, Method C* [3,5]. A plate already prepared without a complex forming agent is developed in a system containing the complex forming agent. This is the method used [3] for the separation of nitroaromatic compounds. The test compounds are applied on the plate and the development carried out by a solvent (cyclohexane/chloroform) saturated with anthracene.

As far as quantitative determination is concerned, some methods have been suggested in the case of PAHs, based on extraction of the spots followed by spectrophotometric determination [6]. They need to be made at $\lambda > 300$ nm to avoid interaction with caffeine. Recovery is very variable, from 80–90% to 20%, according to the hydrocarbons under consideration. In general, as photooxidation of the adsorbed compounds may occur, particularly in the case of PAHs, it is recommended that chromatograms be run in the dark. Some examples of the application of these three methods are provided in Table 2.

2. Selection of Complex Forming Agents

Aromatic hydrocarbons as donors of electrons are the main compounds capable of forming charge-transfer complexes with substances having electron acceptor properties. A wide range of electron acceptors was investigated. The best known are polynitroaromatic compounds: 2,4,7-trinitrofluorenone (TNF) [2,4], 1,3,5- trinitrobenzene (TNB) [2,3], and picric [4,5] and styphnic acids [4]. Other compounds tested were caffeine [4,6,8], urea [4], dimethylformamide (DMF) [4], silver nitrate [4], tetracyanoethylene (TCNE) [8], halogen quinones (chloranil [8,9], bromanil [9], 2,3-dichloro-5,6-dicyanoquinone (DDQ) [9]), *p*-benzoquinone [9], pyromellitic dianhydride [8,10], N-methylated cyclic ureids [8], and bile acids [8].

Among the polynitroaromatic compounds, TNF is the most valuable [4], forming strong complexes with PAHs that have at least three rings [2]. Picric acid used by the precoating method is claimed to be without value [4]. Nevertheless, separation of various PAHs, from naphthalene to perylene, was performed with picric acid using partly coated plates as described before {5]. Among the other compounds cited above, some were without effect—urea [4], DMF [4], and sodium deoxycholate [8]—or not satisfactory in TLC conditions—silver nitrate [4]. TCNE is especially effective for complexing substituted benzenes as compared to PAHs [2]. But in TLC on plates impregnated with TCNE, the R_f values of PAHs were not reproducible [8]. Pyromellitic dianhydride [8,10], tetramethyl ureic acid, and caffeine gave good results [8]. In the latter case, it is well known that the solubilizing

TABLE 2 Thin-Layer Chromatographic Systems Used for the Analysis of PAHs

PAHs	Complexing agent[a] (adsorbent)[b]	Eluant	Methods–Detection	Refs.
Phenanthrene, anthracene, chrysene, and their hydroderivatives	TNF (a)	Benzene– heptane 1:4	A	2
PAHs from biphenyl to 1,12-benzpery-lene (53 derivatives)	TNF (a)	CCl₄	B	Color of complexes
Nitro PAHs	Anthracene (b)	Anthracenic solution (cyclohexane-chloroform)	C UV (365 nm)	3
PAHs (10 compounds)	TNB (3% in benzene) (b)	Cyclohexane alone or with 5% of ethyl acetate	A Color of complexes or spraying with SbCl₅	3
PAHs, from anthracene to coronen (21 compounds)	Picric acid, styphnic acid, TNF, caffeine, urea, silver nitrate, DMF (b, c)	Light petroleum ether (bp 40-50°C) alone or as mixture with ether, pyridine, tetraline, aniline	B and A UV and color of complexes	4, 6
PAHs from biphenyl to perylene (21 compounds	Picric acid (d)	Light petroleum ether	A UV	5
Alkyl benzenes, PAHs from biphenyl to anthanthrene (24 compounds)	TMUA, caffeine, chloranil, TCNE, PMDA (a, e)	Di-n-butyl ether, tetrachloro-ethylene	A UV or spraying with a saturated chloranil solution	8
C₆H₅—(CH₂)n-C₆H₅ (n = 0, 1,2), PAHs (acenes), polyphenyls: C₆H₅—(C₆H₅)n-C₆H₅ (n = 0,1,2)	Quinones (benzo-quinone, chloranil, bromanil, DDQ), TNB (b, d)	Benzene/heptane 1:4	B UV (365 nm and 254 nm), spray-ing with naphtha-cene solution	6

[a]TNF, 2,4,7-trinitrofluorenone; TNB, 1,3,5-trinitrobenzene; DMF, dimethylformamide; TMUA, tetramethylureic acid; PMDA, pyromellitic dianhydride; DDQ, 2,3-dichloro 5,6-dicyanobenzoquinone; TCNE, tetracyanoethylene.

[b]a, silica gel; b, Kieselgel G Merck; c, Al₂O₃ Merck; d, Kieselgel GF254; e, Silifol.

TABLE 3   Effect of Acceptor Structure on R_f Values on Impregnated Silica Gel Plates

Acceptor	R_f values		
	Biphenyl	Naphthalene	Anthracene
None	0.35	0.40	0.33
Chloranil	0.37	0.40	0.32
Bromanil	0.36	0.38	0.33
DDQ	0.32	0.36	0.24
Benzoquinone	0.35	0.34	0.27

Source: Ref. 9.

effect of purines on the aromatics in aqueous solution is due to complex formation and the ability of caffeine in this respect is also well established [4]. Used in a mixture with adsorbents, caffeine has a pronounced influence on the separation of PAH mixtures. However, urea was without effect on the solubilization of PAHs and was also of no value for their separation in TLC [4].

Noteworthy is the case of benzoquinones used by Schenk et al. [9] for the study of multiple-site complexation in CTC-TLC. The effect of acceptor structures on the complexation with donor derivatives such as biphenyl, naphthalene, and anthracene can be seen from the R_f values reported in Table 3.

The evolution of steric hindrance (benzoquinone < DDQ < chloranil < bromanil) agrees with the extend of complexation and the difference in R_f values. The high π acidity of DDQ relative to benzoquinone overcomes its greater steric hindrance, which explains the variations in R_f values in these two cases.

3.   Results and Discussion

Reliable R_f values require a standardized technique. So no tables are presented here, as they would be of limited interest for identification purposes. Nevertheless, comparison of R_f values of a large variety of PAHs, on silica gel alone or impregnated with TNF, proves that the strength of the complexation is a direct function of the extend of the conjugated system—in the present case, the number of aromatic rings [2].

To evaluate the effect of adsorbent impregnation, Harvey and Halonen [2] introduced the binding constant B, defined as

$$B = \frac{\Delta R_f}{R_f} \times 100 \quad \text{with} \quad \Delta R_f = R_f - R_f'$$

where R_f and R_f' are, respectively, the values obtained on non impregnated and impregnated silica gel for the same donor molecule (PAH). The constant B provides a measure of an acceptor's efficiency for decreasing the distance traveled by the test substance ($\Delta R_f > 0$ in all cases).

All the molecular structural features [number of condensed aromatic rings, substitution by methyl groups (+I, +M), planarity], which contribute to the lowering of R_f values (from experimental results), are also factors known to facilitate complex formation. Harvey and Halonen [2] claimed that R_f values of PAHs (on TNF-coated plates) are inversely dependent on the presence of structural features that enhance CT complexation. It is the same for the smaller acceptor TNB, but more discrepancies were found with chloranil. More recently, Schenk et al. [9] considered the case with $\Delta R_f < 0$. They assumed that increasing the number of acceptor molecules on the silica gel plate reduces the total number of hydroxyl sites available per square area. In the case where the donor molecule interacts more with the hydroxyl sites than with the acceptor, the adsorption will be decreased and $R_f{'}$ values will be increased ($\Delta R_f \leq 0$). So a comparison of the B values in a series of compounds would reveal relative strengths with which the molecules bond to the surface at one or more sites, the values of B being positive, zero, or negative.

The authors [9] developed their multiple-sites complexation hypothesis by the TLC of donor molecules on benzoquinone (in these experiments about half of the surface is covered by benzoquinone). The results were discussed considering the planarity of the donors and their effective molecular area. Some results are summarized in Table 4. The negative B values reflect an interaction with benzoquinone that is weaker than the one with the silica gel hydroxyl sites, which is, in turn, a function of the planarity of the benzene rings and their orientation toward the silica gel surface.

4. Conclusions

CT–TLC had been applied to the separation and identification of PAHs. Interest in CT–TLC, well documented in the period 1964–1974, has been supplanted by more powerful modern analytical tools (particularly in the case of complex mixtures such as petroleum and coal products) as GC and HPLC, incorporating automatic

TABLE 4 TLC of Aromatic Hydrocarbons on Benzoquinone-Impregnated Silica Gel Plates

Aromatic hydrocarbons (area in Å)	R_f	R'_f	Constant B
Biphenyl (83)[a]	0.39	0.44	−13
Diphenyl methane (94)[b]	0.36	0.36	0
Dibenzyl (100)[a]	0.40	0.42	−5

Source: Ref. 9.
[a]Almost planar.
[b]Not planar.

injectors and integrators as well as coupling with mass spectrometry. Consequently, no fundamental new works appeared since 1974, as can be seen from the review of Daisey [11] on PAH analysis by complexation TLC. The lack of commercially produced plates could perhaps explain this limitation [11]. Nevertheless, the method retains the advantage of simplicity and is applicable in any laboratory without expensive equipment. It is also applicable to all types of compounds, whatever their volatility. From an analytical point of view, sample size requirements are minimal.

B. Charge-Transfer Liquid Chromatography

The formation of CT complexes had been often employed for the separation of PAHs in liquid chromatography. As early as 1957 a very simple procedure [12] was proposed for the separation of PAHs from their partially hydrogenated derivatives (i.e., pyrene from tetrahydro-, hexahydro-, and decahydropyrenes) using tetrachlorophtalic anhydride alone as adsorbent. The noncomplexed hydroaromatic derivatives were eluted with the solvent (petroleum ether or cyclohexane), and the colored complex PAHs, which remained on the adsorbent, were eluted in a basic medium. But the inconvenience of this procedure (e.g., solubility of the acceptors in the mobile phases) was pointed out by Giehr [13] using other pure classical acceptors.

The application of LC using charge-transfer complexes was developed further in two ways, using adsorbents impregnated with various electron acceptors, and more often, using stationary phases chemically bonded with electron acceptors. The subject has been reviewed by Holstein and Hemetsberger [14,15] in the general framework of donor-acceptor complex chromatography (DACC). The authors [14] classified the strategies applied in CTC in four groups, whatever the procedure (TLC, LC, HPLC) may be: (a) Acceptors are added to the mobile phase and donor compounds are separated. This case is treated in detail in Chapter 4. (b) The inverse case (donors added to the mobile phase) gave rise to few reports, owing to the low solubility of donor compounds in polar solvents. Some examples are given later in the chapter. The last two groups, dealing with CTC using (c) coated or (d) bonded adsorbents, were the only ones reviewed by the authors [14,15]. Consequently, only papers published after 1982 are considered in this chapter. Nevertheless, to provide an overview of the evolution of LC (and HPLC) chromatography, some typical older experiments are reported.

1. LC–CTC with Coated Adsorbents

Tye and Bell [16] described the formation of CT complexes between PAHs and s-trinitrobenzene (TNB) within the chromatographic column. The fractions collected were treated with aqueous NaOH to extract the TNB, and the concentration of each fraction was calculated from its UV spectrum. TNB was preferred to picric acid, which gives instable complexes with four- of five-ring PAHs, and to 2,4,7-trinitrofluorenone (TNF), which gives overly insoluble complexes.

Karger et al. [17] illustrated the combination of selective separation via CTC with the speed and efficiency possible from porous layer beads. The packing material used was Corasil I (37 to 50 μm) coated with 2,4,7-trinitrofluorenone (TNF). By adjustment of TNF percentages (0.06% on Corasil I), mobile-phase water content (water-saturated n-heptane), temperature (25°C), and velocity (2 cm/s given by a pressure of 180 atm), it was possible to optimize the column (3 m $\times$ 1.6 mm i.d.) for the rapid separation of three- to seven-fused-ring PAHs.

More recently, Wozniak and Hites [18-20] described the separation of hydroaromatics from PAHs in coal-derived liquids. A more complete presentation of this study and of the results obtained in our own laboratory by applying this procedure to hydroliquefaction and pyrolysis coal liquids appears in Chapter 6. Nevertheless, a description of the Wozniak procedure and of some results obtained by its application to model compounds [21] are given here. The separation of hydroaromatics, already tried using CTC-TLC with nonpolar solvents by Harvey and Halonen [2] for simple mixtures of a parent PAH and its hydroaromatics, was realized successfully by Wozniak using picric acid as acceptor and alumina as adsorbent. There were two reasons for this choice: the majority of hydroaromatic standards did not form picrates, and alumina cleaved the PAH/picrates during elution to give pure PAHs. The unique characteristic of the procedure is the use of dual-column, picric acid-coated alumina (15% w/w; 4 g) being packed over alumina (3 g) as shown in Scheme 1. The sample is introduced at the top of the column by coating on the picric acid/alumina adsorbent (0.2 g of sample in 2 mL of CH_2Cl_2, adsorbed on 3 g of the picric acid-coated alumina, followed by evaporation of the solvent). The eluting solvents and the corresponding eluted compounds are reported in Scheme 1.

After partial rotating evaporation followed by a stream of dry nitrogen until a remaining volume of 1 mL, the samples were analyzed by GC and GC-MS. By using such a dual column, the PAHs and polar compounds will be strongly retarded by picrate formation and the hydroaromatics slightly retained by alumina. Thus the sequence of elution will be as reported in Scheme 1. The activity of the alumina and the percent of picric acid adsorbed are the two main factors for efficiency of the separation of PAHs and hydroaromatics. A water content of 3% in alumina and the use of 15% picric acid on alumina were the optimum conditions claimed by Wozniak. From his work on model compounds, some dihydro derivatives with a high aromatic character (i.e., dihydropyrene or dihydrobenzopyrene) are eluted with PAHs.

Using almost the same conditions and a more complex model mixture, Diack et al. [21] obtained the results presented in Table 5. In order to follow the fractionation of hydroaromatics, five subfractions eluted with solvent T2 were collected T$_2^i$ (i = 1 to 5). If no hydroaromatics are eluted with mixture T3, except those with a high

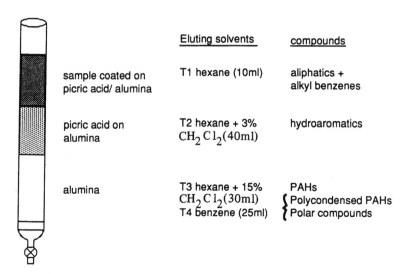

	Eluting solvents	compounds
sample coated on picric acid/ alumina	T1 hexane (10ml)	aliphatics + alkyl benzenes
picric acid on alumina	T2 hexane + 3% CH_2Cl_2(40ml)	hydroaromatics
alumina	T3 hexane + 15% CH_2Cl_2(30ml) T4 benzene (25ml)	PAHs { Polycondensed PAHs { Polar compounds

Scheme 1:   Separation hydroaromatics/aromatics on picric acid column.

TABLE 5   Separation of Model Compounds on Picric Acid–Coated Alumina

Compounds[a]	Index[b]	T_1^c	T_2^1	T_2^2	T_2^3	T_2^4	T_2^5	T_3	T_4
Paraffins (C_{19}–C_{40})	—	+	0	0	0	0	0	0	0
Tetralin	197.69	+	+	+	0	0	0	0	0
Naphthalene	200.00	0	+	+	0	0	0	0	0
OH-1-4,4a,9,10,10a Acenaphthene	254.16	0	0	+	+	+	0	0	0
OH-1-4,4a,9,9a,10 Phenanthrene	273.65	+	+	0	0	0	0	0	0
Anthracene	274.55	+	+	0	0	0	0	0	0
DH-9,10-anthracene	283.88	0	0	0	+	+	+	0	0
DH-9,10-phenanthrene	286.50	0	0	0	+	+	+	0	0
OH-1-8-anthracene	289.76	+	+	0	0	0	0	0	0
OH-1-8-phenanthrene	291.68	+	+	0	0	0	0	0	0
Phenanthrene	300.00	0	0	0	+	+	+	+	0
Anthracene	301.00	0	0	0	+	+	+	+	0
TH-4,5,9,10-pyrene	329.24	0	0	+	+	+	0	0	0
Pyrene (P)	351.50	0	0	0	0	0	0	+	+
HH-1,2,3,6,7,8,P	333.19	0	0	+	+	0	0	0	0
HH-1-6-chrysene	380.54	0	0	+	+	+	0	0	0
Chrysene	400.00	0	0	0	0	0	0	+	+

Source: Ref. 21.
[a]DH, TH, HH, OH = di, tetra, hexa, octahydro.
[b]Retention indexes determined by glass capillary column analysis.
[c]T_2^i as in text.

aromatic character, some PAHs, such as naphthalene, acenaphthene, phenanthrene, and anthracene, were eluted with the T_2 mixture. Furthermore, an increasing alkylation of PAHs (i.e., naphthalenes) decreases the ability of picrate formation or their stability. Indeed, in the case of coal hydroliquefaction liquids, the large amounts of alkylated (C_2 to C_6) two-ring PAHs, more or less complexed with picric acid, gave rise to some difficulties as they were eluted in the T_2 fraction (see Chapter 6).

In all these experiments [18-21] it is noteworthy that using alumina, picric acid was never found in the fractions collected. Nevertheless, Nondek and Malek [22] pointed out that the use of silica gel simply impregnated with electron acceptor has many disadvantages, the primary one being the solubility of the electron acceptor in the usual mobile phases. This difficulty may be overcame by chemical bonding of the electron acceptor, and this procedure was as well developed in liquid chromatography as in high-performance liquid chromatography.

2. LC–CTC with Chemically Bonded Adsorbents

As early as 1964, Ayres and Mann [23,24] used in LC a tetranitrobenzyl polystyrene resin stationary phase for PAH separation. Nondek and Malek [22] described the preparation of a grafted silica gel with 3-(2,4-dinitroanilino)propyl (DNAP) groups and its utilization in LC. The capacity factors of several model PAHs observed on DNAP-Silpearl have higher values than on untreated Silpearl, in agreement with the formation of π-π^* electron transfer complexes. Successful separations of condensed and noncondensed PAHs were obtained in all experiments. The effect of the surface concentration of chemically bonded 2,4-dinitroanilino groups on chromatographic selectivity was also examined [25]. The main characteristics of the LC–CTC examples cited above are summarized in Table 6.

As a complement to the reviews already cited [14,15], an exhaustive study of LC on chemically bonded electron donors and acceptors has been published by Nondek [26]. Besides the preparation of the stationary phases, particularly those used in HPLC, their efficiency was described together with the chromatographic selectivity, temperature, and solvent effects (see Chapter 1).

C. Charge-Transfer High-Pressure Liquid Chromatography

With the appearance of HPLC techniques, bonded silica with electron acceptors or electron donors was intensively developed, particularly in the field of PAHs and their separation in complex mixtures such as petroleum and coal products (see Chapter 6).

Methods developed before 1982 (Table 7). According to extensive studies carried out in the field of CTC in the HPLC mode, since the first example of a chemically bonded stationary phase due to Lochmüller and Amoss [27] in 1975,

TABLE 6 Separation of PAHs by Charge-Transfer Liquid Chromatography

Compounds	Adsorbents	Refs.
Mixture of a given PAH and its hydro derivatives	Powder of tetrachlorophtalic anhydride or other acceptors	12, 13
Fluoranthene/pyrene, petroleum products	Columpack coated with TNB in Carbowax 400	16
PAHs from three to seven fused rings	Corasil I (37–50 μm) coated with TNF	17

Separation of model PAHs from their hydro derivatives, application to coal liquids	Dual adsorbent
	    Alumina coated with picric acid
	    Alumina alone
		18–20, 21

Anthracene/pyrene	Tetranitrobenzyl-bonded polystyrene resin	23, 24
Separation condensed/ noncondensed PAHs	Silica gel (Silpearl 30–40 μm) bonded with DNAP	22, 25

$$\equiv Si-(CH_2)_3-NH-\!\!\!\left\langle\!\!\bigcirc\!\!\right\rangle\!\!-NO_2$$
$$NO_2$$

(DNAP = 2,4-dinitroanilinopropyl)

most of the methods published before 1982 have only been summarized in Table 7, as they have already been reviewed [14,15,26].

Besides the interest in new phase preparation and in the evaluation of their efficiency by comparison with more classical phases [30,34,36–38], the separation and determination of a number of PAHs and of their alkyl derivatives in environmental samples, because of their carcinogenic properties, was another motivation of the studies [27,38]. The silica-bonded groups reported in Table 7 belong to nitro derivatives, except for the phtalimidopropyl groups. Their efficiency and selectivity increase with the number of aromatic rings, in the order aminopropyl < picramidopropyl < trinitrofluorenimine.

Methods developed since 1982. These can be divided into two parts. The first is related to bonded nitroderivatives (summarized in Table 8). Owing to the similarity of CT complex interactions, the subject of chemically bonded phases with donor groups applied to the separation of nitro derivatives are also reviewed in this first part (summarized in Table 10). The second part is related to other π-electron acceptors, such as phtalimidopropyl groups and caffeine, for the main studies (summarized in Table 12).

TABLE 7 Separation of PAHs by Charge-Transfer/High-Pressure Liquid Chromatography[a]

Compounds	Adsorbents[b]	Refs.
Separation of isomeric PAHs and alkyl derivatives	Microporous silica gel (Lichrosorb SI 60) grafted with (a) 3-(2,4,5,7-tetranitrofluorenimino)-propyldiethoxysilane group 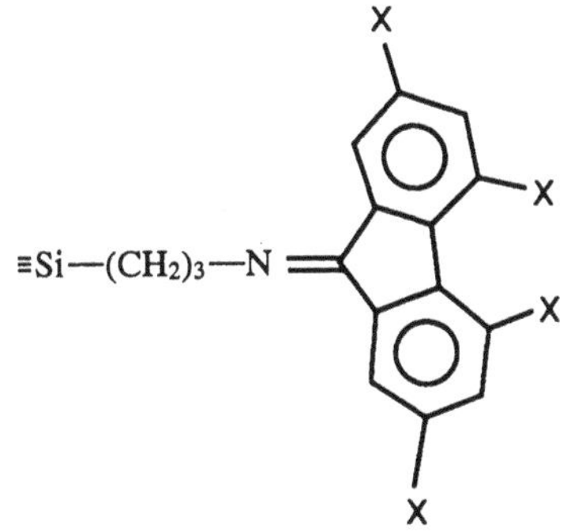	27

$\equiv$Si—(CH$_2$)$_3$—N

| | (b) Other polynitrofluorenimino groups | 28 |
| Acenaphtene/acenaphtylene, alkylbenzenes | Silica (10 μm) grafted with O-propyl-2,4,5,7-tetranitrofluorenone oxime 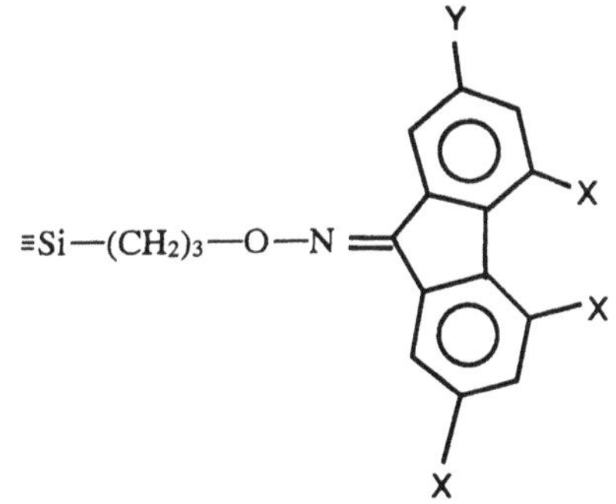 | 29, 30 |

$\equiv$Si—(CH$_2$)$_3$—O—N

| Model PAHs | Lichrosorb Si 60 (8.3–11.5 μm) grafted with 3-(2,4-dinitroanilino)-propyl groups | 31 |

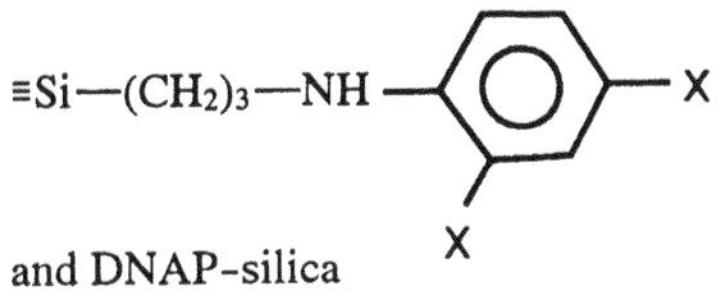

$\equiv$Si—(CH$_2$)$_3$—NH

and DNAP-silica

(Continued)

TABLE 7   (Continued)

Compounds	Adsorbents[b]	Refs.
Model PAHs	Picramidopropylsilica	32, 33

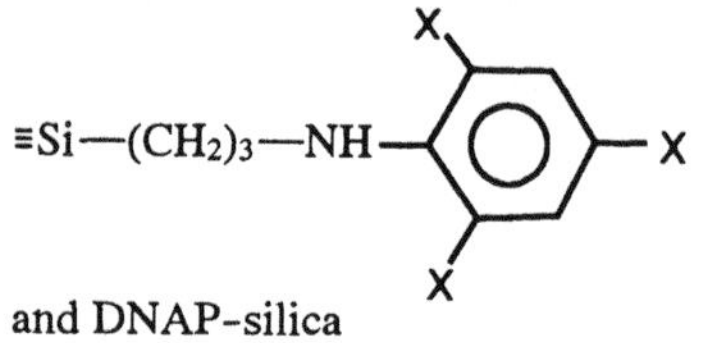

$$\equiv Si-(CH_2)_3-NH-\underset{X}{\overset{X}{\bigcirc}}-X$$

	and DNAP-silica	
Model PAHs, alkyl pyrenes, coal products	Silica grafted with phtalimido groups	34–37

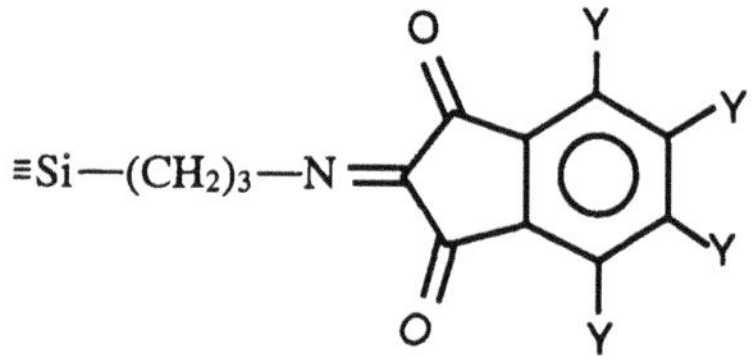

$$\equiv Si-(CH_2)_3-N=\underset{O}{\overset{O}{\bigcirc}}\overset{Y}{\underset{Y}{\bigcirc}}Y$$

	Y = H, Y = Cl (TCI)[c]	
Model PAHs, dust sample extracts	Nucleosil-5 NO₂	38

[a]Methods published before 1982.
[b]In all cases, X = NO2.
[c]Some examples of the application of TCI phase in coal product analysis are provided in Chapter 6.

1.   Chemically Bonded Phases with Nitro Groups (Table 8)

An exhaustive study of chemically bonded phases of the general structure (*1*)

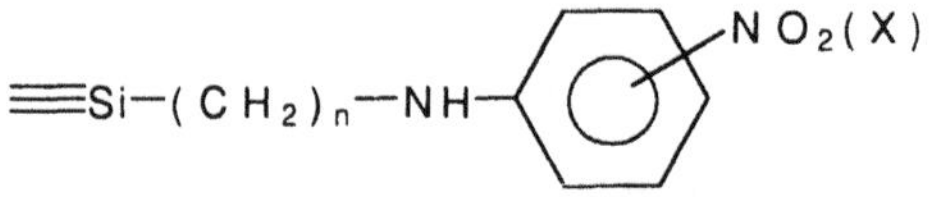

$$\equiv Si-(CH_2)_n-NH-\bigcirc-NO_2(X)$$

with n = 3 and 8
x = (2,4) and (2,4,7)   (*1*)

was made by Grizzle and Thomson [39] and developed further by Thomson and Reynolds [40]. They reported a detailed comparison between these phases for the separation of 86 model compounds, with from one to six aromatic rings, either expected or known to be present in petroleum, coal liquids, and shale oils.

    As gradient elutions or isocratic conditions were used according to the columns, the retention indexes were calculated to obtain comparable results with the following two equations:

$$I_X = I_N + \frac{t'_X - t'_N}{t'_{N+1} - t'_N} \quad \text{(gradient)} \tag{1}$$

$$I_X = \log I_N + \log \frac{t'_X - t'_N}{t'_{N+1} - t'_N} \quad \text{(isocratic)} \tag{2}$$

where I_X is the retention index for compound X compared to standard compounds N and N + 1, and t' corresponds to the corrected retention volumes. For example, the standards used and their I_N values [Eq. (1)] are benzene, 1; naphthalene, 2; phenanthrene, 3; and benzo[a]anthracene, 4. With Eq. (2) these same compounds would have I_N values of 10, 100, and 1000. The results were discussed taking into account (a) the nature of the phase (comparison of columns A, B, and C; (see Table 8), and (b) the influence of alkyl substituents, steric hindrance, and naphthenic substitution on the basic ring structures. The effects of all these parameters (column A) are illustrated in Tables 9a [39], 9b [40], and 9c [40]. As far as model compounds are concerned, the authors found few differences between the phases, with a slight superiority shown by DNAP in the separation of PAHs according to ring numbers [39]. This conclusion was supported in the case of liquid fossil fuel sample analysis (see Chapter 6) [40]. Increasing the distance of the CT groups from the surface of the silica (DNAO and TNAO) offered little advantage in grouping PAHs by ring number but may be of interest in special cases. The following should be noted:

1. It can be seen (Tables 9a and 9c) that alkyl naphthalenes and alkyl biphenyls could interfere with naphthenic compounds, as was reported for LC/picric acid [21].
2. The differences between grafted column A and nonligand acceptor grafted columns B and C (see Table 8) are less than would be expected from the capacity factors obtained by Nondek [26] using LC.

Temperature and eluant effects on the selectivity of 2,4-dinitroanilino (DNA), 2,4,7-trinitrofluorenimino (TNFI), and bis(3-nitrophenyl)sulfone (DNPS) butyl-silica-bonded phases, prepared starting with Nucleosil-10 NH2 10 μm, were studied by Hammers et al. [41]. The first two phases are similar to those reviewed elsewhere [22,27,28] but have a homolog spacer. From this study it can be concluded that:

1. Despite their different structures, TNFI, DNA, and DNPS–silica form complexes of equal stability.
2. Arenes can form CT complexes with electron-acceptor and electron–donor ligands, but in the nitroaromatic phases the interactions are stronger than in the absence of such groups, as in aminobutyl-grafted silica.
3. In all complexing phases, a polyphenyl shows a weaker interaction than is expected for a fused arene with the same number of aromatic carbon atoms.

TABLE 8   Separation of PAHs by Charge-Transfer/High-Pressure Liquid Chromatography on Chemically Bonded Phases with Nitro Groups[a]

Compounds	Adsorbents[b]	Refs.
Model PAHs, alkylated or not, and naphtenic compounds (86 derivatives); application to fossil liquids	*Column A* Lichrosorb Si60 grafted with $\equiv Si-(CH_2)_n-NH$ n y 3 2,4 DNAP 3 2,4,6 TNAP 8 2,4 DNAO 8 2,4,6 TNAO *Column B* Alumina (Woelm) *Column C* Chromagabond $(NH_2)_2$ $\equiv Si-(CH_2)_3-NH-(CH_2)_2-NH-(CH_2)_2-NH_2$ TA = triaminesilica	39, 40
Model PAHs	(a) $\equiv Si-(CH_2)_4-NH$ DNA-silica (b) $\equiv Si-(CH_2)_4-NH$ $\equiv Si-(CH_2)_4-NH$ Bis(3-nitrophenyl)sulfonesilica– DNPS-silica	41
Model PAHs and heteroatomic compounds; application to high-boiling cuts from petroleum origins	Comparison between phases reported in Table 7 (27,31–33) and $\equiv Si-(CH_2)_3-NH_2$ $\equiv Si-(CH_2)_2-CN$	42

(Continued)

TABLE 8   (Continued)

Compounds	Adsorbents[b]	Refs.
Model mixture of benzene, naphthalene, phenanthrene, anthracene	$\equiv Si-(CH_2)_3-NH-Z$ — [ring with X, y substituents]	43
	Comparison of DNAP and TNAP with two new phases:	
Model PAHs	Z X y	43, 47,
	SO_2 NO_2 2,4 DNSP	48
	CO NO_2 3,5 DNBP	
	CO F penta PFBP	
Model PAHs	$\equiv Si-(CH_2)_n-N-SO_2-C_6H_5(5\text{-}m)$	44
	$\quad\quad\quad\quad\quad\mid\quad\quad\quad\quad NO_2(m)$	
	$\quad\quad\quad\quad\quad R$	
	$m = 1\text{-}3,\ n = 1\text{-}4,\ R = H$ or alkyl	
Model PAHs	$\equiv Si-(CH_2)_3-O$ — [ring with X substituents] —X	49
	Compared to Nucleosil-NO_2 and DNBP	
	$\equiv Si-(CH_2)_3-S-Ar$ 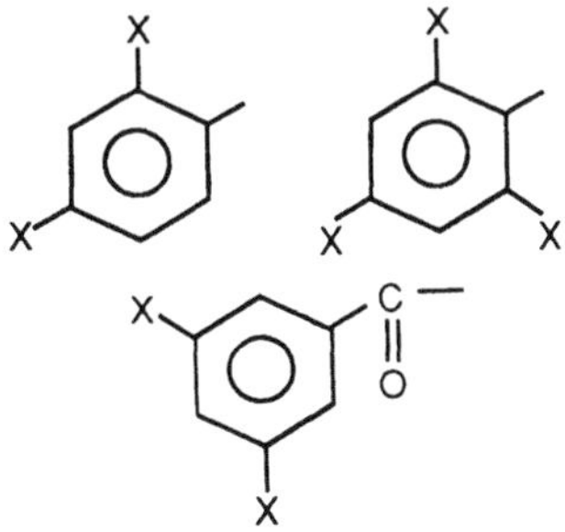	50
	Ar^*: [three aromatic ring structures with X substituents]	

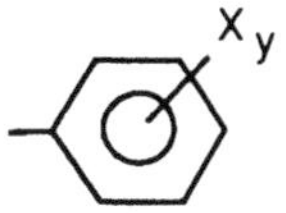

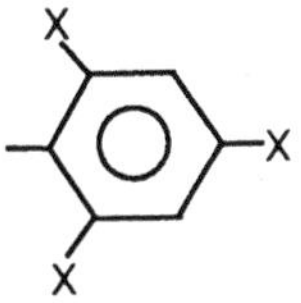

---

[a]Methods published after 1982.
[b]In all cases, $X = NO_2$.

4. The increased selectivity of TNFI phase agrees with the observation of Lochmüller et al. [28] regarding the degree of nitration of the fluorenimine nucleus. Nevertheless, from a practical point of view, the authors [41] do not recommend the fluorenimine phases because of their poor chemical stability and preferred DNA silica in isothermal and isocratic conditions.

The phases studied by Lochmüller and Amoss [27], Nondek et al. [31], and Matlin et al. [32,33] were tested by Eppert and Schinke [42], who made an extensive evaluation of surface adsorption on nitroaryl-bonded phase materials. They discussed the conditions for the separation of PAHs in subclasses, and the influence of the eluant on the selectivity and on the number of theoretical plates.

From a practical point of view, various examples demonstrate the ability of the trinitrophenyl amino groups in the chromatographic analysis of complicated petroleum hydrocarbons (see Chapter 6).

A study of the DNAP phase [29,39,40] was carried on by Nondek amd Ponec [43] considering new phases chosen to test theoretical predictions based on the Hückel molecular orbital theory (HMO). Besides the phases already known, DNAP and TNAP [39,40], two new phases with electron-accepting spacers were synthesized (see Table 8): a 2,4-dinitrophenylsulfamidopropylsilica (DNSP) and a 3,5-dinitro-benzamidopropylsilica (DNBP). They were both prepared from NH_2 silica as described for DNAP [22] and have an electron-accepting spacer instead of an electron-donating one [NH—$(CH_2)_3$] in the other two cases (DNAP and TNAP). The complexing ability increases in the order DNAP ~ DNSP < DNBP < TNAP.

The chromatographic selectivity increases with increasing electron affinity of the ligands. The complexing ability and selectivity depends on the number of electron-accepting substituents on the aromatic nucleus, their positions, and the nature of the spacer. DNSP–silica appeared to be a particularly promising sorbent for HPLC. Indeed, the preparation of six silica gel sorbents of this type (Table 8) was patented [44]. They gave better PAH separation than did the DNA group.

Recently, Nondek [45] described an improved preparation of the picramidopropyl silica (or TNAP) phase from Lichrosorb NH_2 and 2,4,6-trinitrobenzenesulfonic acid, by analogy with the known reactivity of this compound with amine, forming N-substituted picramides. The photoacoustic spectra of TNAP gels prepared in different ways were compared [46].

At the same time, Felix and Bertrand [47] described several preparations of the DNBP phase and its efficiency in model PAH separation and carried on their investigation with a pentafluorobenzamidopropyl silica, prepared in the same way from pentafluorobenzoyl chloride and aminopropylmethylsilica. This new phase can be used in HPLC and RP–HPLC and allowed the separation of model PAHs according to the degree of aromaticity [48].

Although the most common phases consist of nitroaromatics bonded to

TABLE 9 Structural Effects on the Separation of PAHs by CT–HPLC[a]

A. Effect of Naphthenic Substitution on Retention Indexes of Aromatic Hydrocarbons

Compound	I		$\log I$
	Silica-DNAP	Alumina	Silica-R(NH$_2$)$_2$
Cyclohexylbenzene	1.04	1.03	1.27
Indane	1.12	1.07	1.27
1,2,3,4-Tetrahydronaphthalene	1.13	1.09	1.28
2a,3,4,5-Tetrahydroacenaphthene	1.20	1.17	1.31
1,2,3,4,5,6,7,8-Octahydroanthracene	1.15	1.07	1.32
Dodecahydrotriphenylene	1.53	1.20	1.48
9,10-Dihydrophenanthrene	2.17	2.50	2.29
1,2,3,6,7,8-Hexahydropyrene	2.08	2.30	2.00
4b,5,6,10b,11,12-Hexahydrochrysene	2.17	2.63	2.60
9,10-Dihydro-2,4,5,7-tetramethylphenanthrene	1.59	1.64	1.79
5,12-Dihydrotetracene	3.24	3.24	3.62
3,4,4a,5,6,12c-Hexahydrophenanthro[3,4c]-phenanthrene	3.23	2.82	3.87

Source: Ref. 39.

B. Effect of Alkyl Chain Length on Retention Indexes for Aromatic Hydrocarbons

Compound	Chain length	I_x				
		TA	DNAO	TNAO	DNAP	TNAP
Benzene	0	1.00	1.00	1.00	1.00	1.00
Toluene	1	0.92	1.02	1.02	1.04	1.02
Ethylbenzene	2	0.95	0.86	0.92	1.00	0.97
n-Propylbenzene	3	0.87	0.77	0.88	0.97	0.95
n-Butylbenzene	4	0.92	0.75	0.79	0.93	0.95
n-Amylbenzene	5	0.89	0.75	0.84	0.96	0.95
n-Hexylbenzene	6	0.84	0.70	0.72	0.93	0.95
n-Decylbenzene	10	0.86	0.66	0.69	0.97	0.95
n-Pentadecylbenzene	15	0.85	0.63	0.69	0.97	0.93
n-Nonadecylbenzene	19	0.84	0.61	0.69	0.96	0.93
Average for one-ring		0.89 ± 0.05	0.78 ± 0.14	0.82 ± 0.13	0.97 ± 0.03	0.96 ± 0.03
Naphthalene	0	2.00	2.00	2.00	2.00	2.00
2Methylnaphthalene	1	1.99	2.01	2.05	2.04	2.10
2-Ethylnaphthalene	2	1.95	1.91	1.93	1.97	2.03
2-n-Butylnaphthalene	4	1.91	1.83	1.83	1.81	1.95
Average for two-ring		1.96 ± 0.04	1.94 ± 0.08	1.95 ± 0.10	1.96 ± 0.10	2.02 ± 0.06

(Continued)

TABLE 9 (Continued)

Compound	Chain length	I_x TA	DNAO	TNAO	DNAP	TNAP
Pyrene	0	3.35	3.54	3.34	3.70	3.52
1-Methylpyrene	1	3.40	3.71	3.65	3.91	3.76
1-Ethylpyrene	2	3.34	3.43	3.40	3.68	3.52
1-n-Butylpyrene	4	3.18	3.18	3.22	3.52	3.34
Average four-ring		3.31 ± 0.10	3.45 ± 0.22	3.40 ± 0.18	3.70 ± 0.16	3.54 ± 0.17
Average standard deviation		0.06	0.15	0.14	0.10	0.09

Source: Ref. 40.

C. Effect of Steric Hindrance on Retention Index

Compound	I_x TA	DNAO	TNAO	DNAP	TNAP
Benzene	1.00	1.00	1.00	1.00	1.00
Isopropylbenzene	0.92	0.80	0.81	0.96	0.93
p-Diisopropylbenzene	0.83	0.61	0.69	0.86	0.85
1,3,5-Triisopropylbenzene	0.75	0.52	0.69	0.81	0.80
Naphthalene	2.00	2.00	2.00	2.00	2.00
2-Methylnaphthalene	1.99	2.01	2.05	2.04	2.10
2-Isopropylnaphthalene	1.89	1.82	1.90	1.25	1.94
2-n-Butylnaphthalene	1.91	1.83	1.83	1.81	1.95
2-Isobutylnaphthalene	1.90	1.79	1.77	1.61	1.88
1-Phenylnaphthalene	2.40	2.65	2.72	2.19	1.90
Biphenyl	2.05	1.99	1.97	2.01	1.92
2-Methylbiphenyl	1.87	1.82	1.88	1.42	1.77
2-Isopropylbiphenyl	1.82	1.72	1.74	1.24	1.72
4-Methylbiphenyl	2.04	1.96	2.03	2.05	1.98
Benz[a]anthracene	4.00	4.00	4.00	4.00	4.00
9-Phenylanthracene	3.04	2.88	2.92	2.79	2.35

Source: Ref. 40.
[a]The phases are described in Table 8.

aminopropylsilica, new phases were developed to avoid undesirable secondary adsorption. In these cases the electron acceptor group was bonded to silica via ether [49] or sulfide linkages [50].

Ether linkages. The properties of a new *n*-propylpicryl ether phase [49] were compared with those obtained with silica gel Nucleosil–NO$_2$ and 3,5-dinitrobenzamidopropyl phases. The results agree with a retention mechanism by CT. Furthermore, the capacity factors k' of the picryl ether phase were larger than those obtained by Matlin et al. [32,33]. Thus the picryl ether is a better acceptor, due to the greater electronegativity of oxygen compared to nitrogen.

Sulfide linkages. Dinitrophenyl, dinitrobenzoyl, and picrylmercaptopropylsilica (Table 8) were prepared from mercaptopropylsilica [Si(CH$_2$)$_3$—SH] and the corresponding polynitrochlorobenzenes [50]. The capacity factors k' of various PAHs were determined on each stationary phase. A linear relationship between log k' and the number of electrons available for complex formation was shown. Nevertheless, these phases had lower k' values for PAHs than did 2,4-dinitroanilinopropylsilica.

2. Chemically Bonded Phases with Donor Groups Used in the Separation of Nitro PAHs (Table 10)

The preceding examples showed the separation of PAHs, or alkyl PAHs (electron donors) by CTC on grafted silica with nitro groups (electron acceptors). Another method of chromatographic application of D-A interactions was carried on in the "reverse mode," several solutes having π-acceptor properties being separated on chemically bonded donor ligand on silica gel. The most striking examples are summarized in Table 10.

By using a chemically bonded phenyl ether phase for RP–HPLC separation of several nitro compounds, Mourey and Siggia [51] described the formation of weak CT complexes even in a polar solvent system and showed that partitioning into the stationary phase is as much a factor in retention as is CT interaction.

Owing to the relatively small benzene moiety of the phenoxy group, Lochmüller et al. [52] claimed that a high differentiation of large nitro PAHs could not be expected with this phase. They enhanced the separation of nitro PAHs on an end-capped bonded-pyrene silica gel stationary phase [53] used in reversed-phase LC. The authors showed the excellent selectivity and efficiency of the pyrene-bonded phase and its superiority on a phenyl-bonded silica phase, as can be seen from Table 11.

Along the same lines, a new polycyclic aromatic-bonded phase for HPLC, with anthracene as ligand has recently been described by Verzele and Van De Velde [54]. A similar phase, 3-(9′-(10′-methylanthryl))propylsilane, was used in RP–HPLC by Den et al. [55] for the successful separation of nitroaromatic compounds used as classical explosives. Separation of same samples showed only slight differ-

TABLE 10 Separation of Nitro PAHs on Chemically Bonded Phases with Electron Donor Groups by CT-LC and CT-HPLC in Reversed-Phase Mode

Compounds	Adsorbent	Refs.
Nitro derivatives of benzene, aniline, acetanilide, phenol	$\equiv$Si—(CH$_2$)$_3$—O—⬡ RP-HPLC	51
Mono-, di-, and tri-nitro derivatives of benzene, aniline, acetanilide, phenol	$\equiv$Si—(CH$_2$)$_3$—3—pyrene RP-LC	52, 53
PAHs and nitro naphthalenes	$\equiv$Si—(CH$_2$)$_3$—NH—CH$_2$— (anthracene) RP-HPLC	54
Explosive nitro PAHs	$\equiv$Si—(CH$_2$)$_3$— (methylanthryl) 3-(9′(10′-methylanthryl))propylsilane RP-HPLC	55

ences in retention times when a column RP-18 was used with the same mobile phase.

3. Chemically Bonded Phases with Electron Acceptors (Except Nitro Derivatives)

Studies related to tetrachlorophtalimid as acceptor group (Table 12). The tetrachlorophtalimidopropylsilica phase (TCI) carried out by Deymann and Holstein [36,37] was compared to three new phases: pentachloro (PCA), pentafluoro (PFA), and dinitrofluoro (DNFA) anilinopropylsilica by Hellmann [56]. These phases were prepared as usual from the corresponding substituted anilines, condensed with allylbromide and then with trichlorosilane. In the case of pentahalogenated derivatives, mixtures of structures (*1a*) and (*1b*) (Table 12) were obtained, due to a further reaction of the substituted aniline Ar—NH—CH₂— CH=CH₂ with allyl bromide, this reaction being preferred because of a longer reaction time. The efficiency of these phases was tested on model condensed PAHs, with one to seven nuclei, linear catacondensed PAHs, alkyl derivatives of pyrene, methyl derivatives of naphthalene and phenanthrene, and nitrogen derivatives such as carbazoles.

A linear relationship was established between the capacity factors k' and the number of nuclei n: $\ln k' = f(n)$. The slope of the straight line gives an evaluation of the interaction strength between the stationary phase and the solutes.

The effects of temperature (25 to 45°C) and solvent polarity (n-hexane and n-hexane/CH₂Cl₂ mixtures 95:5 or 70:30) were discussed. If a weak temperature effect is observed for all phases, the selectivity increases as the solvent polarity decreases. An interesting point concerns the influence of alkyl substitution on the condensed PAH separation (long-chain alkyl derivatives or polymethyl derivatives).

The influence of chain length was first studied for 1-naphthyl pyrenes with PCA, PFA, and DNFA phases [56] and later with TCI by Holstein [57]. As the results are similar, only the case of TCI is illustrated here (Fig. 1), the relative influences of the three other phases being in the order DNFA > PCA > PFA [56].

The decrease in the capacity factor k' as the chain length increases (except for methyl derivatives) is explained by steric factors masking the donor + I + M effects of the alkyl groups. The study was extended to other condensed PAHs with a typical number of nuclei (two to five) [57]. Two relations were established from experimental results:

$$\ln k' \text{ (methyl PAH)} - \ln k' \text{ (PAH)} = C_{ste} = 0.376 \qquad (3)$$

From Eq. (3) the k' value for an unavailable methyl PAH derivative can be evaluated. By analogy, but only from experimental values in the pyrene series, another relationship was established:

$$\ln k' \text{ (octadecylpyrene)} - \ln k' \text{ (pyrene)} = C_{ste} = -0.800 \qquad (4)$$

From Eq. (4) Holstein estimated the k' values for the unavailable octadecyl derivatives of naphthalene, phenanthrene, and benzopyrene.

When separation of PAHs according to their ring number is considered, as is the case with the TCI phase, the most important consequence of the presence of alkyl derivatives ($n \geq C_4$) is their elution in the preceding group [i.e., an octadecyl pyrene (four nuclei) will be eluted in the phenanthrene group (three nuclei), and so on]. These results do not agree with the first observation, according to which the separation of PAHs is influenced only slightly by alkyl substitution with the TCI phase [15].

As far as methyl derivatives are concerned, particularly in the case of dimethyl-

TABLE 11 Comparison of the Capacity Factor's k' on Phenyl- and Pyrene-Bonded Stationary Phases in 100% Hexane

Compound	Phenyl column k'	Pyrene column k'
Naphthalene	0.08	0.25
Pyrene	0.17	0.51
Chrysene	0.30	0.72
2-Nitronaphthalene	1.05	1.48
Nitrobenzene	1.10	3.10
9-Nitroanthracene	1.14	3.30
4-Nitro-p-terphenyl	1.18	8.56
2-Nitrofluorene	1.44	11.8
1-Nitropyrene	1.52	10.4

Source: Ref. 52.

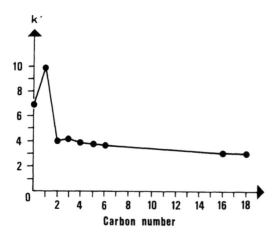

FIGURE 1 Representation of k' values of n-alkyl pyrene versus the number of carbon atoms on the side chain. Phase, TCI; eluting solvent, n hexane/methylene chloride 70:30. (From Ref. 57.)

TABLE 12 Separation of PAHs by Charge-Transfer/HPLC on Chemically Bonded Phases with Acceptor Groups (Except Nitro Derivatives)[a]

Compounds	Adsorbents	Refs.

Model PAHs, coal liquid products	$\equiv$Si—(CH$_2$)$_3$—NH—Ar (*1a*)
	 $\equiv$Si—(CH$_2$)$_3$
	       \
	      N—Ar (*1b*)
	      /
	 $\equiv$Si—(CH$_2$)$_3$ 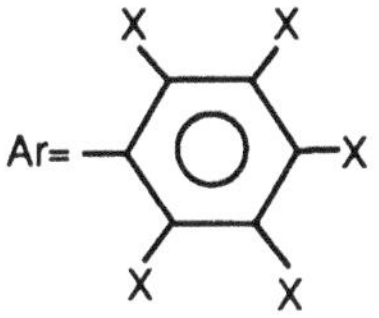
	
	 Ar = (perhalogenated benzene ring, X at positions): X = Cl PCA; X = F PFA 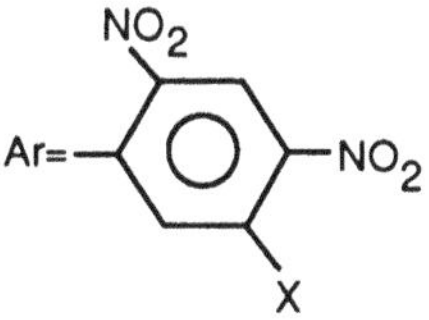
	
	 Ar = (benzene ring with two NO$_2$ and one X): X = F DNFA
		56

Influence of alkyl substitution on PAH retention	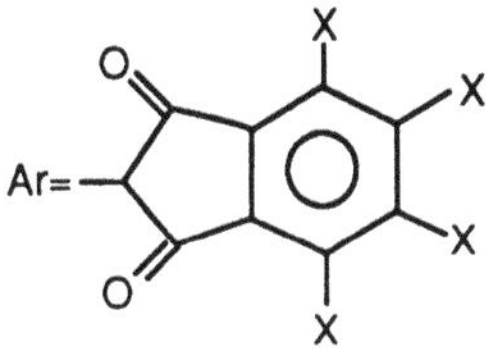 Ar = (tetrahalo-phthalimide/indandione type ring with X substituents) X = Cl TCI	57

Model PAH application to petroleum residues
 Esters of PAHs
	Caffeine-type bonded silica (Lichrosorb Si60 10 μm)
	 $\equiv$Si—(CH$_2$)$_3$—Ar
	   Ar = Theobromine
	   Ar = Theophylline
	   Ar = Theophylline
		
		62
		63–65
		67

Methyl-substituted benzenes, PAHs	Tetracyanoethylene-coated silica gel (Merck Si60 10 μm)	68

[a]Methods published after 1982.

naphthalene, the derivatives with nearby methyl groups (1,2- or 1,8–diMe) have the greatest interaction with the electron acceptors (i.e., with DNFA $k' = 0.18$ against 0.13 to 0.14 for 1,3- and 2,6-diMe). The effect is greater with strong acceptors such as TCI and DNFA, less with weak acceptors such as PFA and PCA [56].

In the case of polymethyl derivatives, it seems that the donor interaction is increased by the +I +M effects of the cumulative methyl groups. Indeed, along the same lines, Stetter and Schroeder [58], studying the retention behavior (in TLC) of nitrotoluenes and quinones on glucose esterified with 3-(pentamethylphenyl)- and 3-(phenanthryl)propionic acids, showed that the pentamethyl phenyl derivative was a much stronger donor than the phenanthrenic derivative.

In the course of coal liquid analysis by CT–HPLC with a TCI phase, the method was applied [59] as described earlier [36,37] on various more or less hydrogenated model PAHs, some of their methyl derivatives (Table 13), and heteroatomic compounds (Table 14). Retention indexes I, based on the Lee and Vassilaros [60] definition used in glass capillary column chromatography (CC), were calculated as follows:

$$I = \frac{tr_x - tr_1}{tr_2 - tr_1} \times 100 + 100z \tag{5}$$

where tr_x, tr_1, and tr_2 are the time elution values of substance x and references 1 and 2 chosen as $tr_1 < tr_x < tr_2$ (naphthalene, phenanthrene, and chrysene as reference compounds) and z is the ring number of substance x.

It is obvious that the separation by TCI follows the number of condensed aromatic nuclei, which is the opposite of the case with glass capillary column chromatography, which follows the total number of rings, saturated or not. In this view the two chromatographies are complementary, as can be seen from Fig. 2 [59]. For some examples of application of these phases to coal and petroleum liquid analysis, see Chapter 6.

Recently, Jadaud et al. [61] showed on an improved TCI-bonded silica that retention is governed only by the interaction between PAHs and the free TCI groups and by their solubility in the mobile phase. They discussed the influence of PAHs structure on the capacity factors.

Studies Related to Caffeine as Acceptor Groups (Table 12).   The ability of caffeine to complex with PAHs is well known, and some examples of applications in CT–TLC have been cited [4,6,8]. More recently, the successful synthesis and use of a bonded caffeine stationary phase for CT–HPLC, with high selectivity for PAHs, were reported by Felix et al. [62–65]. These studies followed an initial investigation of large-scale LC separation (3 g of sample) on an acenaphtene/ acenaphtylene mixture 50:50 on silica gel coated with caffeine (20% in weight) [66]. Preparative HPLC on silica gel coated with caffeine was also studied for the separation of silyl acenaphthenyl mixtures, resulting in high yields of the silylated compounds, with excellent purity [66].

The capacity ratio of the PAHs (from one to four nuclei) measured on the caffeine column compared with those determined on commercial amino and silica columns show the increase in retention due to the strong acceptor caffeine group in the stationary phase. The application to petroleum residues is reported in Chapter 6.

More recently, separation of aromatic esters on a theophylline-bonded phase (as well as 3,5-dinitrobenzamido phase) has been performed [67]. The results were interpreted by a CT mechanism and discussed according to the nature of the aromatic ring substituents.

Although tetracyanoethylene (TCNE) used in TLC on impregnated plates did not give reproducible R_f values of PAHs [8], this electron acceptor has been studied more recently in HPLC. Burger and Tomlinson [68] have attempted to design a simple HPLC system by coating TCNE on to the surface of silica gel. The retention of a series of methyl-substituted benzenes and PAHs has been determined at various temperatures by comparison with nonloaded columns.

D. Gas–Liquid Chromatography

Generally speaking, gas–liquid chromatography (GLC) has been used for the investigation of physicochemical phenomena, particularly for the study of charge-transfer complexation. So it is not surprising that workers using this technique were more interested in theoretical studies on CT complexation than in true analytical applications. This is a striking difference from the topics developed earlier for TLC, LC, and HPLC by CT complexation.

The subject was reviewed in 1979 by Laub and Pecsok [69; 107 references therein]. These authors proposed the solvent dependence of the complex formation constants K_f between various donor and acceptor compounds obtained by various spectroscopic methods (UV, NMR) as explaining the differences between GLC and spectroscopic data. They attributed these differences to solvation interactions arising in spectroscopic determinations. They have tried to show that the chromatographic technique was valid for studying CTC and even gave much more information regarding the determination of vertical ionization potential and electron affinities, molecular substituent and out-of-plane deformation effects, and steric hindrance to charge transfer.

Complementing the review of Laub and Pecsok [69] are some other works [70–73]. A gas chromatographic method has been investigated by King and Quinney [70] to study CT interactions between alkyl benzenes and molten CBr₄. They showed that the CT interactions are temperature dependent (decreasing rapidly with increasing temperature). Eon et al. [71] established new general equations to calculate by GLC the equilibrium constants for complexing reactions between a volatile solute (furan, thiophen, 2-methyl thiophen) and a nonvolatile reagent

TABLE 13   Retention Indexes *I* on TCI Phases of Model PAHs and Hydroaromatic
Derivatives

Structure	*I*
1,2,3,4,4a,9,9a,10-Octahydroanthracene	140
1,2,3,4,5,6,7,8-Octahydroanthracene	143
4,5,9,10-Tetrahydropyrene	165
4-Methylbiphenyl	165
Biphenyl	166
1-Phenylnaphthalene	177
9,10-Dihydrophenanthrene	182
1,2,3,6,7,8-Hexahydropyrene	185
1,2,3,4,5,6-Hexahydrochrysene	194
Naphthalene	200
1,2,3,4,4a,11,12,12a-Octahydrochrysene	202
1,2,3,10b-Tetrahydrofluoranthene	206
2-Methylnaphthalene	219
1-Methylnaphthalene	220
Fluorene	236
1,5-Dimethylnaphthalene	237
2,7-Dimethylnaphthalene	236
2,6-Dimethylnaphthalene	240
1,2,3,3a,4,5-Hexahydropyrene	246
Acenaphtene	252
1,2,3,4,7,8,9,10-Octahydrochrysene	256
1,2,7-Trimethylnaphthalene	257
Acenaphtylene	267
4,5-Dihydropyrene	276
Phenanthrene	300
Anthracene	305
Benzo[a]fluorene	337
1,2,3,4-Tetrahydrochrysene	336

(Continued)

TABLE 13 (Continued)

Structure	I
Benzo[b]fluorene	345
4H-Cyclopenta[d,e,f]phenanthrene	342
Fluoranthene	372
Pyrene	399
Chrysene	400
9,10-Dimethylanthracene	400
11-Isopropylbenzo[a]fluorene	400
5,6-Dimethylchrysene	410
5-Methylchrysene	412
1-Methylpyrene	417
Benzo[d,e]fluoranthene	421
1-n-Butylpyrene	424
Picene	428
Benzo[a]pyrene	443
Benzo[e]pyrene	443
Perylene	448
(1,2,3-cd-Indeno)pyrene	460
Dibenzo[a,h]pyrene	482
Dibenzo[ghi]perylene	493
Coronene	500
4-Methyl benzo[ghi]perylene	515

Source: Ref. 59.

(dibutyl tetrachlorophthalate), dissolved in a stationary liquid phase (squalane). On the basis of these equations, equilibrium constants and thermodynamic functions of reactions between dibutyl tetrachlorophthalate and five-membered heterocyclic derivatives were calculated [72]. Excellent agreement was observed between the experimental results and the theoretical predictions. In the same way, Castells [73] used GLC for studying complexing equilibrium at 60°C between aromatic hydrocarbons and TNB, dissolved in dinonylphthalate. The thermodynamic stability

TABLE 14 Retention Indexes *I* on TCI Phase of Model Heteroatomic Compounds

Structure	*I*
Benzo[*b*]thiophene	186
p-Cresol	218
Dibenzofuran	222
N-Methylcarbazole	242
Dibenzothiophene	265
Benzo[*b*]naphto[*d*]furan	304
Indole	332
3-Methylindole	347
5-Methyl-1,2,3,4-tetrahydrocarbazole	347
1,2,3,4-Tetrahydrocarbazole	355
2,6-Dimethylnaphthol	384
Carbazole	411
β-Naphthol	438
2-Methylquinoline	461
Benzo[*c*]carbazole	475
Hydroquinone	489

Source: Ref. 59.

constants for the complexes were calculated by means of a series of relations and the values compared with spectrometric data.

The same topics continued to be developed after 1975. For example, summarizing the results obtained until 1976 in the GLC studies of electron donor–acceptor systems and the problems associated with the interpretations of the results, Crowne et al. [74] tried to separate the "chemical" from the "physical" interactions involved in the complexation. In this regard they studied the complexes of several naphthalene derivatives with picric acid using GLC and spectroscopic techniques. GLC was also used to study molecular associations involving 1-iodo 1-dodecyne and several Lewis donors [75].

Although CT–GLC is considered as an analytical tool for the separation of complex mixtures of PAHs or other derivatives, such as heterocyclic and heteroatomic compounds, very few applications are quoted in the literature after 1970. The de-

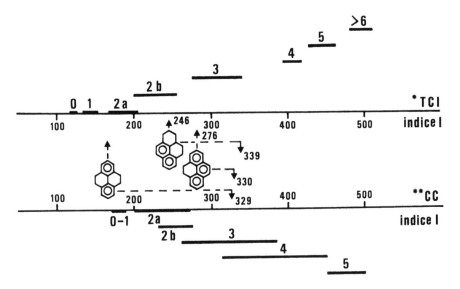

FIGURE 2 Comparison of TCI and CC chromatographies. The figures with an asterisk give a number of aromatic cycles (e.g., 0, paraffins; 1, alkylbenzenes; 2a, biphenyls; 2b, naphthalenes; 3, phenanthrenes; etc.). The figures with a double asterisk give the total number of cycles. (From Ref. 59.)

velopment of more powerful gas chromatographic methods, such as glass capillary column chromatography, can easily explain this loss of interest. In fact, it is only in the first development of GLC that some examples can be found. They involved the separation of isomeric nitrotoluenes [76], hydrocarbons, and amines [77,78] by complexation with TNF, of several alkyl benzenes complexes with tetrahalophthalate esters (i.e., separation of *m*- and *p*-xylenes) [79]. An attempt has also been made to use CT–GLC for the separation of olefins [80]. The TNB–olefin complexes exhibit progressive increased stability with increasing double-bond substitution, but the steric effects are less pronounced than for olefin–Ag$^+$ complexation. Nevertheless, the differences in retention times among structural isomers are sufficiently characteristic to provide a basis for assigning major structural features.

E. Application of Charge-Transfer Complexation to the Resolution of Racemic Polycyclic Aromatic Hydrocarbons

A general view of the utilization of chiral stationary phases (CSPs) for the separation of enantiomers is given in Chapter 5. Some aspects of the mechanism of chiral recognition relevant to the liquid chromatographic separation of enantiomers have been reviewed by Pirkle and Pochapsky [81]. In this section only the resolution of optical isomers on CSPs acting mainly by π-π interactions (DA–CSPs) is considered, as it is a logical continuation of the studies reported in this chapter.

As early as 1955, Newman et al. [82] described the synthesis of 2-(2,4,5,7-tetranitro-9-fluorenylideneaminooxy)propionic acid (2) (X = CH3) [chiral R(-) and S(+) TAPA], a useful agent for the optical resolution of some PAHs [e.g., phenanthro(3,4-c)phenanthrene]. The dextro and levo forms were made by condensing 2,4,5,7-tetranitrofluorenone with (-) and (+) isopropylidene aminooxypropionic acid. This work was the starting point for the direct resolution of enantiomers by HPLC on CSP.

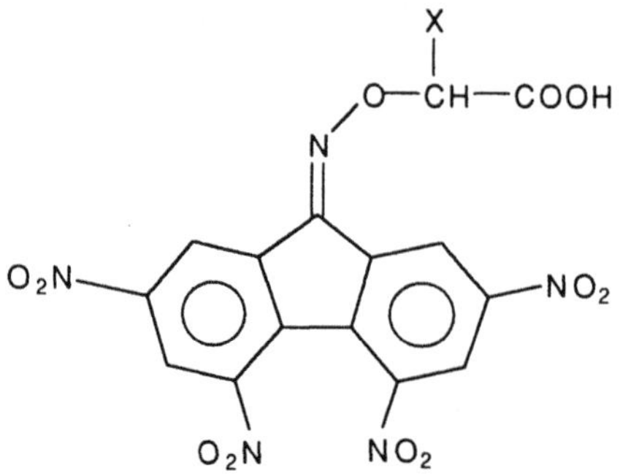

X = CH3	TAPA
X = C2H5	TABA
X = i-C3H7	TAIVA
X = C4H9	TAHA

(2)

Klemm and Reed [83] described the use of columns of silicic acid impregnated with (2) (+) or (-) for the partial optical resolution of 1–naphtyl-2-butyl ether and 3,4,5,6-dibenzo-9,10-dihydrophenanthrene. The method was extended to the partial resolution of 9-sec-butylphenanthrene [84].

But it is in the field of helicenes that the method was developed for the separation of P and M enantiomers. [Depending on whether the identified helix is left- or right-handed, the isomer is designated by "minus" (M) or "plus" (P) with the prefixes (+) and (-) indicating the sense of optical rotation.]

Mikes et al. [85] used TAPA and its homolog TABA (2: X = C2H5), TAIVA (2: X = iso-C3H7), and TAHA (2: X = C4H9) derived, respectively, from butyric, isovaleric, and hexanoic acids as chiral charge-transfer complex forming stationary phases, in situ coated on silica microparticles. They showed that the bulkiness of the X group at the chiral center of the CT acceptor is critical for the resolution of the CT donor. For example, if the [6] and [14] helicenes were completely resolved on R(-) TAPA, the [5] helicenes could be separated only on R(-) TABA, but after 10 recycling steps. R(-) TAPA was also used as a bonded solid phase (microsilica particles—Partisil 7).

The selector–selectand interactions were discussed on a molecular basis. Assuming that the selector molecules are oriented parallel to the surface of the support, the selectivity of the resolution depends on the one hand on the bulkiness of the X group, and on the other hand, on the shapes of the semicavity of the helicenes. It was suggested that the interpenetration of one molecule into a cavity of another, or topographical complementary convexity and concavity of the selector and selectand, may be necessary for chiral recognition. (The terms *selector* and *selectand* were introduced by Mikes instead of ligand–substrate, solvent–solute, and so on.) In the TAPA–helicene system, the former possesses chiral convex and the later chiral concave topography [86]. The same group [86] also realized the resolution of two heterohelicenes: dinaphtho[1,2-*d*; 1′,2′-*d*′]benzo[1,2-*b*; 4,3-*b*′]dithiophene and dithieno[3,2-*e*; 3′,2′-*e*′]benzo[1,2-*b*; 4,3-*b*]bis(1)benzothiophene. On a linked TAPA column, the resolution factors were similar to those of [6] and [5] carbohelicenes, respectively.

The chiral selectivity of TAPA was compared to that of another optically active selector, binaphthyl-2,2′-diyl hydrogen phosphate (binaphthyl phosphoric acid— BPA; (*3*) [86], covalently linked to microporous silica particle. Whereas all carbo

R = NH—(CH2)3—
R = NH(CH2)2—NH—(CH2)3

(*3*)

and heterohelicene selectands can be resolved on TAPA columns, only helicenes containing heteroatoms or electron-donating substituents could be separated easily by BPA. For example, significant resolution effects were observed with 8,20-dibromodiphenanthro[4,3-*a*; 4′,3′-*j*]chrysene and 9,10-diaza[7,8; 11,12]dibenzohepta helicene, whereas no resolution effect was observed with [12] helicene, in contrast to the good resolution of these three helicenes on a TAPA column.

It must be noted that recently a total conversion of racemic [5] thiaheterohelicene (thieno[3,2-*e*; 4,5-*e*′] di[1] benzothiophene) into a single enantiomer was obtained by crystallizing the charge-transfer complex with *S*-TAPA [87].

Another chiral sationary phase was described by Lochmüller and Ryall [88] using L-alanine as the chiral group (*4*). The HPLC separation of racemic 1-aza[6]-helicene and heptahelicene can be achieved with this simple bonded phase. As this CT agent could hardly participate in the type of interaction that has been proposed to occur with TAPA [86], the authors suggested that enantiomeric resolution may require only one significantly strong CT interaction near a center of asymmetry.

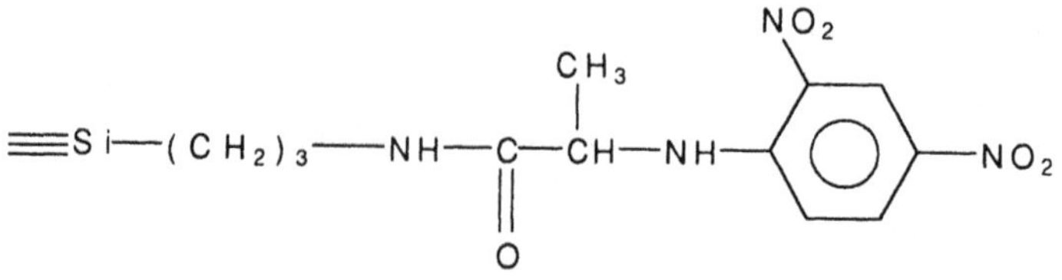

$$\equiv Si\!-\!(CH_2)_3\!-\!\!NH\!-\!\underset{\underset{O}{\|}}{\overset{\overset{CH_3}{|}}{C}}\!-\!CH\!-\!NH\!-\!\!\left\langle\bigcirc\right\rangle\!-\!NO_2 \qquad (4)$$

Some biological molecules, such as flavins, have certain general structural features recalling those of TAPA. For example, Kim et al. [89] showed that both riboflavin (5) and TAPA (2) have a tricyclic system suitable for CT complexation with PAHs and a side chain containing one (or more) asymmetric center(s). Indeed, they reported the resolution of [6] to [14] carbohelicenes on silica gel coated with riboflavine.

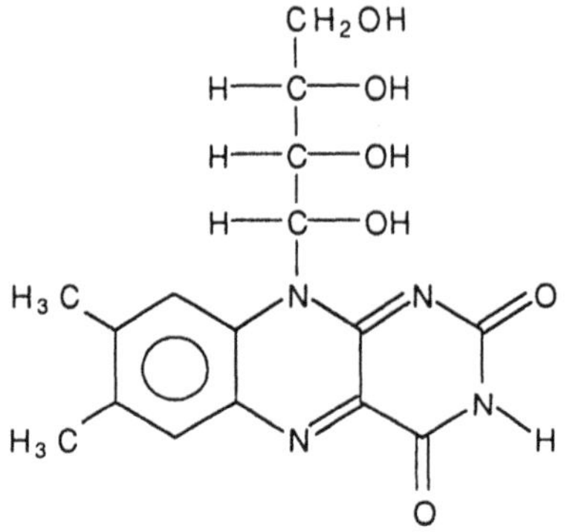

(5)

III.   AROMATIC AMINES AND AZAARENES

As azaarenes and other nitrogen bases (e.g., aromatic amines) are efficient donors of π- and n-electrons, several chromatographic methods based on charge-transfer interactions with acceptors were used.

A.   Aromatic Amines

Aromatic amines form weak charge-transfer complexes with aromatic nitro compounds. The equilibrium constants of the complex formed with a given acceptor are usually different, depending on the donor strength of amines.

1.   Charge-Transfer/Thin-Layer Chromatography

The formation of weak charge-transfer complexes by CT–TLC was carried out about 20 years ago in two ways. In one study Parihar et al. [90] described the separation of explosive nitroaromatic compounds, such as 2,4,6-trinitrotoluene (TNT), m-dinitrobenzene (m-DNB), 2,4,6-trinitrophenyl N-methyl nitramine (tetryl), and

2,4-dinitrochlorobenzene (DCNB). They used Kieselgel G containing 3% α-nap-thylamine and a mixture of toluene/ethylene dichloride (9:1) with 3% α-naph-tylamine as eluting solvent.

In a second study using adsorbents (Kieselgel G, cellulose benzoate, or acetate, for example) impregnated with tetryl or DCNB [91], picryl chloride, or 1,3,5-trinitrobenzene [92], the same authors described the separation and identifi-cation of various aromatic amines (substituted anilines, toluidines, anisidines, naphtylamines). The mobilities of the complexes were almost in order of the basicities of the corresponding amines. It was possible to characterize distinctly the complexes formed from 1 to 2 μg of individual amines. A two-dimensional TLC was also described for the separation of solid and liquid amines, using silica gel G plates impregnated with m-DNB and a set of solvents (n-heptane saturated with ammonia, carbon tetrachloride saturated with water for liquid amines, and chloro-form saturated with water and carbon tetrachloride, as above, for solid amines) [93]. A similar technique was developed by Snyder and Welch [94,95]. They re-ported that a number of aromatic donors were adsorbed by DNB cotton cellulose [94], adsorption being attributed to CT complex formation. Then the authors em-ployed DNB paper for the chromatographic separation of substituted anilines [95]. They found a relationship between the donor strengths of aromatic anilines with electron-releasing substituents and the R_f values (i.e., the R_f values could be corre-lated with the σ Hammett constant only in the case of electron-releasing sub-stituents).

2. Charge-Transfer/Gas–Liquid Chromatography

Gas–liquid chromatography (GLC) was also tried by Crowne et al. [96] for the separation of substituted anilines on TNF added to tritolyl phosphate and polypropylene sebacate as stationary phases.

3. Charge-Transfer/Liquid Chromatography (HPLC)

Using the already described phases 3-(2,4-dinitroanilino)propylsilica (DNAP) [31] and 3-(2,4-dinitrobenzenesulfamido)propylsilica (DNSP) [43], Nondek and Chvalovsky [97] studied the separation of azaaromatics, anilines, and alkyl aro-matic amines by liquid chromatography in the HPLC mode. The aim of the study was the establishment of on-line preconcentration of azaarenes in view of their identification in petroleum products and their separation from other nitrogen bases. The capacity factors k' of several selected azaaromatics, anilines, and alkyl aro-matic amines were determined in HPLC experiments, using the phases described above and CH_2Cl_2 (+0.5% isopropanol) as the mobile phase. A precolumn was used for on-line preconcentration [98]. As in the case of PAHs, DNSP–silica seems to be the most efficient acceptor (all solutes having higher retention on DNSP–sil-ica than on DNAP–silica).

From k' values it can be suggested that the retention is influenced by the basicity of the solutes: weak bases such as anilines and pyrroles are retained less than are azaaromatics having a pyridine ring. Alkylaromatic amines such as benzylamine and phenylethylamine are very strongly retained, as are azaarenes with several pyridine rings. PAHs with up to three or four rings are eluted before the weak nitrogen bases, which permits a group separation. The on-line preconcentration was used in combination with CT–HPLC for the trace analysis of nitrogen bases in petroleum products from gasoline to diesel fuels (see Chapter 6).

In their HPLC analyses, iodine–amine CT complexation was applied by Clark et al. [99] to enhance the UV detectability of amines. For example, an HPLC analysis is described for *N,N*-dimethylbenzylamine which includes direct chromatography of the free amine, on-line formation of the complex, and the detection of amine in the complexed form with a 20-fold increase in peak area. The case of biological amines is treated in Section IV.

B. Azaarenes

Independent of the study above, a high-speed liquid chromatographic method for analysis of azaarenes, found in airborne particulate matter, was described as early as 1972 by Ray and Frei [100]. (We recall that this group also contributed to the separation of azaarenes by argentation chromatography (see Chapter 4, Section V and Ref. 168). The authors [100] prepared the first chemically bonded nitroaromatic phase for LC by bonding aromatic isocyanate to the surface silanol groups of porous glass beads (corasil I). The column was specifically developed for the separation of polynuclear azaheterocycles. The mixture of model compounds is the same as that used in argentation chromatography. The elution order (similar to that reported in Fig. 13 of Chapter 4) was discussed in terms of steric and electronic effects. The columns can be used for a large number of separations in routine quantitative analysis of azaarenes at the nanogram and even the subnanogram concentration level.

Holstein [37] applied HPLC with the TCI phase on both PAHs (see Table 7) and on various nitrogen heterocycles. In the latter case the interpretation of the results is more difficult, as several factors take part in the complexation: adsorption on the free silanol groups, heteroatom position, and alkylation (chain length and its position).

Nitrogen heterocycles were also studied by Hellmann [56] in HPLC experiments on DNFA, PCA, and PFA silica-bonded phases (Section II.C). Their behavior was different according to their acid–base properties. All derivatives with acidic character (pyrrole, indole, carbazole, 2,3-benzocarbazole) are eluted with a 70:30 *n*-hexane/CH_2Cl_2 mixture. A linear relation was found in all cases between the capacity factor k' and the number of aromatic rings condensed to the pyrrole nucleus. Then these heterocycles can interfere with the corresponding PAHs. All derivatives with basic character (pyridine, quinoline, acridine) are retained on the column. In case of the neutral *N*-alkyl carbazoles (*N*-methyl, ethyl and butyl carbazoles) the influence of the chain length is the same as described earlier for alkyl pyrenes (Section II.C).

More recently, Bertrand et al. [101] described analytical and preparative charge-transfer liquid chromatographies on 3,5-dinitrobenzamidosilica gel [47] for the separation of 1,4- and 1,5-bis(trimethylsilyl)indole isomers as well as 4- and 5-trimethylsilylindole isomers. The separation of various alkyl pyrazines was reported by Vasundhara and Parihar [102].

IV. BIOLOGICAL COMPOUNDS: AMINO ACIDS, PEPTIDES, NUCLEOSIDES, AND NUCLEOTIDES

The existence and characteristics of charge-transfer complexes in the field of biomolecules had long been known and the subject was reviewed by Slifkin in 1971 [103] but without chromatographic applications. But soon after, the charge-transfer chromatography of various biological compounds was studied exhaustively by both liquid chromatography (LC and HPLC) and thin-layer chromatography (TLC).

A. Charge-Transfer/Liquid Chromatography

In the case of liquid chromatography, CTC on highly cross-linked dextran derivatives (Sephadex G-25) and agarose (Sepharose 6B) was developed by Porath's group [104–108, 110, 111]. As the method is performed under mild conditions, it is suitable for the fractionation of substances where the use of GC is not feasible. Aromatic amino acids (His, Phe, Tyr, and halo derivatives), aromatic biological amines (e.g., tryptamine, tyramine, dopamine, adrenaline), oligopeptides, nucleosides, and nucleotides were investigated on hydrophilic gels coupled with electron-acceptor or electron-donor ligands. Nevertheless, water close to the matrix may participate in hydrogen bonding or in dipole–dipole or ion–dipole interactions with the solutes, and the adsorption of aromatic and heterocyclic compounds cannot be caused by electron charge transfer alone [106].

The general structure of the gels is the following:

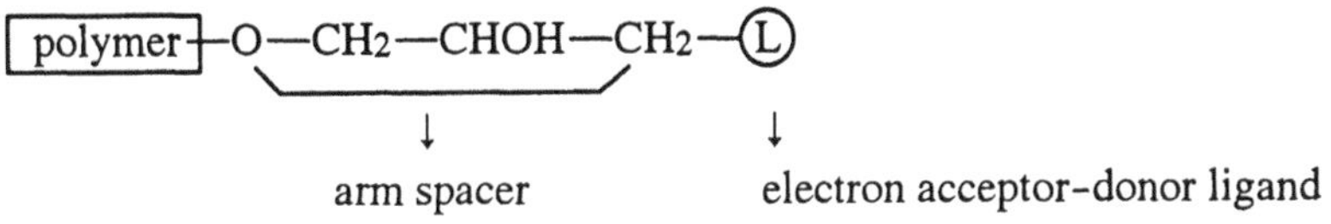

$$\boxed{\text{polymer}}\!\!-\!\!O-CH_2-CHOH-CH_2-\boxed{L}$$

↓ arm spacer   ↓ electron acceptor–donor ligand

where polymer = Sephadex or Sepharose. The main phases are obtained according to the following scheme:

epoxypropyl Sephadex (EP-Sephadex G-25)

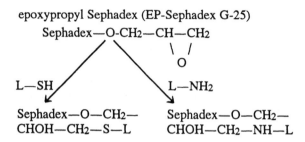

The main ligands are summarized in Table 15. The results were expressed as a function of the reduced elution volumes V_E/V_T, V_E being the sample elution volume and V_T the total bed volume.

The first study with DNP as ligand and various model substances (amino acids, biologic amines, oligopeptides) provided convincing evidence that adsorption was due primarily to the electron-acceptor capacity of the ligands and the electron-donor properties of the solutes [104]. The latter point is illustrated by the relevant examples of Table 16.

Various factors affecting adsorption were studied: temperature (generally, the adsorption increased as the temperature is decreased from 20°C to 4°C), pH, salt adsorption to the gel matrix, and adsorption to the acceptor gels (nature of the ligand acceptors) [106].

Acriflavin–Sephadex gel was used for large-scale fractionation of oligonucleotides [107,108]. Indeed, most studies on biological charge transfer have been carried out on purines and pyrimidines, owing to the interest on nucleic acids in molecular biology. The authors described the chromatographic behavior of adenosine and AMP on different adsorbent gels with DNP, DDQ, chlorpromazine, PCP, and acriflavin as acceptors. As acriflavin–Sephadex yielded the largest differences in V_E/V_T, it was selected for the chromatographic study of nucleotides, nucleosides, and their corresponding bases.

Several observations can be made:

1. The separation is a function of the size or the aromatic ring numbers. Indeed, it was observed that the V_E/V_T ratio fall into two categories, the purine compounds being more retarded than the pyrimidine ones.
2. Adenosine is less retarded than adenine, as the ribose residue affects the adsorption by its hydrophobic character and/or by steric hindrance.
3. When several adsorption centers are introduced into the solute (e.g., phosphate groups), cooperative CT will increase the strength of the adsorption considerably, as illustrated in Fig. 3.

Thus mono-, di-, and trinucleotides can easily be separated on the basis of the number and character of nucleotide units [107]. Temperature, ionic strength, and

TABLE 15 Main Ligands Used in CTC on Gel Matrices (Sepharex G-25 and Sepharose 6B)

Electron-acceptor ligands
  DNP–S–polymer
  (DNP = 2,4-dinitrophenyl) (104–106)

DDQ–S–polymer
(DDQ = 2,3-dicyano-5,6-dichloroquinone)
(105, 106)

ANA–polymer
(ANA = aminophthalamide) (106)

PCP–polymer
(PCP = pentachlorophenyl) (106)

Electron-donor ligands
  Chlorpromazine Sephadex G-25 (105)

*Acriflavin gels: both electron-donor
  and electron acceptor ligands* (107)

TABLE 16 Some Examples of Experiments on DNPS Sepharose
6B (Amino Acids and Oligopeptides)

	V_E/V_T		V_E/V_T
Trp	1.29	Tyr	1.11
Trp–Trp	5.27	Tyr–Tyr	1.41
		Tyr–Tyr–Tyr	2.08

Source: Ref. 104.

solvents have been found to have wide effects on the CT interaction. This fact can be exploited to obtain better resolution [108].

The procedure was applied by Arus et al. [109] to the separation of brominated nucleotides from nonbrominated nucleotides. The authors showed the great efficiency of the combination of CT and affinity chromatographies for the preparative purification of mono- and dinucleotide mixtures.

The effect of urea on the retention of adenosine on acriflavin and/or pentachlorophenyl Sephadex was discussed by Ochoa et al. [110]. It is well known that urea interacts with dyes, aromatic amino acids, and the bases of nucleotides by way of CT complexation. Then the decreasing effect of urea on the retention of adenosine has been interpreted as being due to its electron donor–acceptor properties, acting competitively against the tendency of adenosine to complex on grafted Sephadex or Sepharose. Interestingly, the addition of urea to saline buffer allows differentiation of the adsorption strength of different homopolynucleotides on acriflavin–Sepharose in the order

poly G > poly A > poly C > poly U

The utilization of tryptophan and its dipeptide as ligand, coupled to epoxypropyl Sephadex (or Sepharose), was tried by Vijayalakshmi and Porath [111]. The adsorption of tyrosine, trytophan, and related compounds (proteins) on these gels has been attributed mainly to CT interactions leading to π-bonding, in addition to hydrogen bonding and hydrophobic adsorption.

The separation of aromatic amino acids (Tyr, Trp, Phe) on DNP-Sephadex was studied by Ishihara et al. [112], who discussed the parts of CT and hydrophobic interactions. Comparing the values of the hydrophobicity of the energy of the highest occupied molecular orbital (HOMO) and the V_E/V_T ratio for the amino acids tested, the authors concluded that the effect of the nitro groups observed in the case of DNP-Sephadex is actually based on CT interaction.

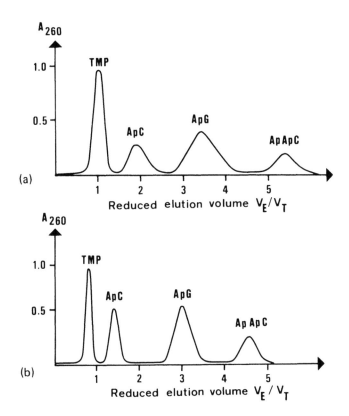

FIGURE 3 Chromatographic resolution of oligonucleotides: thymidine monophosphate (TMP), adenylcytidine (ApC), adenylguanosine (ApG), and adenylyladenylcytidine (ApApC). (a) 0.1 M ammonium acetate buffer (pH 6), 0.2 M ethanolamine at 20°C; (b) 0.1 M ethylmorpholine buffer (pH 7), 2 M NaCl at 20°C. (From Ref. 107.)

B. Charge-Transfer/High-Performance Liquid Chromatography

The first attempt to unite the technique of HPLC and the CT chromatography using microparticulate silica-bonded acriflavin was described by Small et al. [113] in 1982. This technique was named HPCTC (high-performance charge-transfer chromatography) [114]. The similarity of this adsorbent with the acriflavin–Sephadex (Table 15) is shown here (6).

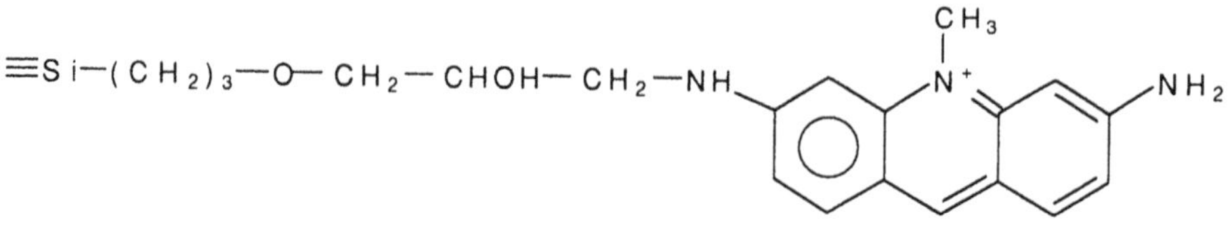

$$\equiv Si-(CH_2)_3-O-CH_2-CHOH-CH_2-NH$$

(6)

The results were expressed in terms of V_E/V_0, with V_E the elution volume of the sample and V_0 the void volume of the column. As anticipated from the results obtained with LC [107,108] the resolution of purines and pyrimidines on acriflavin-silica in 0.1 M N-ethylmorpholine/acetate buffer (pH 7.0) showed that adenine was the most retarded base, followed by guanine and cytosine. Similarly, adenine was more retarded than adenosine, owing to steric hindrance due to the sugar moiety. Separation of the adenosine phosphate in the order AMP < ADP < ATP was obtained by linear gradient elution, with 0.01 to 0.1 M N-ethylmorpholine/acetate buffer (pH 7.0) in less than 8 min, against hours in LC on acriflavin/Sephadex.

In conclusion, HPCTC found wide application in the separation of base nucleosides, nucleotides, oligonucleotides, polynucleotides, as well as coenzymes, vitamins, and drugs. The same procedure allowed purification of homo- and hetero-oligonucleotides [114]. For example, purification of oligonucleotides up to a heptadecamer, using HPCTC, succeeded in about 40 min with high purity and reasonable yields, as can be seen from Fig. 4.

C.   Charge-Transfer/Thin-Layer Chromatography

The application of CT–TLC to biological systems was reported by Slifkin et al. [115,116] for purines, pyrimidines, nucleotides, and amino acids. Three kinds of experiments have been carried out with regard to plate preparation: (1) plates are made using a slurry of silica gel and acceptor; (2) chromatography of mixtures of donor and acceptor was carried out on standard silica gel plates; and (3) grafted silica gel [85] or cellulose [117], with riboflavin as acceptor, was used as adsorbent. In all cases, Kieselgel G F254 was used. The results were expressed in terms of the B constant [2].

The differentiation between purines and pyrimidines reported in LC [107,108] and HPLC [113] was also observed in TLC experiments performed as in experiment type 2 above on mixed solutions of bases with different electron acceptors (chloranil, bromanil, TNB, and TCNE). If no complexation occurred in the case of pyrimidines, in almost all cases of purines the R_f values of the mixtures showed retardation compared with the individual components.

R_f and B values of various compounds (purines, pyrimidines, and amino acids) were determined in impregnated TLC plates with p-benzoquinone, chloranil, and

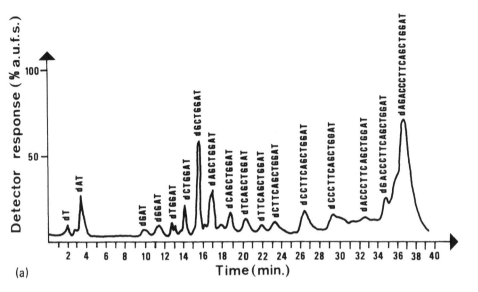

(a)

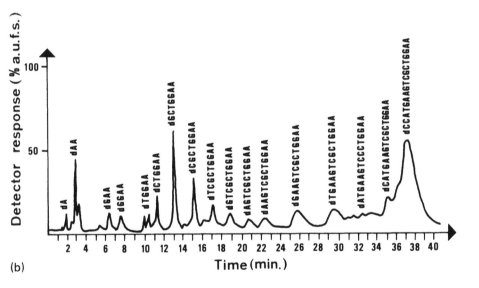

(b)

FIGURE 4   (a) Elution profile of hetero-oligonucleotide CATI (10 g). (b) Elution profile of hetero-oligonucleotide DABI (10 g). Buffer conditions: A and B 20 mM potassium phosphate, 20% acetonitrile (v/v), pH 6.5; B, 20 mM potassium phosphate, 20% acetonitrile, 1 M potassium chloride, pH 6.5. Gradient, 0–30% B in 50-min concave curve: flow rate, 1 mL/min; pressure, 13.8 MPa (2000 psi); A.U.F.S., 0.2; temperature, ambient (18–22°C). (From Ref. 114.)

TNB (see experiment type 1 above). The polarity of the solvents appeared to be a determining factor acting on the B values, which can be positive or negative.

Concerning this point, already explained by Schenk et al. [9] as being due to "masking" of active hydroxyl sites of silica gel by the acceptor (see Section II.A.3), the authors [115] suggested another hypothesis, according to which some of the impregnated acceptor itself moves up the plate, with the solvent dragging the donor with it. However that may be, the main results showed little interaction for adenosine, uracil, cytosine, and adenine, unlike AMP, ADP, ATP, and c-AMP. But the major interaction in the case of nucleotides occurs through the phosphate group and there is competition between the acceptor (riboflavin) and silica gel. To avoid the drawback of acceptor migration, adsorbents with riboflavin grafted to silica gel (or to cellulose) (see experiment type 3 above) were tried. Effectively with bound acceptors, there were no negative B values, reinforcing the author's hypothesis.

Nevertheless, other results obtained by the same group [116,118] in CT–TLC of nucleic acid bases and amino acids with polycyclic aromatic hydrocarbons (PAHs) suggested exclusively the presence of "masking silica gel effects." In these new experiments, the interaction of nucleic bases [116,118] and amino acids [118] with PAHs (pyrene, phenanthrene, anthracene, naphthalene) was studied by CT–TLC as follows: TLC of PAH donors was performed on silica gel plates coated with the nucleic acid bases acting as acceptors, detection being made by observation with a UV lamp (254 nm). For example, the results obtained from pyrene are reported in Table 17. From these results the authors concluded that the interaction of PAHs with purines correlates well with the electron-accepting ability of purines based on their structure (i.e., uric acid having three carbonyl groups, adenine and hypoxanthine having no electron-withdrawing groups but electron-donating amino or hydroxyl groups).

It is noteworthy that in this experiment purines act as electron acceptor against PAHs as electron donors. This is contrary to the behavior of the purines (as electron donors), with riboflavin acting as the electron acceptor [115]. In the case of pyrimidines (having good acceptor properties) the results are less obvious, owing to the negative B values (thymine, uracil), which now appear to be due to masking rather than to migration of the bases with the solvent.

Finally, the interaction of PAHs with adenosine nucleotides, correlated directly with the number of phosphate groups, was attributed to electrostatic forces in addition to CT complexation. The role of the masking effect was also pointed out during CT–TLC of amino acids with pyrene [116], as negative values of B were obtained without any observable migration of the impregnant (amino acids) with the solvent, whatever the impregnant concentration may be.

Interaction between pairs of amino acids influences the structure and association capacity of proteins and also has biological consequences. In recent research Cserhati and Szögyi [119] tried to show this interaction using CT–TLC in reversed-phase mode. For this purpose, the effects of D and L amino acids were de-

TABLE 17 *B* values of Pyrene from TLC Impregnated Plates
(Solvent = Chloroform/Heptane 1:99)

Impregnant	*B* values
Purines	
Adenine	7
Guanine	16
Hyproxanthine	28
Uric acid	36
Pyrimidines	
Thymine	33
Uracil	-12
Cytosine	-5
Nucleoside-nucleotides	
Adenosine	3
AMP	20
c-AMP	26
ADP	34
ATP	46

Source: Ref. 116.

termined separately on the retention behavior of TRP on Polygram UV 254 plates (Macherey-Nagel, Düren, RFA). The results were expressed in terms of R_M values (see Section V.A), characterizing molecular lipophilicity in RP-TLC [120]. Whereas some amino acids, such as Ala, Ser, and Gly, did not influence the R_M value of Trp, others increased or decreased the retention of Trp. The authors suggested that those amino acids that interact with Trp gave either more lipophilic or more hydrophilic molecular complexes than did uncomplexed Trp. They concluded that free amino acids interact with each other in aqueous media, the strength of interaction depending on the structure of the particular amino acid pair. Nevertheless, this strength is very weak compared with the strengths of other interactions determined under similar RP-TLC conditions between nonionic surfactants and phospholipids, as will be shown later. The same authors described adenine–amino acid interactions by CT-TLC [120].

V. MISCELLANEOUS

Some examples of chromatographic applications of CT complexation in different fields not already discussed are provided below.

A. Study of the Interaction between Phospholipids and Nonionic Surfactants

Cserhati et al. used CT chromatography to study the interaction between phospholipids and nonionic surfactants. Several papers have been published [121,122] on this subject with the following main results. Nonionic surfactants, present in pesticide formulations, modify membrane permeability and this effect depends on their structural characteristic.

It is assumed that nonionic surfactants of general formula

$$Q—O—(CH_2—CH_2—O)—H$$

(with Q = hydrophobic part of various structures) interact with phospholipids, forming organic molecular complexes. As CT chromatography is suitable for studying such complex formation, the authors applied RP-TLC to investigate the interaction of dioleoyl phosphatidyl choline (DOPC) with some nonionic surfactants. They were mixed in the eluant at various concentrations, and the dependence of the DOPC lipophilicity (evaluated by the parameter R_M) on surfactant concentration was considered to be related to the interaction forces:

$$R_M = \log\left(\frac{1}{R_f} - 1\right)$$

The degree of lipophilicity decrease corresponds to the stability of the complex. As the authors pointed out, their data do not prove that the interaction DOPC/nonionic surfactant is due to CT complexation [121]. Therefore, they considered CT chromatography to be a "method suitable for detecting any type of weak interaction between two molecular species" (although there are many types of weak interactions having nothing to do with charge transfer or EDA interactions). In this method a chloroform solution of DOPC is spotted on the plates (DC cellulose Alufolien, Merck), with water/ethanol as the eluant. Surfactants were dissolved in the eluant in a concentration depending on the strength of interactive forces. After development and drying (105°C), detection is made by iodine vapor [121]. In this way, the interactions between several synthetic phospholipids and the nonionic surfactant nonylphenylnonylglycolate were studied [122]. The method proved to be a versatile and simple procedure to use to detect minor differences between the complex-forming capacities of similar compounds and does not require complicated instrumentation.

B. Study of the Interaction between Fungicides and Thiol Amino Acids

In the same way, the interaction of fungicides [123] such as 1-phenyl-2-nitro-3-acetoxyprop-1-ene and its precursor, 1-phenyl-2-nitro-propane-1,3-diol acetate, with thiol amino acids and glutathione was studied by RP–TLC. The activities of various bioactive α,β-unsaturated nitrocompounds are assumed to be due to their reactivity toward sulfhydryl groups of biomolecules, and it is assumed that the fungicides block the sulfhydrydryl-containing systems, essential to normal metabolism of fungi. From the R_f values obtained in the additive-free eluant and the R_f' values obtained in eluants containing various additives, three parameters characterizing the strength of interaction were calculated: (1) $R_M - R_M'$, (2) a binding constant B, and (3) a parameter

$$B' = \frac{R_f' - R_f}{1 - R_f} \times 100$$

The third relation was proposed to avoid negative B values, as was often the case. The main results of this study confirm that the fungicide can act by blocking the sulfhydryl groups of essential proteins in fungi and that the interaction with thiol amino acids involves molecular complex formation. The interaction of amino acids and some peptides with the herbicides diquat and paraquat was also determined by reversed-phase charge-transfer chromatography by Czerhati et al. [124].

C. Identification of Pharmaceuticals in TLC

Many natural or synthetic compounds of pharmaceutical interest can form colored CT complexes with π-acceptor molecules. Then π-acceptors as spray reagents [e.g., chloranil, fluoranil, 2,5-dichloro(p-benzoquinone)] were often reported in the identification of pharmaceuticals on TLC plates. Some examples are detection of alkaloids [125–127], penicillins [128], diuretics and oral hypoglycemics [129], triterpenoids and steroids [130], and drugs with a tertiary N-ethyl group [131]. Another example concerns the separation by TLC of CT complexes between iodine (as acceptor) and phenothiazines derivatives on Kieselgel GF254 Merck [132].

D. Detection of Sulfur-Containing Pesticides on TLC

Various sulfoxides, sulfones, and sulfides (all chemical groups that could be present in pesticides), separated on silica gel, were detected by spraying with various acceptor reagents (e.g., TCNE, DDQ, chloranil) [133]. As anticipated, the sulfone groups are weaker bases and required development at 80°C to be detected.

E. Separation of Conjugated Derivatives

A lot of practical applications using CT–TLC were published by Parihar et al. [134]. They concerned the separation of conjugated compounds acting as donors (sorbic acid; vitamin A; α-, β-, and γ-carotenes; lycopene; β-citraurin) on plates impregnated with acceptor nitroaromatic compounds such as tetranitro (TNDPS) and hexanitro (HNDPS) diphenylsulfides. Cinnamic acid and its 4-methoxy and

3,4-dimethoxy derivatives were separated by liquid chromatography with silicic acid + a 12% solution of 7-(2,3-dihydroxypropyl)theophylline as stationary phase (135). By using 11 stationary phases and among them, six charge-transfer stationary phases, and after a correspondence factor analysis (CFA), Walczak et al. [136] claimed the superiority of DNAP phase for the separation of E-s-cis and Z-s-cis isomeric chalcones.

REFERENCES

1. N. Kucharczyk, J. Fohl, and J. Vymetal, *J. Chromatogr., 11*: 55 (1963).
2. R. G. Harvey and M. Halonen, *J. Chromatogr., 25*: 294 (1966).
3. M. Franck-Neumann and P. Jössang, *J. Chromatogr., 14*: 280 (1964).
4. A. Berg and J. Lam, *J. Chromatogr., 16*: 157 (1964).
5. H. Kessler and E. Müller, *J. Chromatogr., 24*: 469 (1966).
6. J. Lam and A. Berg, *J. Chromatogr., 20*: 168 (1965).
7. R. Foster, *Organic Charge Transfer Complexes*, Academic Press, New York, 1969, p. 378.
8. V. Libickova, M. Stuchlik, and L. Krasnec, *J. Chromatogr., 45*: 278 (1969).
9. G. H. Schenk, G. L. Sullivan, and P. A. Fryer, *J. Chromatogr., 89*: 49 (1974).
10. G. D. Short and R. Young, *Analyst, 94*: 259 (1969).
11. J. M. Daisey, in *Handbook of Polycyclic Aromatic Hydrocarbons* (Alf Bjorseth, ed.), Marcel Dekker, New York, 1983, p. 407.
12. N. P. Buu-Hoï and P. Jacquignon, *Experientia, 13*: 375 (1957).
13. A. D. Giehr, Diplom Arbeit TU Clausthal, Clausthal Zellerfeld (RFA), 1974.
14. W. Holstein and H. Hemetsberger, *Chromatographia, 15*: 186 (1982).
15. W. Holstein and H. Hemetsberger, *Chromatographia, 15*: 251 (1982).
16. R. Tye and Z. Bell, *Anal. Chem., 36*: 1612 (1964).
17. B. L. Karger, M. Martin, J. Loheac, and G. Guiochon, *Anal. Chem., 45*: 496 (1973).
18. T. J. Wozniak, *The Chemical and Biological Characterization of Hydroaromatic Compounds in Coal Derived Synfuels*, Ph.D. thesis, 1984; UMI Dissertation Information Service, Bell and Howell Information Company, Ann Arbor, Mich., 1988.
19. T. J. Wozniak and R. A. Hites, *Anal. Chem., 55*: 1791 (1983).
20. T. J. Wozniak and R. A. Hites, *Anal. Chem., 57*: 1320 (1985).
21. M. Diack, R. Gruber, D. Cagniant, H. Charcosset, and R. Bacaud, *2nd International Rolduc Symposium*, The Netherlands, May 22–27, 1989; *Fuel Process. Techn. 24*: 151 (1990).
22. L. Nondek and Y. Malek, *J. Chromatogr., 155*: 187 (1978).
23. J. T. Ayres and C. K. Mann, *Anal. Chem., 36*: 2185 (1964).
24. J. T. Ayres and C. K. Mann, *Anal. Chem., 38*: 861 (1966).
25. L. Nondek and M. Minarik, *J. Chromatogr., 324*: 261 (1985).
26. L. Nondek, *J. Chromatogr., 373*: 61 (1986).
27. C. H. Lochmüller and C. W. Amoss, *J. Chromatogr., 108*: 85 (1975).
28. C. H. Lochmüller, R. R. Ryall, and C. W. Amoss, *J. Chromatogr., 178*: 298 (1979).
29. H. Hemetsberger, H. Klar, and H. Ricken, *Chromatographia, 13*: 277 (1980).
30. H. Hemetsberger and H. Ricken, *Chromatographia, 15*: 236 (1982).

31. L. Nondek, M. Minarik, and Y. Malek, *J. Chromatogr., 178*: 427 (1979).
32. S. A. Matlin, W. J. Lough, and D. G. Bryan, *J. High Resolut. Chromatogr. Chromatogr. Commun., 3*: 33 (1980).
33. S. A. Matlin and J. S. Tinker, *J. High Resolut. Chromatogr. Chromatogr. Commun., 2*: 507 (1979).
34. D. C. Hunt, P. J. Wild, and N. T. Crosby, *J. Chromatogr., 130*: 320 (1977).
35. N. T. Crosby, D. C. Hunt, L. A. Philp, and I. Patel, *Analyst, 106*: 135 (1981).
36. H. Deymann and W. Holstein, *Erdoel Kohle Erdgas Petrochem. Brennst. Chem., 34*: 353 (1981).
37. W. Holstein, *Chromatographia, 14*: 468 (1981).
38. E. P. Lankmayr and K. Müller, *J. Chromatogr., 170*: 139 (1979).
39. P. L. Grizzle and J. S. Thomson, *Anal. Chem., 54*: 1071 (1982).
40. J. S. Thomson and J. W. Reynolds, *Anal. Chem., 56*: 2434 (1984).
41. W. E. Hammers, A. G. M. Theeuwes, W. K. Brederode, and C. L. De Ligny, *J. Chromatogr., 234*: 321 (1982).
42. G. Eppert and I. Schinke, *J. Chromatogr., 260*: 305 (1983).
43. L. Nondek and R. Ponec, *J. Chromatogr., 294*: 175 (1984).
44. L. Nondek, Czech. C.S. 233,990 (C1C01B33/157) Dec. 1, 1986; Appl. 83/8660, Nov. 22, 1983; see *Chem. Abstr., 107*: P137053s.
45. L. Nondek, *J. High Resolut. Chromatogr. Chromatogr. Commun., 8*: 302 (1985).
46. L. Nondek, *J. High Resolut. Chromatogr. Chromatogr. Commun., 11*: 217 (1988).
47. G. Felix and C. Bertrand, *J. High Resolut. Chromatogr. Chromatogr. Commun., 7*: 160 (1984).
48. G. Felix and C. Bertrand, *J. High Resolut. Chromatogr. Chromatogr. Commun., 8*: 362 (1985).
49. G. Felix and C. Bertrand, *J. High Resolut. Chromatogr. Chromatogr. Commun., 7*: 714 (1984).
50. K. J. Welch and N. E. Hoffman, *J. High Resolut. Chromatogr. Chromatogr. Commun., 9*: 417 (1986).
51. T. H. Mourey and S. Siggia, *Anal. Chem., 51*: 763 (1979).
52. C. H. Lochmüller, M. L. Hunnicutt, and R. W. Beaver, *J. Chromatogr. Sci., 21*: 444 (1983).
53. C. H. Lochmüller, A. S. Colborn, M. L. Hunnicutt, and J. M. Harris, *Anal. Chem., 55*: 1344 (1983).
54. M. Verzele and N. Van De Velde, *Chromatographia, 20*: 239 (1985).
55. T. G. Den, K. C. Ho, C. H. Lee, and T. C. Chang, *HRC & CC, 9*: 409 (1986).
56. G. Hellmann, Dissertation zur Erlangung des Doktorgrades für Ingenieur wissenschaften der TU Clausthal., Clausthal-Zellerfeld (RFA), 1984.
57. W. Holstein, *Erdoel Kohle Erdgas Petrochem. Brennst. Chem., 40*: 175 (1987).
58. H. Stetter and L. Schroeder, *Angew. Chem. Int. Ed. Engl., 7*: 948 (1968).
59. Ph. Cléon, M. C. Fouchères, D. Cagniant, D. Severin, and W. Holstein, *Chromatographia, 20*: 543 (1985).
60. M. Lee and D. L. Vassilaros, *Anal. Chem., 51*: 768 (1979).
61. P. Jadaud, M. Caude, and R. Rosset, *J. Chromatogr., 439*: 195 (1988)
62. G. Felix and C. Bertrand, *J. Chromatogr., 319*: 432 (1985).
63. G. Felix, C. Bertrand, and F. Van Gastel, *Chromatographia, 20*: 155 (1985).

64. C. Bertrand, G. Felix, C. Biran, and M. Fourtinon, *J. Liq. Chromatogr.*, *10*: 853 (1987).
65. G. Felix, E. Thoumazeau, J. M. Colin, and G. Vion, *J. Liq. Chromatogr.*, *10*: 2115 (1987).
66. G. Felix, C. Bertrand, and A. Fevrier, *J. Liq. Chromatogr.*, *7*: 2383 (1984).
67. G. Felix, A. Rahm, and C. Bertrand, *Chromatographia*, *25*: 451 (1988).
68. J. J. Burger and E. Tomlinson, *Anal. Proc.*, *19*: 126 (1982).
69. R. J. Laub and R. L. Pecsok, *J. Chromatogr. Chromatogr. Rev.*, *113*: 47 (1975).
70. J. W. King and P. R. Quinney, *J. Chromatogr.*, *49*: 161 (1970).
71. C. Eon, C. Pommier, and G. Guiochon, *Chromatographia*, *4*: 235 (1971).
72. C. Eon, C. Pommier, and G. Guiochon, *Chromatographia*, *4*: 241 (1971).
73. R. C. Castells, *Chromatographia*, *6*: 57 (1973).
74. C. W. P. Crowne, M. F. Harper, and P. G. Farrell, *J. Chromatogr. Sci.*, *14*: 321 (1976).
75. R. Queignec and M. Cabanetos-Queignec, *J. Chromatogr.*, *209*: 345 (1981).
76. R. O. C. Norman, *Proc. Chem. Soc.*, 151 (1958).
77. A. R. Cooper, C. W. P. Crowne, and P. G. Farrell, *Trans. Faraday Soc.*, *62*: 2725 (1966).
78. A. R. Cooper, C. W. P. Crowne, and P. G. Farell, *Trans. Faraday Soc.*, *63*: 447 (1967).
79. S. H. Langer, C. Zahn, and G. Pantazoplos, *J. Chromatogr.*, *3*: 154 (1960).
80. R. J. Cvetanovic, F. J. Duncan, and W. F. Falconer, *Can. J. Chem.*, *42*: 2410 (1964).
81. W. H. Pirkle and T. C. Pochapsky, *Chem. Rev.*, *89*: 347 (1989).
82. M. S. Newman, W. B. Lutz, and D. Lednicer, *J. Am. Chem. Soc.*, *77*: 3420 (1955).
83. L. H. Klemm and D. Reed, *J. Chromatogr.*, *3*: 364 (1960).
84. L. H. Klemm, K. B. Desai, and J. R. Spooner, Jr., *J. Chromatogr.*, *14*: 300 (1964).
85. F. Mikes, G. Boshart, and E. Gil-Av, *J. Chromatogr.*, *122*: 205 (1976).
86. F. Mikes and G. Boshart, *J. Chromatogr.*, *149*: 455 (1978).
87. H. Nakagawa, K. I. Yamada, and H. Kawazura, *J. Chem. Soc. Commun.*, 1378 (1989).
88. C. H. Lochmuller and R. Ryall, *J. Chromatogr.*, *150*: 511 (1978).
89. Y. H. Kim, A. Tishbee, and E. Gil-Av, *J. Am. Chem. Soc.*, *102*: 5917 (1980).
90. D. B. Parihar, S. P. Sharma, and K. K. Verma, *J. Chromatogr.*, *31*: 120 (1967).
91. A. K. Dwivedy, D. B. Parihar, S. P. Sharma, and K. K. Verma, *J. Chromatogr.*, *29*: 120 (1967).
92. D. B. Parihar, S. P. Sharma, and K. K. Verma, *J. Chromatogr.*, *29*: 258 (1967).
93. J. P. Sharma and S. Ahuya, *Z. Anal. Chem.*, *267*: 368 (1973).
94. S. L. Snyder and C. M. Welch, *J. Polym. Sci. Polym. Lett. Ed.*, *11*: 695 (1973).
95. S. L. Snyder and C. M. Welch, *J. Chromatogr.*, *105*: 157 (1975).
96. C. W. P. Crowne, M. F. Harper, and P. G. Farrell, *J. Chromatogr.*, *61*: 7 (1971).
97. L. Nondek and V. Chvalovsky, *J. Chromatogr.*, *312*: 303 (1984).
98. L. Nondek and V. Chvalovsky, *J. Chromatogr.*, *268*: 395 (1983).
99. C. R. Clark, C. M. Darling, J. L., Chan, and A. C. Nichols, *Anal. Chem.*, *49*: 2080 (1977).
100. S. Ray and R. W. Frei, *J. Chromatogr.*, *71*: 451 (1972).
101. C. Bertrand, G. Felix, C. Biran, and M. Fourtinon, *J. Liq. Chromatogr.*, *10*: 853 (1987).
102. T. S. Vasundhara and D. B. Parihar, *J. Chromatogr.*, *194*: 254 (1980).

103. M. A. Slifkin, *Charge Transfer Interactions of Biomolecules*, Academic Press, London, 1971.
104. J. Porath and K. D. Caldwell, *J. Chromatogr., 133*: 180 (1977).
105. J. Porath, *J. Chromatogr. Chromatogr. Rev., 159*: 13 (1978).
106. J. Porath and B. Larsson, *J. Chromatogr., 155*: 47 (1978).
107. J. M. Egly and J. Porath, *J. Chromatogr., 168*: 35 (1979).
108. J. M. Egly, J. L. Ochoa, C. Rochette-Egly, and J. Kempf, in *Protides of Biological Fluids 27th Colloquium, 1979* (H. Peteers, ed.), Pergamon Press, New York, 1979, p. 747.
109. C. Arus, M. V. Nogues, and C. M. Cuchillo, *J. Chromatogr., 237*: 500 (1982).
110. J. L. Ochoa, J. Porath, J. Kempf, and J. M. Egly, *J. Chromatogr., 188*: 257 (1980).
111. M. A. Vijayalakshmi and J. Porath, *J. Chromatogr., 177*: 201 (1979).
112. K. Ishihara, T. Iida, N. Muramoto, and I. Shinohara, *J. Chromatogr., 250*: 119 (1982).
113. D. A. P. Small, T. Atkinson, and C. R. Lowe, *J. Chromatogr., 248*: 271 (1982).
114. P. A. D. Edwardson, C. R. Lowe, and T. Atkinson, *J. Chromatogr., 368*: 363 (1986).
115. M. A. Slifkin, W. A. Amarasiri, C. Schandorff, and R. Bell, *J. Chromatogr., 235*: 389 (1982).
116. M. A. Slifkin and S. H. Liu, *J. Chromatogr., 269*: 103 (1983).
117. C. Arsenis and D. B. McCormick, *J. Biol. Chem., 241*: 330 (1966).
118. M. A. Slifkin and H. Singh, *J. Chromatogr., 303*: 190 (1984).
119. T. Cserhati and M. Szögyi, *J. Chromatogr. Biochem. Appl., 434*: 455 (1988).
120. T. Cserhati and M. Szögyi, *J. Liq. Chromatogr., 11*: 3067 (1988).
121. T. Cserhati, M. Szögyi, and L. Györfi, *J. Chromatogr., 349*: 295 (1985).
122. T. Cserhati and M. Szögyi, *J. Biochem. Biophys. Methods, 14*: 101 (1987).
123. G. Gullner, T. Cserhati, A. Kis-Tamas, and G. Mikite, *J. Chromatogr., 355*: 211 (1986).
124. T. Czerhati, M. Szoyi, and Z. Szigeti, *Chromatographia, 26*: 305 (1988).
125. G. Rucker and A. Taha, *J. Chromatogr., 132*: 165 (1977).
126. H. J. Huizing, F. De Boer, and T. M. Malingré, *J. Chromatogr., 195*: 407 (1980).
127. S. P. Agarwal and M. A. Elsayed, *Planta Med., 45*: 240 (1982).
128. S. P. Agarwal and J. Nwaiwu, *J. Chromatogr., 323*: 424 (1985).
129. S. P. Agarwal and J. Nwaiwu, *J. Chromatogr., 351*: 383 (1986).
130. S. P. Agarwal and J. Nwaiwu, *J. Chromatogr., 295*: 537 (1984).
131. A. M. Taha and M. A. A. El-Kader, *J. Chromatogr., 177*: 405 (1972).
132. M. R. Gasco and A. Bodrato, *Farmaco Ed. Prat., 26*: 337 (1971).
133. L. Fishbein and J. Fawkes, *J. Chromatogr., 22*: 323 (1966).
134. D. B. Parihar, O. M. Prakash, I. Bajaj, R. P. Tripathi, and K. K. Verma, *J. Chromatogr., 59*: 457 (1971).
135. E. Sondheimer and I. E. Pollak, *J. Chromatogr., 8*: 413 (1962).
136. B. Walczak, M. Dreux, J. R. Chretien, L. Morin-Allory, M. Lafosse, and G. Felix, *J. Chromatogr., 464*: 237 (1989).

4

Argentation Chromatography: Application to the Determination of Olefins, Lipids, and Heteroatomic Compounds

D. Cagniant
University of Metz, Metz, France

I. INTRODUCTION

Some metals, acting as electron acceptors, form molecular complexes with certain organic compounds acting as electron donors. The properties of these complexes were found to be so favorable for use in chromatographic processes that a large number of papers described various techniques adapted to the separation of a great variety of organic compounds. All types of chromatography are involved: thin-layer and paper chromatographies, gas and liquid chromatographies, and more recently, high-performance liquid chromatography.

In the beginning of this chapter, it is necessary to define in what cases the chromatographic separation based on complexes formed between metal ions and organic compounds must be considered as *charge-transfer complex chromatography* (CTC) or as *ligand-exchange chromatography* (LEC). The distinction is not always obvious and is sometimes grounds for discussion in the literature, argentation chromatography and related techniques being the most conspicuous.

For Helfferich [1–3], who was the first to propose the term *ligand exchange* applied to chromatography, the technique allows the separation of "ligands" that would coordinate with metal ions loaded on cation-exchange resin and would be displaced selectively by the mobile phase. For example, using copper or metal ions coordinated at first with ammonia, amines could displace ammonia from the metal ions and in the turn would be displaced down the column by passing more ammonia.

A similar definition was proposed some years after by Muzzarelli et al. [4]: "LEC is based on the principle that a molecule or ion in part of a complex fixed on a support can be released because a different molecule enters to form a more stable complex, or because the complex collapses when the medium is altered." For these authors, argentation chromatography and related techniques are not LEC because the complexing agent is merely added to silica or alumina and is not bound.

For Davankov and Semechkin [5], such a definition of LEC is not satisfactory, as it ignores the multiplicity of LC processes in which various ligand exchanges undoubtedly take place and also as it rejects gas-LEC, which requires a simple addition of complex-forming ion salts to a standard support. Then the authors [5] suggest that LEC should be defined as a process in which interaction between the stationary phases and the molecules to be separated occurs during the formation of coordination bonds inside the coordination sphere of the complex-forming ion. It is the exchange of ligands bound to the central ion of the metal that suggests the term *ligand-exchange chromatography*. The marked tendency of Ag^+ ion to bind different complex-forming molecules is the main distinguishing feature of silver-containing sorbent. It is reasonable, therefore, to include thin-layer argentation chromatography in the definition of the LEC above.

On the other hand, Walton [6] considered that the ligand concept can be interpreted very broadly and claimed that "there is no fundamental reason why, for example, the binding of aromatic hydrocarbons by pyromellitic anhydride adsorbed on silica gel, which occurs by formation of *charge-transfer complexes*, and has been applied in thin-layer chromatography, should not be regarded as ligand exchange. To perform chromatography, the bound substances must of course be displaced or removed somehow, and the displacement must occur selectively" Commenting on the view of Muzzarelli et al. [4], Walton said: "We shall use a broader definition that includes argentation chromatography and related techniques, but we shall stick to cases in which the fixed partner in the complex formation is a metal ion, and generally, though not always, consider the metal ion to be bound by an ion-exchanging material." In other words, all types of chromatography in which the stationary phase contains metal ions and in which the substances to be separated form complexes with these metal ions can be considered as LEC. The distinction between ion-exchange and ligand-exchange reactions is shown in the following typical reactions [7]:

Ion exchange:

$$Zn^{2+} + 2R\!-\!NH_4 \longrightarrow R_2\!-\!Zn + 2NH_4^+ \tag{1}$$

Ligand exchange:

$$R_2Zn(NH_3)_4 + 4H_2N—(CH_2)_2—CH_3$$
$$\longrightarrow R_2Zn(H_2N—(CH_2)_2—CH_3)_4 + 4NH_3 \quad (2)$$

Simultaneously, Davankov et al. [8] defined LEC "as a chromatographic technique in which the formation of coordination bonds is the dominant mechanism responsible for the separation of solute species."

It is noteworthy that the term *ligand exchange* implies the displacement of one type of coordinated ligand by another [see Eq. (2)]. This is particularly the case for LEC separation of amino acid enantiomers. But in other complexation between metal ions acting as electron acceptors and organic compounds acting as electron donors, the ligand exchange is not as clear.

So, to conclude the discussion, the definition put forth by Muzzarelli et al. will be adopted subsequently as it seems to be without ambiguity, although more restrictive. Consequently, argentation chromatography will be classified as charge-transfer complex chromatography. It will be the same for several studies published under LEC, related, for example, to the separation of organic sulfides acting as electrons donors and stationary phases loaded with metal ions acting as electron acceptors.

II. HISTORICAL DEVELOPMENT

Argentation chromatography is a well-known method for the separation of unsaturated compounds, particularly in the case of geometrical isomers, limited at the beginning of its development (1960s) to gas chromatography (GLC or GSC), liquid chromatography (LC), and thin-layer chromatography (TLC). The instability of the silver-loaded stationary phase at elevated temperature ($t > 65°C$) limits the use of the GC methods to low-boiling substances. This drawback is at the origin of the exhaustive utilization of TLC until now, where recent applications involve separation in the field of fatty acids, lipids, prostaglandins, or steroids, for example.

During the last decade, high-performance liquid chromatography (HPLC) with silver-loaded support (Ag-HPLC), or in reversed-phase mode (RP-HPLC), has been applied to most of the separations once dominated by TLC. Nevertheless, with many convenience products on the market (i.e., commercial plates impregnated with silver nitrate), TLC is less operator dependent and is also relatively inexpensive compared with HPLC. Moreover, many methods involve classical argentation TLC used in conjunction with other chromatographic separations (GC, HPLC, RP-HPLC).

The topics treated in this chapter will give many examples of argentation chromatography classified into two main parts: olefinic compounds and lipids. In addition, some other examples will concern miscellaneous applications in the field of heteroatomic compounds.

III. OLEFINIC COMPOUNDS

According to Dewar [9], the bond between an olefinic ligand and a metal ion, such as Ag$^+$, is formed by donation of electrons from the filled π-orbital of the olefin to the vacant s orbital of Ag$^+$ and back donation of d electrons from the metal ion to the antibonding π^*-orbital of the olefin. The bonding in the complex will be affected by the availability of electrons in the filled π-orbital and the ease of overlap of these orbitals. So both polar and steric factors determine the stability of the silver ion–olefin complexes. In fact, the steric effects have been found to be strong and in many cases are sufficient to explain the influence of structure on the stability constants.

According to these basic principles the chromatographic applications of silver–olefin complexing can be classified as follows:

1. Chromatographic separations of hydrocarbon mixtures (saturates, olefins, aromatics)
2. Selective separations of olefins:
    a. Monounsaturated hydrocarbons
    b. Polyunsaturated hydrocarbons: dienes
    c. Terpenes
    d. Pheromones
3. Complexation of olefins with other metals

A. Chromatographic Separations of Hydrocarbon Mixtures

Bradford et al. [10] were the first to report that by dissolving silver nitrate in polyethylene glycol (stationary phase), unsaturates were retained more strongly than saturates in gas chromatography (GC). Unfortunately, the columns tend to decompose from 85°C [11]. The optimum working conditions are within 0 to 30°C, as the complex stability at $t > 50$ to 60°C drops to a value that is no longer significant for the separation. Further the stability of silver nitrate solutions in the organic solvents used is poor at room temperature. The use of glycerol or polyethylene glycol afforded no change in the stability [11]. Therefore, only very volatile olefins can be separated in this way.

To perform the separation of higher olefins from paraffinic, cyclic, and aromatic hydrocarbons, Janák et al. [12] used LC with Porapak Q (a styrene/ethylvinylbenzene copolymer, which has a high sorption affinity to hydrocarbons) as a stationary phase and a complex-forming solution (various concentrations of silver nitrate and propanol/water ratio) as mobile phase. This allows the separation of olefins from hydrocarbon mixtures by LC and/or TLC, avoiding the serious drawbacks of the classical separations by GC on columns containing silver nitrate solutions (i.e., in ethylene glycol).

In the conditions described [12] the olefin elution is more rapid than in systems in which the mobile liquid does not contain Ag⁺, while the retention of aromatic and paraffinic hydrocarbons does not change, as schematized in Fig. 1. In this way the separation of olefins is not longer restricted to the very volatile ones.

Using columns with aqueous solutions of complexing metal ion (Ag⁺ and Hg²⁺) as the stationary phase, Wasik and Tsang [13] demonstrated the feasibility of separating compounds by groups such as olefins, alkanes, and aromatics by GLC. Mercuric ions do not complex with alkanes or with aromatics, while silver ions complex weakly with aromatics. Using, for example, 50% w/w of an aqueous solution of 5.0 M AgNO₃ and 0.05 M Hg(NO₃)₂ as the stationary phase (solid support-60 to 80-mesh Chromosorb P acid wash), aromatics are eluted longer times relative to alkanes while olefins are retained completely. This study was applied only to model compounds.

As the application of silver nitrate phases in GC is limited at elevated temperatures, the development of *high-performance liquid chromatography* (HPLC) afforded a great improvement as well in the separation of olefinic, saturated, and aromatic hydrocarbons as in the separation of geometric isomers (see Section III.B.1). As these new procedures essentially found application in the structural group analysis of petroleum products, they will be developed with more details in the special topic devoted to this subject. In the dual-column technique described by McKay and Latham [14], aromatic hydrocarbons were separated from olefinic and saturated hydrocarbons by HPLC on the silica gel column, and these two groups were then separated by HPLC on the silver–coated silica gel column. Applied to high-boiling distillates and residues of shale oil, the time required for the separation is about 2 h, with recovery of material better than 90%.

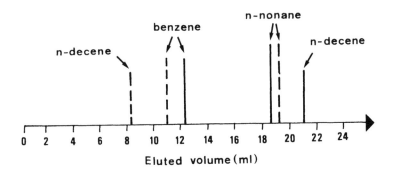

FIGURE 1 Separation of paraffinic, olefinic, and aromatic hydrocarbons by LC with silver nitrate solution as mobile phase. Solid line, liquid chromatography: solvent, n-propanol/water 2:1; flow rate, 0.3 mL/min. Dashed line, liquid chromatography: solvent, n-propanol/water 2:1 + AgNO₃ (0.08 g/mL); flow rate, 0.24 mL/min.

A similar method was also applied by Matsushita et al. [15] to a rapid hydrocarbon analysis of gasoline (25 min). Under their conditions the order of elution was saturates, olefins, and aromatics, with CCl4 as the mobile phase and an infrared (IR) detector.

More recently, Chartier et al. [16] carried out an HPLC analysis offering a significant improvement in resolution between saturated and olefinic compounds. Silver-coated silica gel, or silver chemically bonded to silica gel, is used, allowing separation of the three groups (saturated, olefinic, and aromatic compounds) on the same column, using backflushing. As the method concerns gasoline analysis, a detailed description is reported in Chapter 6.

Note: To increase column stability, another method of silver complex utilization, as adsorbents in GSC, was carried out by Cook and Givand [17] for the separation of saturated, olefinics, and aromatic hydrocarbons (model compounds only). In this method the solid support (Chromosorb W) is coated by a complex of general form AgL_2NO_3, where L represents the ligand molecule (pyridine, picolines), helium being used as the carrier gas. Nevertheless, the column temperature is limited to 40°C to ensure thermal stability of all complexes.

B. Selective Separation of Olefins

1. *Monounsaturated Hydrocarbons*

The potential selectivity of Ag^+-unsaturated compound complexes, according to the number, geometry, and position of $C=C$ doubled bonds, accounts for the significant development of chromatographic separations of olefins. Up to 1979 the literature is well covered by leading articles, such as those of Guha and Janák in 1972 (180 references covered) [18] and of Szczepaniak and Nawrocki in 1975 (127 references covered, Polish) [19]. In 1979 these authors published a review paper in English on the same subject (140 references covered) [20]. It is noteworthy that these papers [19,20], involving application of transition metal complex formation in GC, are not limited to Ag^+ complexes. So after recalling the main results summarized in these fundamental papers and several others of the same period, we shall focus our attention on work published after 1979.

Gas chromatography (GLC–GSC) analysis. By GLC with silver nitrate/ethylene glycol as the stationary phase, Shabtai et al. [21–25] studied alkyl cyclenes, from cyclobutene to cycloheptene derivatives, and the corresponding isomeric alkyl cyclanes. From their exhaustive work the following general rules can be drawn:

1. Substitution at the double bonds decreases the retention volume. This behavior of 1-alkyl cyclenes was attributed to the decrease in complex stability caused by steric hindrance of the alkyl group at the double bond [24].
2. Cyclic (like open-chain) olefins having a substituent in the 3-position show the highest retention volume among the isomers. This fact was correlated with an

increase in the complex stability by inductive effect of the methyl substituent in the 3-position [24,26].

3. The size of the cycle: cyclobutenes have less tendency to form complexes than the corresponding five- and six-membered cycloolefins [23]. Correlations between complex stability and olefin structure were deduced by comparing the relative retention volumes of the isomers, taking into consideration conformation of the seven-membered ring system [24]. The stability constants [26] of olefin/silver ion complexes in ethylene glycol agree with the observations above: lower stability of 1-alkyl isomers, higher tendency for complex formation of the 3-alkyl isomers and a lower stability constant of 1-methyl cyclopentene. These facts were corroborated at the same time by Muhs and Weiss [27], also measuring the equilibrium constants for the silver complex reactions. These constants are found to be strongly affected by the olefin structure, and their variations were discussed in terms of steric strain and electronic effects. The conclusions agreed well with Dewar's model: "The bonding in the complex will be affected by the availability of electrons in the filled orbital and the ease of overlap of these orbitals as is determined by steric factor." As the influence of electronic effects on complex formation is difficult to observe, steric hindrance and strain are more straightforward and most of the effects observed are rationalized using steric arguments. It is noteworthy that by GLC on silver nitrate/ethylene glycol phases, mixtures of volatile isomers whose boiling points differ by only 0.1° are easily separated and analyzed quantitatively.

Separation of cis-trans isomers is another well-studied example [21,22] of the usefulness of argentation chromatography. By GC, using a column packed with a saturated solution of silver nitrate in diethylene glycol on celite and helium as eluant, mixtures of cis-trans olefines were separated by Cope et al. [28–31]. Equilibration of *cis*-and *trans*-cyclenes (C_9 to C_{12}) was also described using a solution of silver nitrate in tetraethylene glycol but at an elevated temperature (88 to 100°C) [31], which is difficult to explain according to the well-known instability of the column above 65°C.

Macroreticular cation-exchange resins are highly adsorption packing for GSC. In the silver form, the resin retains aromatic and olefin compounds strongly and separates geometric isomers. Some examples were given by Hirsch et al. [32] for butenes and pentenes.

Liquid chromatography (LC) analysis. LC was also applied for the separation of cis and trans isomers of high-molecular-weight olefins produced by the isomerization of 1-olefins (C_{12} to C_{20}) (columns of alumina impregnated with 30% silver nitrate, 99:1 pentane/ethyl ether eluting *trans*-olefins and 97:3 pentane/ethyl ether eluting *cis*-olefins) [33]. By high-speed LC, Mikes et al. described the separation of isomeric cyclolefins with 0.8% $AgNO_3$/1.7% ethylene glycol on Corasil II or 5% $AgNO_3$ on silica Woelm 32. The correlation of retention times with known argentation constants was discussed [34]. The system was also applied to the separation of some unsaturated natural compounds such as fatty acids, prostaglandins and pentacyclic triterpenes.

High-performance liquid chromatography (HPLC) analysis. Since their appearance, HPLC techniques were carried out in the case of unsaturated compounds. Many olefinic isomers were rapidly analyzed using HPLC columns packed with silver nitrate–coated silica gel [35]. As these columns are not commercially available, Heath and Sonnet [36] carried out a method for the in situ coating of silver nitrate onto silica gel using commercial HPLC columns. The ability of these columns to resolve geometrical isomers was tested using decenes, octadecenes, and 9-tetradecen-1-ol acetate (*E* and *Z*) (see Section III.B.4).

In the same way the development on *reversed-phase high-pressure liquid chromatography* (RP–HPLC) afforded great improvements in the application of argentation chromatography. They were discussed by Vonach et al. [37], who pointed out some drawbacks of the preceding methods: instability of the column above 120°C in the argentation GC; several restrictions in the case of argentation LC (including TLC), such as difficulty in adaptation of the silver concentration to the complexation strength of the solute; too long retention times of solutes undergoing the strongest complexation; and so on. The usefulness of RP–HPLC on chemically bonded phases, by using mobile phases containing silver salts at various concentrations, were described [37,38] in the case of olefins, sesquiterpenes, fatty acid esters, and triglycerides as well as heterocyclic compounds. Several advantages of the RP separations were claimed: the more polar solutes are eluted first; low Ag^+ concentrations (ca. 10 to 20 mol/L) in the mobile phase are sufficient for the common separation; good reproducibility of retention times; no direct influence of the mobile phase on the chromatographic properties of the support. As far as olefins are concerned, Schomburg and Zegarski [38] described the separation of *cis*- and *trans*-2-hexenes, octenes, and decenes. The method was applied in an RP mode on a 10-μm Lichrosorb RP 8 column and water/isopropanol mixtures of variable ratio with silver nitrate (concentrations between 0.5 and 2.5%) as the mobile phase. The elution order of cis and trans isomers, the cis compounds being eluted before the trans isomers, is opposite to the separations performed with a silver nitrate–containing stationary phase, as the better complexed species is preferentially found in the mobile phase in which the transport of the solute molecules takes place. Cis-trans separation can be carried out not only on group geometry but also according to size. Other examples of olefin separations were done [37] using $AgClO_4$ as the silver salt, in the case of homologous (*Z,E*)-2 alkenes (C_5, C_8, C_{10}).

Another interesting feature of argentation chromatography, well summarized by Guha and Janǎk [18], is the application of a secondary isotope effect to the separation of deuterated and tritiated olefins. An example of the influence in electron density around the double bond on the *Rf* values was afforded by Fuggerth [39] in the case of substituted stilbenes, separated into their cis and trans isomers by TLC and HPLC.

X=F, Cl, Br, OCH3, OH $Y = C_6H_5-$

If TLC on silica gel alone did not give satisfactory separation, Ag–TLC [silica gel HF254 + 2% aqueous solution silver nitrate (5% w/w)] gave excellent separation. In the same way, if HPLC without silver nitrate is ineffective, an excellent separation can be done with Ag–HPLC (Partisil 10 impregnated with 5% w/w silver nitrate in ethyl acetate).

Spots with higher Rf values (TLC) and peaks of shorter retention time (HPLC) are assigned to Z structures. Reversed-phase chromatography (methanol/water 200:50) gave direct separation of Z and E isomers except with X = Cl, Br (broad peaks). In these cases, two separate peaks were obtained with silver nitrate in the eluant. What happens in the case of polyunsaturated compounds is discussed below.

2. Polyunsaturated Hydrocarbons: Dienes

In argentation GC of dienes, complex formation with silver nitrate differs markedly in ethylene glycol compared with aqueous medium, where it was found that conjugation may reduce the stability of the complex drastically [23]. In ethylene glycol no such effect is apparent, as can be seen by examination of Table 1, where some characteristic examples of dienic compounds are reported with, for comparison, values of K_1 (the equilibrium constants [27]) and of r (the retention volumes [23].

The following conclusions can be outlined:

1. A cumulative double bond has a lower forming capacity than that of a single double bond.
2. Unconjugated dienes coordinate strongly in ethylene glycol. 1.4-Pentadiene ($n = 1$) has a retention volume 3.6 times higher than calculated on the basis of additivity (1-pentene $r = 0.19$) and 1,5-hexadiene ($n = 2$) is irreversibly retained by the column. The 1,5-diene system appears to possess the optimum configuration for chelation, the diolefin acting as bidentate ligands [23,27], the greater stability ($K_1 = 28.8$) being explained by lesser strain in the chelate ring.

TABLE 1 Relation between Equilibrium Constants K_1 and Retention Volumes r for Some Dienic Compounds

Dienes	K_1[a]	r[b]
Cumulenes		
Allene	0.8	0.036
Methylallene	0.8	—
Unconjugated dienes		
$CH_2{=}CH{-}(CH_2)_n{-}CH{=}CH_2$		
$n = 1$	10.2	1.34
$n = 2$	28.2	Retained irreversibly by the column
$n = 3$	14.7	
$n = 4$	11.3	
$n = 5$	10.4	
$n = 6$	7.8	
$CH_2{=}CH{-}(CH_2)_2{-}CH{=}CH_2$	28.8	
2-Me	22.1	
2,5-DiMe	13.3	
Conjugated dienes		
Butadiene	4.2	
Isoprene	3.1	0.32
1,3-Pentadiene		
cis	4.4	0.63
trans	3.5	0.46
Cyclenes		
Cyclopentadiene	4.6	0.97
Cyclooctadienes		
$\triangle_{1,3}$	3.2	
$\triangle_{1,4}$	14.4	
$\triangle_{1,5}$	75	
$\triangle_{1,3,5,7}$	91	

[a]K_1, equilibrium constant for the following reaction in the solvent L (in this case, ethylene glycol):From Ref. 27. [olefin]L + [Ag(NO$_3$)]L ⇌ [olefin-AgNO$_3$]L

[b]r, retention volumes at 30°C relative to toluene (30 g of AgNO$_3$/100 mL of ethylene glycol). From Ref. 23.

3. Cyclooctadiene series. When the 1,5-diene system is held rigidly in a ring, such as $\triangle 1,5$, the value of K_1 is extremely high ($K_1 = 75$) and the chelate ring is stable even in aqueous solution at 90 to 100°C [27]. It is the same for $\triangle 1,3,5,7$, but the possibility of reaction with a second silver ion is not excluded in this case.

4. Conjugated dienes. They are somewhat poorer ligand than the corresponding monoolefins (i.e., butadiene $K_1 = 4.2$, 1-butene $K_1 = 7.7$), the weakening of the bonding being the result of delocalization. Even if cyclopentadiene ($r = 0.97$) has a higher retention volume than that of conjugated open-chain dienes (effect of double bonds in a five-membered ring), in this case also, the conjugation seems to weaken the coordination capacity since its retention volume is much lower than the additive value calculated from cyclopentene ($r = 0.90$).

By RP-HPLC Vonach and Schomburg [37] and Schomburg and Zegarski [38] described the separation of polyunsaturated hydrocarbons, such as 1,4- and 1,7-oc-tadienes, 1,9-decadiene, 1,4,9-decatriene, several mono- and polyunsaturated cyclooctenes, and the four geometrical isomers of 1,5,9-cyclododecatrienes (using silver nitrate [38] or silver perchlorate [37]) (Fig. 2).

3. Terpenes

Argentation chromatography was introduced in the field of terpenoïds in the early 1960s principally by TLC. Several examples of separation of monoterpene mixtures as well as sesquiterpenes, tetracyclic terpenes, and mono, sesqui, and diterpenic alcohols were reviewed by Morris [40] in 1966 and Guha and Janǎk [18] in 1972. Until very recently no significant modifications to these TLC techniques were introduced and the original procedures involving the use of silica gel layers containing silver nitrate and gypsum are still used. The detection is performed by spraying with a solution of chlorosulfonic acid in acetic acid, phosphomolybdic acid in ethanol, or antimony perchlorate in chloroform.

The superiority of silver perchlorate to silver nitrate in TLC of terpenes was claimed by Prosad et al. [41]. The silver salt needs to be acid-free; otherwise, a catalyzed rearrangement of certain terpenes may occur. The same authors showed that gypsum in the adsorbent decreased the resolution.

In 1985, Kohli and Badaisha [42] claimed that silver iodate was superior to silver nitrate for certain terpenoid separations, without being able to explain the differences in the behavior of the studied terpenoids toward these two complexing agents.

From all these results very few conclusions were drawn in relation to structure. According to Lawrence [43], some some possible conclusions are the following:

1. The R_f values of a monoterpene hydrocarbon varies with the impregnation percentage of silver nitrate.

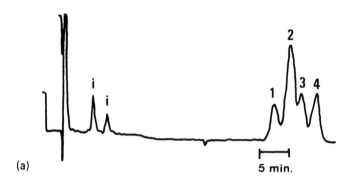

(a)

5 min.

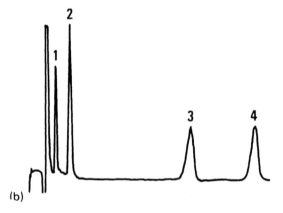

(b)

FIGURE 2 Separation of the four isomeric 1,5,9-cyclododecatrienes. Column, 150 × 4 mm i.d.; stationary phase, 5-μm Nucleosil 5 C_{18}; mobile phase: (a), methanol/water 3:1 v/v; (b),(a) + 3 × 10^{-3} N AgClO$_4$; pressure, 80 bar; flow rate, 0.7 mL/min; temperature, 24°C; detector, RI; attenuation, ×4. Peaks: 1, *Z,Z,Z;* 2, *Z,Z,E;* 3, *Z,E,E;* H, *EEE*; i, impurity. (From ref. 29.)

2. Cyclic terpenes with single internal double bond do not readily form complexes,
3. Cyclic or acyclic terpenes with two nonterminal double bonds do not readily form complexes unless the double bonds are cis conjugated,
4. Cyclic or acyclic terpenes with exocyclic or terminal double bonds do form complexes.

Along the same lines, it was found in the field of allylic and propenylic derivatives of benzene or cyclohexene that only allylic isomers readily formed complexes with silver cations [44].

For gas chromatographic separations we shall return to the papers of Shabtai et al. [21] and Muhs and Weiss [27]. The P-menthenes studied [21] appeared as examples of semi- and endocyclic isomers in the cyclohexene series. In Table 2 are reported several values of the equilibrium constant K_1 [27].

The high value of K_1 ($K_1 = 33.7$) for 2,5-norbornadiene agrees with its appropriated conformation for chelation. The bridging group accounts for the greater value of K_1 in α-pinene compared to 1-methyl cyclohexene ($K_1 = 0.5$), for the greater similarity of K_1 in bridged six-membered rings (2-methylene norborane, β-pinene, camphene) with methylene cyclopentane ($K_1 = 4.0$) rather than with methylene cyclohexane ($K_1 = 6.0$). The greatest value of K_1 found in 2-norbornene is ascribed to the large amount of strain at the double bond in this compound [27].

Another example of the influence of stereochemistry on the behavior of isomers is given [45] by the changes in retention times of *syn-* and *anti*-7-carbomethoxynorbornenes in the presence or absence of silver nitrate on a vapor-phase chromatographic column of diethylene glycol on celite; the presence of silver nitrate had no effect on the retention time of the synisomer (with the hindered double bond), but under identical conditions, the antiisomer was slowed down by a factor of 2, due to the complexing of the silver ion with the unhindered double bond.

R2=H R1=COOMe(anti)
R2=COOMe R1=H (syn)

An excellent example of the ability of argentation chromatography to separate positional isomers differing in location of the double bond (and consequently, in steric hindrance at the double bond) was provided by Hara et al. [46] in the field of sesquiterpene alcohol isomers. This work is also a rare example of utilization of LC in the terpenoid series. Using a silica gel column impregnated with silver nitrate in methanol (see Heftmann et al. [47]), they succeeded in the separation of eudesmol isomers (*e*: α,β,γ) and hinesol (h). These compounds were identified in two fractions isolated from the bark of *Magnolia obovata* Thumn and *Atractylodes lancea* D. C., mixtures of α, β, γ-*e* (fraction 1) and of β-*e*, h (fraction 2). The adsorption sequence $\gamma < \alpha < \beta$ observed was opposite to the order of steric hindrance at the double bond. By analytical silica gel LC, without Ag, using binary solvent:

TABLE 2   Values of K in the Terpenoid Series

Compounds		K_1	Compounds		K_1
	α-Pinene	1.1		α-Phellandrene	5.1
	Camphene	3.1		2,5-Norbornadiene	33.7
	β-Pinene	3.7		2-Norbornene	62
	2-Methylene norbornane	4.3			

Source: Ref. 27.

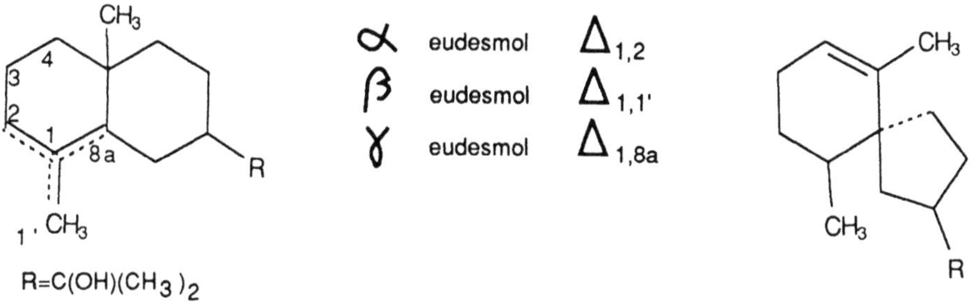

α eudesmol $\Delta_{1,2}$

β eudesmol $\Delta_{1,1'}$

γ eudesmol $\Delta_{1,8a}$

R=C(OH)(CH$_3$)$_2$

systems containing n-hexane, the separation of the α, β isomers was unsuccessful (Table 3).

But if by means of impregnation with silver nitrate, an increase in system reactivity allowed the separation of α, β isomers, it is difficult to remove the silver nitrate from the silica gel surface and to recover the original column material. So a high-efficiency preparative silica gel column was tested (column 3).

The separation of β-eudesmol and hinesol was improved [48] using preparative HPLC on silver nitrate–impregnated silica gel column prepared by a modification of the procedure cited [47]. In this way the silver salt was almost quantitatively retained by the silica gel surface, avoiding the drawback cited earlier. Silver nitrate was gradually removed by use of a strongly polar solvent; however, the columns can be reactivated repeatedly by applying the impregnation procedure described.

Using RP–HPLC, as in the case of olefins(section III.B.1), Vonach and Schomburg [37] described the separation of sesquiterpenes (δ and γ cadinenes (Fig. 3)), owing to the much stronger complexation of the exocompound (1).

An interesting case of effect of silver ions on RP–HPLC separation of retinyl esters was described by de Ruyter and de Leenheer (49). Using octadecyl silica column with mobile phases consisting of different amounts of AgNO3 dissolved in

TABLE 3 Retention Behavior of α, β, γ-Eudesmol on Silica Gel Columns Using Binary Solvent Systems (n-Hexane/Acetone 97:3)[a]

Column[b]	Capacity ratio k'			Separation factor	
	γ	α	β		
				α/γ	β/α
1	5.32	5.71	5.71		
2	6.58	7.24	8.42	1.10	1.16
3	7.94	8.74	8.9	1.10	1.02

Source: Ref. 46.

[a]See also Ref. 46.

[b]1, silica gel; 2, Ag–silica gel; 3, high-efficiency preparative silica gel.

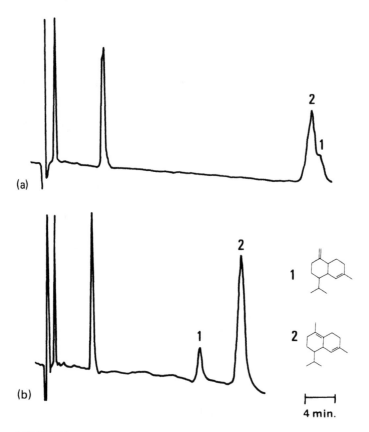

FIGURE 3 Separation of δ and γ cadinenes; conditions as in Fig. 2. (a) methanol/water
5:1 v/v; (b) A + 8 × 10⁻³ *N* AgClO₄; attenuation, ×1. (From ref. 29.)

methanol, the authors were able to separate retinyl esters on the basis of chain
length and unsaturation. The most important esters are retinyl palmitate (R = C_{16})
stearate (R = C_{18}), oleate (R = $C_{18:1}$), and linoleate (R = $C_{18:2}$):

$$CH{=}CH{-}C{=}CH{-}CH{=}CH{-}C{=}CH{-}CH_2{-}O{-}\underset{O}{\overset{}{C}}{-}R$$

Where R is the number of atoms and unsaturation in the fatty acid chain of the
retinyl esters.

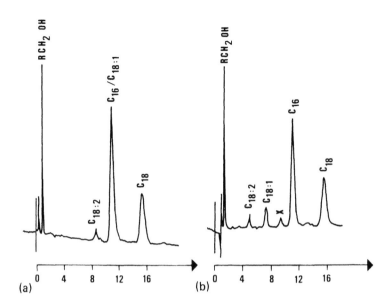

FIGURE 4 HPLC record of a serum extract. Column, 15 × 0.32 cm, 10 μm RSIL-C$_{18}$ HL; flow rate, 1 mL/min; mobile phase, (a) CH$_3$OH, (b) CH$_3$OH/58.9 × 10^{-3} M (Ag$^+$); detection, 330 nm; x, unidentified peak. (From ref. 49.)

By varying the silver-ion concentration of the mobile phase, it was possible to control the retention characteristic of the unsaturated compounds. Silver nitrate did not alter the retention of esters with a saturated fatty acid despite unsaturation in the retinyl moiety. An example of the applicability of the method to biological sample is done in Fig. 4, which records HPLC analysis of a blood sample after ingestion of vitamin A.

A method described as *argentation medium-pressure liquid chromatography* (Ag–MPLC), both cheap and efficient for the preparative separation (on a g = gram scale), was proposed by Evershed et al. [50]. The phase, a silica gel support (Kieselgel 60) loaded with 20% of its weight of silver nitrate (dissolved in acetonitrile), was used for making glass columns (100 cm × 1.5 i.d.). The columns were prepared for reuse by backflushing with diethyl ether and can be stored dry if not needed for an extended period. Mixtures of Z and E isomers such as heptadecene and nonadecene were separated, as were ZE and ZZ α-farnesene (*1*) and Z and E 7-methyl 6-nonen-3-one (*2*) (not completely separated in similar conditions). E isomers eluted before Z, and ZZ before ZE.

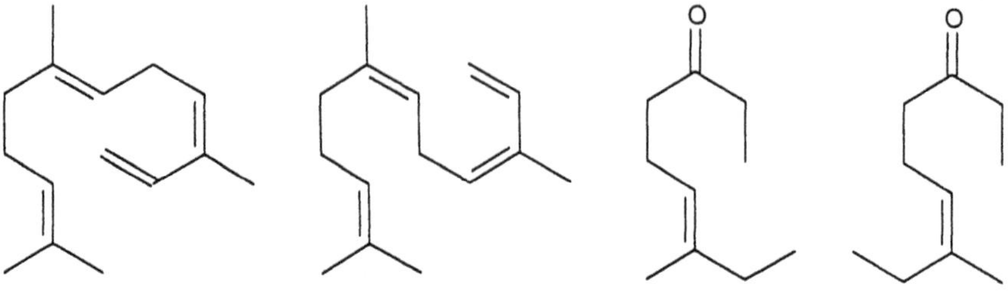

(*1ZE*)　　　　(*1ZZ*)　　　　(*2E*)　　　　(*2Z*)

Another preparative separation has been described [51] in the case of two terpene alcohols differing only in a $\triangle^1$ double bond, cafestol and kahweol, extracted from green coffee beans in the form of esters of fatty acids. In the structure of kahweol below, the C_1-C_2 bond is saturated in cafestol.

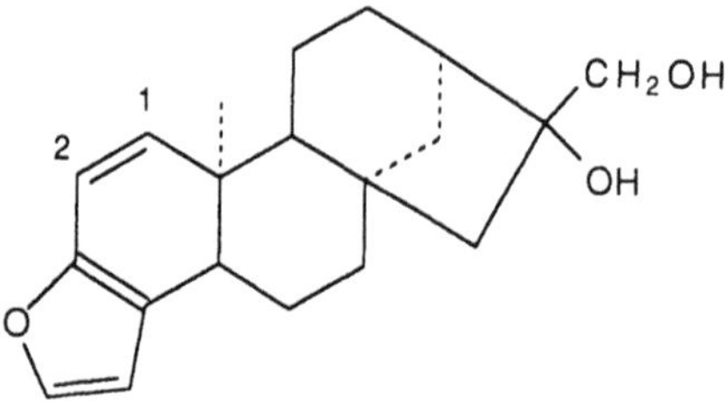

Analytical samples of the palmitate esters were separated by Ag–TLC [52]. This technique, extended by preparative liquid chromatography on silica cartridges impregnated with 10% silver nitrate, allowed complete separation of gram quantities of these diterpene esters.

Carrying on their work on diterpene resin acids and esters developed since 1964, as well by Ag–LC, Curran and Zinkel [53] described the advantages of the use of silver ion as bound salts of macroreticular sulfonic acid ion-exchange resins.

4.　Pheromones

Many sex attractants, "insect pheromones," used for monitoring insects were found to be acetates of geometrical isomers of long-chain unsaturated alcohols (with 12, 14, or 16 carbon atoms), such as, for example, cis-9-tetradecen-1-ol acetate (*cis*-9-TDA) and cis-11-tetradecen-1-ol acetate (*cis*-11-TDA). As attractancy depends on the geometrical structure, several improved methods were tried for the purification and analysis of synthetic products. If small amounts of pure isomers can be collected by Ag–TLC [54], several authors attempted to carry out large-scale (1 to 600 mg) procedures particularly by Ag–HPLC, using commercially available columns. Houx et al. [55,56] described a low-pressure liquid chromatographic system (20 to 50 atm) using columns packed with macroporous cation-exchange resin commercially available, treated with silver nitrate. The columns are eluted with methanol. About 1000 chromatograms have been made without any deterioration of the columns. The best resolution of the geometrical isomers of mono- and diunsaturated acetates (or fatty acid methyl esters) was obtained at $t < 10°C$ and for trienes at 40°C. This method avoids the drawbacks of HPLC on AgNO3-coated silica gel, used by Heath et al. with benzene as the mobile phase [36, 57] (which also required that the eluant be washed with NaCl to remove the silver that bled from the column) and also the drawbacks of RP–HPLC on Lichrosorb RP 8 and silver nitrate-containing aqueous isopropanol as the mobile phase used by Schomburg and Zegarski [38]. In this case the mobile phase contained up to 2.5% of silver nitrate, which is corrosive. In a similar way, Heath et al., using their technique for in situ coating of Ag^+ onto silica gel in HPLC commercial columns [35, 36, 57, 58] (cf. section III. B.1) compared the ability of commercial and prepared columns to separate tetradecen-1-o1 acetates [36].

We conclude this chapter, devoted to olefinic compounds, with an example illustrating the ability of Ag–chromatography to resolve complex cases. It was afforded by Heftmann et al. [47], who succeed the separation by Ag–HPLC of pairs of gibberellins (plant growth factors), differing from each other only by one double bond. They were separated in the form of their p-nitrobenzyl esters on Porasil A impregnated with silver nitrate; elution was made with several mixtures of solvents according to the structure of the compounds. The same authors reported that closely related gibberellin analogs are also easily resolved on Ag–TLC in the form of their p-nitrobenzyl esters, detected by spraying the plates with dilute sulfuric acid [59].

R1=OH	R2=H		R1=OH	R2=H	Δ 3.4
R1=H	R2=OH		R1=H	R2=OH	Δ 2.3
R1=R2=OH			R1=R2=OH		Δ 3.4

C. Complexation of Olefins with Other Metals

In the preceding section we presented the potential selectivity of Ag^+ unsaturated complexes, according to the number, geometry, and position of unsaturation. To complete this chapter we shall recall that weak complexes between transition metals [e.g., Co(II), Ni(II), Zn(II), Mn(II), V(II)] and several organic compounds containing π-bonds or free electron pairs (N, O, S. halogens) have also found application in chromatographic separation and in methods permitting the calculation of stability constants by GC. The subject was well reviewed by Guha and Janăk in 1972 ("on charge transfer complexes of *metal* in the chromatographic separation of organic compounds" [18], by Szczepaniak et al. in 1979 ("application of transition metal complex formation in GC"[20]), and also by this last group in 1984 [60]. In this paper the authors were able to draw conclusions from research performed during the decade 1974–1984.

It is clear that none of the methods involving metal complexes other than those described with silver ions have reached a development comparable to argentation chromatography, particularly in the field of olefinic compounds and lipids. Nevertheless, as far as olefinic compounds are concerned, we have summarized some of the most striking examples in Table 4.

If GLC and GSC were the main methods in all these studies [61–65] Mikes et al. [34] tested the application of complex–forming stationary phases in high-speed liquid chromatography. Independent of their work in Ag–chromatography already cited [34], they applied the phase prepared with Rh(II) acetate deposited on Corasil II, introduced by Schurig et al. [65], to the separation of butenes. These compounds are eluted in the following order: butene-1, *trans*-butene-2, isobutene, and *cis*-butene-2.

IV. LIPIDS

It is perhaps in the field of separation and identification of lipids, fatty acids (FAs), and their methyl esters (FAMEs) that argentation chromatography was found to be very suitable and had known the greatest development. A great deal of investigation was performed at Unilever Research Laboratory in the 1960s. Most papers published before 1966 were covered in the excellent review of Morris [40] completed in 1972 by Guha and Janăk [18], in 1979 by Scholfield [66], and as far as TLC is concerned, by Morris and Nichols in 1970 [67] and by Kirchner [68] in 1978.

Examination of the literature of the last decade (1978–1988) convinced us that argentation chromatography remains a fundamental method in the field of lipids and related compounds, with beneficiation of recent developments of HPLC and new modes of detection. The method was carried out by chromatography on silver nitrate–impregnated silicic acid, Florisil, or ion-exchange resin column, by TLC on silicic acid impregnated with silver nitrate solution in one or two dimensions, by paper chromatography, and by reversed-phase partition chromatography, even in HPLC mode. It was used alone or in conjunction with other types of chromatography (gas–liquid chromatography, liquid chromatography, reversed-phase TLC, etc.) (18).

TABLE 4 Examples of Utilization of Transition Metal/Olefin Complexes in Chromatographic Separations

Compounds	Metals	Conditions	Results	Refs.
Monoolefins	Rhodium coordination compounds; Rh(I) (CO)$_2$ β-Diketonates (in squalane solution) derived from 3-trifluoroacetyl-d-camphorate	GC	Marked selectivity in their interaction with olefins; interesting reagent for difficult problems in olefin analysis.	61, 62
Alkanes, alkenes, alkynes	Anhydrous chlorides Co(II), Mn(II), V(II)	GSC on columns packed with chlorides	Best utilized for compounds having isolated π-bonds	63
Olefinic and aryl compounds	Anhydrous chlorides Co(II), Mn(II)	GSC on columns packed with chlorides	Determination of heats of sorption (they vary directly with the π-electron density of adsorbate and inversely with the number of $3d$ electrons of the adsorbent)	64
Monoolefins, cycloolefins, dienes	Diphenylphosphines complexes with CoCl$_2$ and CoBr$_2$ chemically bonded to silica (Porasil C)	GC	The interaction with electron-donating compounds is stronger in the case of CoBr$_2$ than in the case of CoCl$_2$; the packings containing bonded metal complexes have much better ability to separate different compounds	60
Monoolefins, cycloolefins, prostaglandins, terpenes	Solid dimeric Rh(II) acetate/Corasil II	HPLC	The reference concerns essentially Ag–HPLC with an example of Rh–HPLC in the butene series.	34

For papers published before 1972, we send readers back to the reviews mentioned above [18,40] and focus our attention on those published after that date,

A. Fatty Acids (FAs) and Methyl Esters (FAMEs)

1. General Considerations

Fatty acids as well as resin acids are generally separated as their methyl esters. It is well known from the older studies [18,40,66–68] that they are separated according to:

1. The degree of unsaturation
2. The geometry of the double bond (i.e., separation of cis and trans monoenoic esters). This was particularly fruitful in the elucidation of the structure of polyunsaturated fatty acids.
3. The position of the double bond (i.e., separation of positional isomers of unsaturated fatty acids). In dienoic esters, the effect of silver complexing increased with increasing separation of the double bonds. As separation is basically affected according to the degree and the type of unsaturation, with little if any separation of different chain lengths, argentation chromatography was often used in conjunction with other partition methods (i.e., GLC and LC) or in two-dimensional TLC (reversed-phase TLC in the first dimension and argentation in the second, for example [18]).

Except for some classical separations of positional and geometric isomers of FAs and FAMEs [69–71], the main developments occurring in the last decade concern essentially the separation of polyunsaturated FAs and FAMEs and of hydroxy acids.

2. Separation of Positional or Geometric Isomers of Monoenoic FAs and FAMEs (C_x: 1 $\triangle^y$ c and C_x: 1 $\triangle^y$ (t) (cis = c, trans = t)

International restrictions on the content of erucic acid (C_{22}:1 $\triangle$ 13c) in edible oils have caused renewed interest in the determination of positional and geometrical FA isomers by argentation TLC combined with GC. In 1983 the results of an international collaborative study for the determination of erucic acid in four samples of edible fats and oils with differing levels C_{22}:1 $\triangle$ isomers other than erucic acid were published by Ackman et al. [69]. Two methods were compared by nine laboratories: an open-tubular (capillary) GLC method and an $AgNO_3$-TLC. The pros and cons of each method were reported in detail. The means and repeatabilities of the two methods were similar, but on one hand, the GLC method showed better reproducibility, and on the other hand, the $AgNO_3$-TLC method took a much longer time per analysis and involved several additional steps, including recovery of a fairly small amount (<1 mg) of methyl ester from a TLC plate for subsequent

GLC measurement. Nevertheless, the official EEC method for determining erucic acid [70] is based on this low-temperature TLC technique.

To increase the speed of the TLC analysis, Breuer et al. [71] used silver-impregnated alumina sheets, with toluene as the developing solvent (2°C or -20°C, 40 to 60 min). The spots visualized under UV light (366 nm) after spraying with dichlorofluorescein were scraped from the plates, and after extraction, the FAMEs were submitted to GC or CC. Ag–TLC with alumina sheets also allows separation of FAs without preliminary derivatization, but in this case a double development is required with intermediate drying.

A new procedure of sorbent impregnation, separation, and detection used in high-performance thin-film chromatography (HP–TFC) was described [72]. Separation is performed on Kieselgel 60 thin films impregnated by submerging the plates vertically into acetonitrile solution containing silver nitrate, phosphotungstic acid, and cobalt nitrate. The method is claimed to provide highly selective resolution of lipidic derivatives such as cis and trans monounsaturated fatty alcohols, fatty acids, and their methyl esters. It is possible to couple with quantitative densitometry.

Qualitative and quantitative analysis of trans FAs, isolated from complex FA mixtures, were performed by Heckers and Melcher [73]. Combining preparative Ag–TLC (silica gel H plates impregnated with silver nitrate and dichloromethane as developing solvent) and reversed-phase high-performance TLC (HP–TLC), this method allows the separation of *trans*-hexadecenoique and *trans*-octadecenoique FAMEs. In the same way we find examples of analysis of fish oils based on argentation chromatography in conjunction with GLC [74] or RP–HPLC [75].

To separate the two FAMEs of a "critical" pair, Ozcimder and Hammers [75] compared the utility of RP–HPLC (on Lichrosorb RP-18, with acetonitrile as eluant) and Ag–HPLC. The separation of cis and trans isomers, in complex esters mixtures in different RP fractions (although restricted to monoenoic esters with chain length $\geq$, C_{15}), further facilitates the analysis of the RP fractions by means of capillary GLC, including positional isomers (not characterized by RP–HPLC alone). Ag–HPLC appears to be less suitable as a prefractionation method in total FAMEs analysis. However, it seems to be a good method for the fractionation of unsaturated species with three to six double bonds.

Lam and Grushka [76] separated cis-trans isomers of p-bromophenacyl esters of C_{16} and C_{18} FAs ($C_{18:1}$ $\triangle^{11}$ t and c; $C_{18:1}$ $\triangle^{9}$ t and c; $C_{16:1}$ $\triangle^{9}$ t and c) and *cis* and *trans*-permethin (a synthetic pyrethroid insecticide) using a silver-loaded aluminosilicate as a support for HPLC. Silica gel was treated with sodium aluminate to form a polyanionic surface and the counterions were then exchanged for silver ions. Trans isomers always eluted before cis isomers, except with permethrin, where the opposite order was observed, owing to steric factors. This new column did not suffer from the Ag bleeding problems often met in other methods.

Some years later Chan and Levett [77] applied Ag RP–HPLC [octadecyl silane-bonded 10-μm (Partisil-10-ODS) and 5-μm (Spherisorb S5-ODS) silica particles] for the separation of *p*-bromophenacyl esters of oleic and elaidic acids, and of erucic and brassidic acids, the cis isomers eluting first, which is the reverse of the example above by HPLC.

3. Separation of Polyunsaturated FAs and FAMEs and Oxygenated FAs

Besides separations performed by application of well-known techniques to samples of various origins, the general trends observed in this particular field of FA analysis relate to research on new procedures for impregnation, new phases, new techniques, or adaptation of older ones to special cases, with the aim of succeeding in a rapid and improved separation of unsaturated FAMEs according to their unsaturation.

New procedures for impregnation. Reviewing the different impregnation processes (by either dipping or immersing the precoated plates in an aqueous solution of 5 to 40% of silver nitrate or silver sulfamate [78], Inomata et al. [79] developed a simple, rapid, and convenient procedure for impregnating precoated silica gel plates by spraying 40% aqueous silver nitrate solution. This method gives reproducible analysis of polyunsaturated FAMEs. Radioautography and radioscanning were carried out with ^{14}C-labeled FAMES in addition to localizing spots with dichlorofluorescein. Identification and quantification of FAMEs were also made by GLC after Ag–TLC.

The preparation of silica gel and plates with the sorbent layer firmly fixed with silica sol allows the dipping of the plates in various reagents without destroying the layers. Impregnation was made by dipping in methanol saturated with silver nitrate, and further the use of two-dimensional TLC is improved. Svetashev and Zhukova [80] described by this procedure the separation of labeled FAs and FAMEs, whereas GLC with special radioactivity detectors lacks sufficient sensivity and is quite expensive. The separation was performed on the silver nitrate-impregnated plates, and following the separation, the silver nitrate was washed off with a saturated sodium thiosulfate solution. After washing and drying a reversed-phase chromatography (methanol/water 95:5 saturated with decane) was carried out in the second direction. Over 90% of the radioactivity was recovered and the resolving power is close to that of GLC on a packed column.

Simple and rapid separations of FAMEs containing zero to six double bonds were proposed by several [81–83]. The experimental procedure described by Dudley and Anderson [81] consists in Ag–TLC (the silica gel G-AgNO$_3$-impregnated plates are developed twice in a solvent system containing hexane/diethyl ether/acetic acid 94:4:2) followed by GC–CC carried out on the methanol extracts of the scraped bands. The method, utilized on a model mixture, was applied to the separation of retinal FAMEs. Similarly, FAMES containing from one to six double bonds were separated by Ag–TLC (silica gel G coated with 7.5% AgNO3 aqueous solution) and a multidevelopment procedure in three solvent systems [82]. Shukla and Srivastava [83] used ammonia-complexed silver ion in TLC. Ammonia silver ion plates give better separation of FAMEs than do aqueous-prepared silver ion plates. The method was also employed for the FAME separation of cod liver oil triglycerid.

New phases. New developments of column chromatography on silver ion-saturated resins arose in the field of polyunsaturated FA [84,85] and mono- or dihydroxy fatty esters [86]. Scholfield has used column chromatography with silver ion bound to macroreticular sulfonic acid resins since 1964 (see Ref. 18). More recently [84,85] he developed improved separations of FAMES using, on the one hand, newer resins of greater surface area, and smaller particles (53 to 54 μm), obtained by grinding the resin [84], and on the other hand, using HPLC [85] with, if necessary, temperature programming (25 to 70°C). Retention volumes of a number of esters are consistent with the observations mentioned above regarding factors influencing retention in argentation analysis of these compounds: that retention volumes of saturates are nearly the same and that the retention volume for a long-chain saturate corresponds to that for an unretained substance. Retention volumes for monoenes are much less than those for dienes and depend mainly on the configuration of the double bond. They decrease when the double bond is conjugated with the carbonyl and increased by a terminal double bond. With dienes, retention volumes depend on the number of methylene groups between double bonds as well as on configuration. The following example illustrates the advantages of this Ag-HPLC in the case of polyunsaturates. If 30 h was required to remove methyl linoleate from a column suitable for monoene separation [84], only 30 min was required from a new HPLC column, but methyl linolenate gave a discerned peak only by increasing the temperature. Further improvement of the method is foreseen using 5- or 10-μm particles.

As some of the macroreticular resins used are not always readily available, Scholfield [87] investigated HPLC with silver nitrate silicic acid and benzene (see Heath and Sonnet [36]) and compared the results with those given by LC on the sulfonic acid method [84]. Both methods gave good separations of FAMEs. The HPLC procedure provides for faster separations on an analytical scale; the LC procedure does not require high-pressure apparatus and is suitable for larger samples.

Other ways to overcome the tenacity with which polyunsaturated are retained with silver resin chromatography were described by Emken et al. [88], who removed linolenate from a silver resin column with 10% 1-hexene in methanol, and by Adlof et al. [89–92], who used partial argentation resin chromatography (PARC) in the case of methyloctadecanoate isomers and mono, di, and tri-, and tetraenoic FAMEs. In this technique only a part of the total sulfonic acid protons of the resin (Amberlyst XN 1010) is replaced by silver ions. The variations in reten-

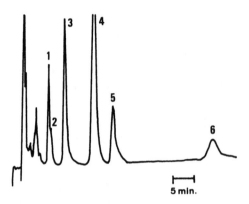

FIGURE 5   Separation of unsaturated C₁₈ fatty acid methyl esters. Conditions as in Fig. 2. Mobile phase, methanol/water 5:1 v/v + 10^{-2} N AgClO₄; temperature, 2°C. Peaks of methyl esters: 1, linolenic acid; 2, α-linolenic acid; 3, linoleic acid; 4, oleic acid; 5, elaidic acid; 6, stearic acid. (From ref. 37.)

tion of unsaturated compounds on various partially silvered resins (54, 61, 72, and 91%) were summarized. *cis*-Trienoic and tetraenoic fatty esters were eluted with methanol on PARC columns containing 36 and 17% silver. Partial silvering of the resin improved peak shapes, while sample elution time and elution volume were reduced [89]. For example, a 36% PARC column gave a selective separation of all-*cis* dienoic, trienoic, and tetraenoic fatty esters, these last compounds being eluted with 1-hexene (see Ref. 88). In general, the column can be reused indefinitely.

Rakoff and Emken [86] extended the use of silver-saturated Amberlyst XN 1010 to the separation of geometrically isomeric methyl mono and dihydroxy fatty esters, such as methyl 12-hydroxy *cis* (and *trans*)-9-octadecenoates (for the monohydroxy compounds) and methyl threo-12, 13-dihydroxy *cis* (and *trans*)-9-octadecenoate (for the dihydroxy compounds). It is apparent from the retention volumes that the hydroxy groups have a retarding effect. It should be recalled that Morris [40] earlier pointed out that by argentation TLC, dihydroxy esters were separated not only on the basis of unsaturation but also according to the threo and erythro configurations of the glycol group.

New techniques or adaptation of older ones to special cases.   RP–HPLC on chemically bonded phases (5-μm Nucleosil-5C 18) with a methanol/water mixture + 10^{-2} N AgClO₄ under the conditions mentioned earlier for olefins and terpenes (Section III.B.3) allowed the separation of unsaturated C₁₈ FAMEs, such as methyl esters of linolenic, γ-linolenic, linoleic, oleic, elaidic, and stearic acids (see Fig. 5) [37]. It is seen that separation of the two linolenic esters, which differ from each

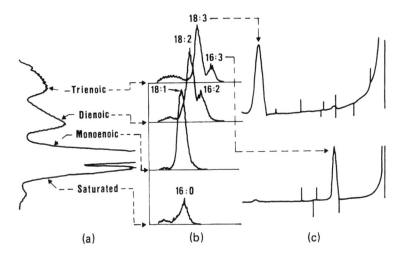

FIGURE 6 Separation of uniformly labeled fatty acids by TLC. (a) Typical radiochromatogram scan showing the fractionation of the methyl esters into groups with different degrees of unsaturation. Silica gel G containing 10% silver nitrate; hexane/diethyl ether 80:20 v/v. (b) Typical radiochromatogram scan showing the fractionation of groups of methyl esters with the same degree of unsaturation into individual compounds. Silica gel G impregnated with undecane: acetic acid/water 90:10 v/v. (c) Typical gas chromatograms showing the purity of individual compounds isolated by consecutive TLC in systems (a) and (b). (From ref. 95.)

other only by the position of one of the three double bonds, is still incomplete (peaks 1 and 2).

Sebedio et al. [93,94] used silver nitrate–impregnated chromarod with a FID Iatroscan (see Section IV.C.1) to separate FAMEs according to the double-bond configuration. This technique was developed by this group of workers for the separation of FAMEs with different degrees of unsaturation, particularly for the separation of cis-trans isomers. An advantage of the AgNO₃–Iatroscan method over the GLC method is the small sample size (10 μg) and the short time needed for an analysis.

An example using conjunction of several chromatographic methods has been shown [95]. Labeled uniformly FAMEs were prepared by methanolysis of the total lipids of *Chlorella pyrenoidosa* that had been grown in the presence of $^{14}CO_2$. The FAMEs were fractionated by TLC on silica gel impregnated with 10% silver nitrate using hexane/diethyl ether (80:20) as the developing solvent. The plates were dried and scanned for radioactivity (Fig. 6A). After elution (diethyl ether) of the zones of sorbent carrying the various fractions of labeled methyl esters, the FAMEs (in hexane) were submitted to reversed-phase chromatography on silica gel G plates impregnated with undecane (solvent fractionating: acetic/water 85:15 or

90:10) (Fig. 6B). The FAMES zones were scrapped off and eluted with Et2O, then tested for purity by GC (Fig. 6C).

Lee Ken Jee and Lam [96], studying the isolation of unsaturated and oxygenated FAMEs, reported TLC properties of several dimethylene interrupted *cis-cis* (c/c) and *trans-trans* (t/t) methyl octadecadienoates and methyloctadecadiynoates on impregnated silver nitrate silica, the t/t compounds being less polar than the corresponding c/c isomer. An interesting case of oxygenated FAMEs concerns the THF and THP derivatives, (*1*) and (*2*), separated on thin layers of silica impregnated with 10% silver nitrate [97].

$$CH_3-(CH_2)_x \underset{O}{\diagdown} (CH_2)_y-COOCH_3 \qquad x+y=12 \quad x=0-11$$

(*1*) $\qquad\qquad\qquad\qquad\qquad\qquad y = 1 - 12$

$$CH_3-(CH_2)_n \underset{O}{\diagdown} (CH_2)_m-COOCH_3 \qquad n+m=11 \quad n=0-10$$

(*2*) $\qquad\qquad\qquad\qquad\qquad\qquad m = 1 - 11$

Other cases of oxygenated FAMEs (*cis*-9,10-epoxy-*trans*-3 *cis*-12-octadecadienoic acid and 9-hydroxy-*trans*-3,*trans*-10,*cis*-12-octadecatrienoic and 13-hydroxy-*trans*-3,*cis*-9,*trans*-11-octadecatrienoic acid) were separated [98], combining a first TLC (silica gel G; solvent: hexane/ether 80:20), which allows the separation of epoxy esters from hydroxy esters, followed by Ag–TLC on each of these classes of oxygenated compounds for their separation according to the number of double bonds (solvent: benzene/chloroform/ether 50:50:2 in the case of epoxy esters and 50:50:15 in the case of hydroxy esters).

Separation of conjugated trienoic FA isomers from each other as well as from oleic, linoleic, linolenic, and saturated FAs was investigated [99] on Ag–TLC (silica gel G plates) on the basis of their methyl esters prepared from various oils. A 30% ratio of impregnation is suitable for the best resolution.

Another separation of linoleic and linolenic acids by Ag–TLC is due to Bielen and Turina [100]. Mycolic acids (α-alkyl β-hydroxy polyunsaturated FAs) from several *Norcadia* were fully separated and characterized by a combination of Ag–TLC and GC–MS. The GC–MS analysis of each subclass of the mycolic acids, after Ag–TLC, revealed that the major species consisted of the even-C-numbered mycolic acids, ranging from C38 to C68 [101].

B. Arachidonic Acid, Prostaglandins, and Leukotrienes

With the separation by Ag–TLC of all of the known prostaglandins (PGs) in 1964, the paper of Green and Samuelsson [102] marks the beginning of the use of argentation chromatography in the field of oxygenated FA derivatives, metabolites of arachidonic acid (AA): prostaglandins and leukotrienes. Ag–TLC and Ag–HPLC are currently used for the identification of this class of compounds, as each of these two methods offers specific advantages in comparison with the same methods carried out without silver ion.

As far as Ag–TLC is concerned, Green's method was applied by Davies et al. [103] in conjunction with LC on silicic acid and GS–MS for the identification of PGs in prevertebral blood. Many other examples are described in the literature and the subject was reviewed by Daniels [104] in 1976. Some years later, Greenwald et al. [105] pointed out that Ag–TLC (plates precoated with Whatman LK 6D silica gel dipped in a 10% solution of $AgNO_3$ in 86% aqueous ethanol) gave baseline separation of arachidonic acid and its metabolites isolated from human platelets, in contrast to the use of nonimpregnated silica gel plates. If argentation results in decreased counting efficiency with time, when quantitation is done by liquid scintillation methods, this drawback is eliminated by precipitating silver with sodium chloride. The efficiency of the method is shown in Fig. 7, compared with other TLC conditions, for thromboxane (TxB2), arachidonic acid (AA), 12-hydroxyeicosatetraenoic acid (HETE), and 12-hydroxyheptadecatrienoic acid (HHT). Separation of the same hydroxy FA metabolites had been achieved by Bills et al. [106] but they did not succeed in the separation of HETE and AA.

In the course of the synthesis of PGs via the general method of Corey, Ag–TLC was used as a purification procedure at different steps of the synthesis (i.e., purification of 15-epi-PGF2 from 15-epi-13, 14-dihydro corresponding alcohols). Interestingly, the resolution of PGs by saturation class can also be accomplished by HPLC using a column of a pellicular silica support (Corasil II) without the inclusion of silver ion [107].

More recently, Bomalaski et al. [108] described an Ag–TLC system that separates PGE1, PGE2, and other major eicosanoids, including dihomo-γ-linolenic acid, the immediate FA precursor of PGE (solvent: chloroform/methanol/acetic acid/H_2O 90:7.5:5:0.8).

The HPLC methods were developed concurrently with the Ag–TLC methods. HPLC on silicic acid and RP–HPLC on octadecylsilyl silica have proved to be very useful for the separation of PGs and monohydroxy FA metabolites of AA. Nevertheless, as Powell [109] pointed out, in some cases HPLC does not give good separations and RP–HPLC is time consuming, as it necessitates the removal of aqueous solvents. So Ag–HPLC was carried out by several authors. Since 1973 an example of separation of PGF2γ and PGF2β on a silica column impregnated with $AgNO_3$ (solvent: toluene/dioxane/acetic acid/2-methyl butanol 150:150:10:5) was re-

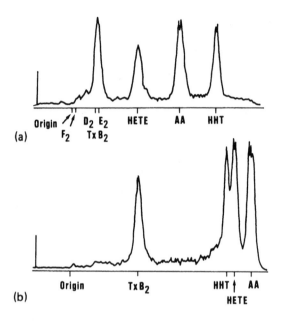

FIGURE 7 Product separation of platelet arachidonic acid metabolites. Approximately 40,000 cpm of authentic standards were spotted on TLC plates and developed as follows: (a) ethyl acetate/2,2,4-trimethylpentane/acetic acid/water 110:50:25:100 (AgNO$_3$-impregnated silica gel); (b) ethylacetate/2,2,4-trimethylpentane/aceticacid/water 110:50:25:100 (silica gel). (From ref. 105.)

ported by Mikes et al. [34] in comparison with Ag-TLC. Merritt and Bronson reported in 1977 [110] the separation of mixtures of *p*-nitrophenylacyl derivatives of PGs A$_2$, B$_2$, E$_1$, E$_2$, F$_1$, and F$_{2\alpha}$, and some of their synthetic isomers and analogs, by Ag-HPLC on a silver ion-loaded cation-exchange column of R Sil CAT (5 μm). with UV detection. According to Powell [109,111], this method may not be very suitable for the analysis of PGs from biological sources, due to the UV adsorption of unrelated compounds. Moreover, it needs a derivatization step, and the not very volatile *p*-nitrophenylacyl derivatives render subsequent GC/MS analysis difficult. In several papers, Powell [109,111–113] described Ag-HPLC separations applied to underivatized metabolites of AA and 8, 11, 14-eicosatrienoic acid using the same column as before and various solvent mixtures. Their choice followed mechanistic considerations [111] related to the interaction of solutes with the stationary phases. Retention times on Ag column appear to be determined by a combination of interactions: on one hand, between the silver ions of the stationary phase and double bonds of the solute (these interactions decrease by the addition of acetonitrile, which competes with olefinic groups for the silver ion) and on the other hand, between polar groups of the stationary phase and polar groups of the solute (these interactions decrease by addition of methanol, which competes for polar sites). Using various proportions of methanol or acetonitrile added to the eluting solvents (chloroform/acetic acid) the selectivity of the stationary phase can be optimized for the separation of any given group solutes. Applications are described concerning the separation of a variety of underivatized eicosenoic acids, monohydroxyeicosenoic acids, and PGs as well as the corresponding isotopically deuterium and tritium metabolites [112,113] of AA. Separation of unlabeled metabolites of AA from their deuterium and tritium analogs is based on the increased affinity of isotopically substituted double bonds for the silver ions of the stationary phase.

Accuracy and precision of the analytical methods are necessary in the pharmaceutical industry to provide a high level of confidence in the amount of PGs determined. A silver-modified normal-phase HPLC system was developed by Kissinger and Robins in 1985 for prostaglandin bulk drugs and triacetin (glycerine triacetate) solutions [114]. Silver nitrate present in the mobile phase results in high selectivity for cis-trans isomers with conventional silica columns (Dupont Zorbax SIL). To provide high detectability at 254 nm, PGs were esterified with α-bromo-2'-acetonaphthone prior to chromatography. The analytical technique was applied to triacetin solutions containing as little as 10 μg/mL of 15-(R)-methyl PGE2 (arbaprostil).

C. Neutral and Polar Lipids

1. General Considerations

In the following discussion regarding argentation chromatography applied to lipids, we distinguish the case of neutral lipids (triglycerides or triacylglycerols) from that of polar lipids (phospholipids principally). A great many papers on the subject, have been published particularly between 1960 and 1978, and most of them were analyzed in the leading reviews already cited [18,40,67,68]. As Ag–TLC is one of the most useful techniques, alone or in conjunction with other chromatographic methods, the literature up to 1978 is well covered by Kirchner [68], so we will discuss the principal developments during the last decade.

2. Neutral Lipids

Quantitative analysis of food lipids relies on accurate setup and calibration of instrumentation [115]. For example, in the analysis of triglycerides by Ag–TLC, errors in detection by densitometry are aggravated by the difficulty in quantitatively charring the triglycerides to carbon, although Dallas and Padley (116) have demonstrated this possibility. The most recent developments in triglyceride argentation chromatography were made in two fields: (a) research into more accurate detectors in Ag–TLC, and (b) utilization of HPLC on silver-impregnated silica, on silver-loaded resins, and in reversed-phase mode.

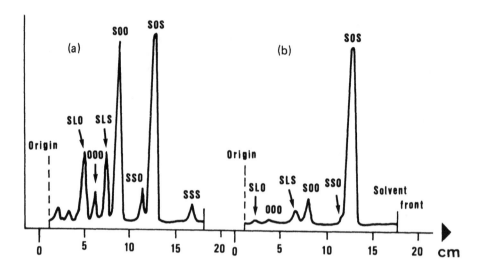

FIGURE 8 The 5% silver nitrate TLC separation of (a) palm oil triglycerides and (b) cocoa butter triglycerides. Detection by fluorescence scanning using a Zeiss KM3. S, saturated; O, oleate; L, linoleate. Solvent, two developments in 75% chloroform in cyclohexane. (From ref. 115.)

As far as Ag-TLC is concerned, Hammond (115) used a detector by fluorescence scanning with a Zeiss KM 3 apparatus. An example of separation of palm oil and cocoa butter triglycerides is reported in Fig. 8.

The use of TLC on Chromarod with an Iatroscan flame ionization detector, a technique cited earlier in connection with FAME analysis [93,94], is actually well developed for lipid separations. In 1967, Padley [117] pioneered a silica-coated quartz rod passed directly through FID. These rods were developed commercially in Japan as Chromarods and Iatroscan. Itoh et al. claimed in 1979 [118] that the procedure on Ag-Chromarod was powerful for the separation of triglycerides and phosphatidyl choline. More recently, Jee and Richtie [119] tried to achieve separation of monounsaturated triglyceride isomers (i.e., POS and PSO, with P = palmitoy, O = oleoyl, S = stearoyl), separations previously carried out only by Ag-TLC. Quantitatively, the method is inferior to TLC with scanning densitometry, but it is more rapid.

As far as HPLC methods are concerned, the resolution of glyceride isomers differing only in the position of the acyl groups [i.e., 1,3-distearoyl-2-oleoyl glycerol (SOS) and 1,2-distearoyl-3-oleoyglycerol (SSO)] cannot be achieved by HPLC on C_{18} columns; it is only possible on silver nitrate–impregnated columns at subambient temperature.

Using silver-loaded silica HPLC, Hammond [115] and Smith et al. (120) ob-

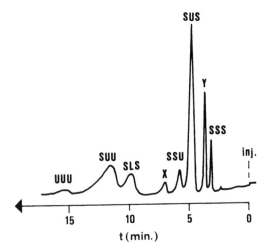

FIGURE 9   Separation of triglycerides at 6.8°C using a benzene mobile phase. Y = methyl santalbate. Column, 25 × 0.4 cm i.d., glass-lined steel; packing, Partisil 5 with a 10% AgNO$_3$ loading; flow rate, 1 mL/min; detector, refractive index; sensitivity, S × 8; injection volume, 30 μL; temperature, ambient. (From ref. 120.)

tained a good resolution of triglyceride mixtures [trisaturated (SSS), 2-mono and 3-monounsaturated (SUS and SSU), 2,3-diunsaturated (SUU), and 2-linoleo (SLS) triglycerides] (see Fig. 9). The effect of temperature and silver nitrate concentration (2.5 to 15%) on the separation, as well as on the quantitation by internal standard (Y = methyl santalbate), were discussed [120].

The separation of three authentic unsaturated triglycerides by Ag–RP–HPLC [37] is illustrated in Fig. 10, which shows (a) strong interaction in the mobile phase due to the nine double bonds of trilinolenin (peak 1), and (b) good separation between triolein (peak 2) and trielaidin (peak 3), which contains the same number of double bonds but with a different configuration. The method was applied to the identification of triolein and diolein–palmitin in a Spanish oil [37].

Bezard et al. [121] investigated the possibility of fractionating pure triglyceride fractions by RP–HPLC after preliminary separation by Ag–TLC. Two methods were tried: (a) Ag–TLC (Kieselgel G plates impregnated with 20% (μ-Bondapack (C$_{18}$) and (b) Ag–HPLC (column packed with silicic acid coated with silver nitrate (see Refs. 120 and 122), followed by RP–HPLC as before.

Another combination of two chromatographic methods has been published by Takano and Kondoh [123]. In a previous work [124] these authors desribed the use of a new post column reactor detector (PCRD) for the analysis of triglycerides with high sensitivity. By using this glyceride-selective PCRD (GS–PCRD), triglycerides of natural fats and oils were separated and quantitatively analyzed

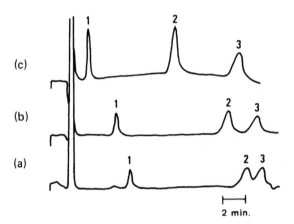

FIGURE 10   Separation of unsaturated triglycerides with and without Ag⁺. Column, 150 × 4 mm i.d., stationary phase, 5-μm Nucleosil 5 C_{18}; mobile phase: (a) methanol/isopropanol 3:1 v/v; (b) A + 10^{-2} N AgClO₄; (c) A + 5 × 10 N AgClO₄. Pressure, 80 bar; flow rate, 0.9 mL/min; temperature, 24°C; detection RI; attenuation, ×4. Peaks: 1, trilinolenin; 2, triolein (Z); 3, trielaidin (E). (From ref. 37.)

with nonaqueous reversed-phase (NARP) chromatography (octadecyl chemically bonded silica column). Nevertheless, complete separation of all triglyceride species could not be accomplished (i.e., separation of SUS and SSUtriglycerides). So the authors combined Ag–HPLC (infrared detector) with subsequent NARP chromatography of the trapped peak. They were able to separate all triglyceride species except positional isomers of SSS triglyceride.

A quantitative determination of triglycerides was performed in two steps by Kalo et al. [125]. The triglycerides were separated by level of unsaturation on silica gel G plates impregnated with 25% silver nitrate. Then the fatty acid composition of the fractions isolated by Ag–TLC was determined by the sodium methoxide method [126] followed by CC analysis.

Some practical applications of Ag–TLC can be considered only as coarse but rapid analytical tests. Argentation TLC was applied to the detection of mineral oils (liquid paraffin and transformer oil) in edible oils [127]. Up to 3% of adulterant mineral oils in various vegetable oils has been detected using silica gel G, sprayed with silver nitrate solution, and eluted with benzene. In the same way, adulteration of butter fat by 5% hydrogenated peanut oil, mohua oil, or tallow was detected by Ag–TLC (solvent: C₆H₆ /HOAc 99.5:0.5; UV detector) [128].

Very recently, a process useful for quality control and adulteration, blending, and hydrogenation detection in edible fats and oils was published [129]. The complex triglyceride mixtures are analyzed by two on-line chromatographic methods: adsorption on Ag-impregnated Kieselgel minicolumns connected to a RP–HPLC Hypersil ODS (5μm) column, with propionitrile as eluant. The triglycerides are separated according to saturation and chain length.

3. Polar Lipids

Two approaches to the analytical problem of separation and identification of phospholipids can be distinguished.

1. Because of the difficulties caused by the polarity of the phosphobase group, a first set of successful chromatographic separations have been achieved by first removing this polar group with transformation in neutral lipids. Several methods were tried: enzymatic hydrolysis [130], chemical reduction [131], and conversion of the polar group to a nonpolar derivative [132,133]. These types of chromatographic analysis were reviewed in 1974 by Viswanathan [134]. This approach is in common use, as is clear from the following examples.

   Interest in the phospholipid classes present in a cellular membrane, to determine their influence on the functional aspects of such membranes, prompted Das et al. [135] to improve a suitable analytical procedure. One of the essential steps involved in the determination of phospholipids is their transformation in the corresponding diglycerides and their resolution as monoacetylderivatives, according to the number of double bonds in the molecule, by Ag–TLC (silica gel G, aqueous solution silver nitrate 25%). The plates were developed twice up to 22 and 24 cm in light petroleum ether (bp 60 to 80°)/diethyl ether/acetic acid (75:35:1) at 25°C. The pool extract was freed from silver nitrate by washing with water. The composition of the different mono acetyl diglycerides (MADGs) in each fraction was obtained either by direct GC–MS or by GLC. The efficiency of the procedure depends to a large extent on the Ag–TLC, and the difficulty lies in the choice of the solvents for TLC.

   Kennerly [136] recently described an improved analysis of various species of phospholipids with Ag–TLC. Phosphatidyl choline, serine, and phosphatidyl ethanolamine were converted to phosphatidic acids (PA) by enzymatic hydrolysis, then transformed in their dimethyl esters (DMPAs). A technique is described for the separation on a single chromatogram, using two successive developments (first with chloroform/methanol 96:4; second, after drying, with the same solvents 98.5:1.5). Diacylglycerol (DAG) was also converted to [32_p]PA in the presence of [32_p]-ATP by *E. coli* DAG kinase. An autoradiograph of Ag–TLC of these labeled PAs transformed with diazomethane in DMPAs is shown in Fig. 11.

   Successive developments were used by Michalec et al. [137,138] in their microchromatographic Ag–TLC method on Silufol R UV254 sheets impregnated with 5% silver nitrate aqueous solution. The method, first described for

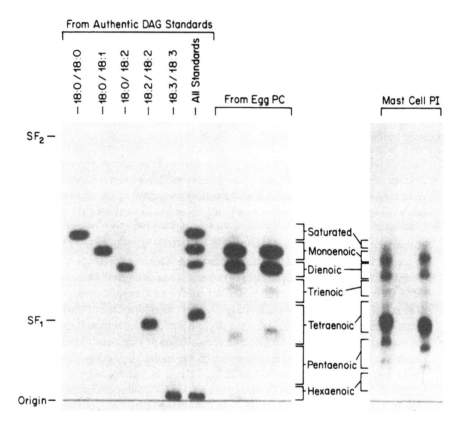

FIGURE 11 Autoradiograph of argentation TLC analysis of defined species of PA, egg PC, and mast cell PI. Left-hand panel: egg lecithin was hydrolyzed by *Cl. welchii* phospholipase C and the resultant diacylglycerol converted to ^{32}P-labeled PA. Authentic DAG species were similarly converted to their (^{32}P) PA derivatives. After treatment with diazomethane, the ^{32}P-labeled DMPA product was isolated by TLC, extracted from the plate, and spotted at the origin of a plate previously dipped in 5% silver nitrate and activated as described in the text. Sequential development was accomplished using chloroform/methanol mixtures (96:4 to 5 cm above origin followed by 98.5:1.5 to 16 cm above origin). Right-hand panel [^{32}P] PI was acetylated and methylated as described in the text and subclasses separated by argentation TLC as described above except that chloroform/methanol 98.75:1.25 was used in the second development 18:0 = stearic acid; 18:1 = oleic acid; 18:2 = linoleic acid; 18:3 = linolenic acid. PA = phosphatidic acid; PC, phosphatidyl choline; PI, phosphatidyl inositol. (From Ref. 136.)

the separation of saturated, monoenoic and dienoic long-chain bases extracted form a human brain sphingolipid mixture [137], was extended to FAMEs (trans 22:1 $\triangle^9$ and trans 18:1 $\triangle^9$) and lipids extracts of several biological sources [138].

2. The second approach considers the separation of molecular species of intact phosphatides. Morris [40] pointed out that it is more difficult to separate these polar compounds on the basis of unsaturation than it is to separate neutral lipids. The first successful separations were made by Ag-TLC (i.e., lecithins [139-141], phosphatidyl ethanolamine [142], phosphatidyl inositols [143], plasmalogens (alken-1-enylacyl phosphatides) [144]). This method, always used, is improved by utilization in conjunction with other chromatographies. For example, Salem et al. [145] described the separation by Ag-TLC of brain phosphatidyl serines on the basis of the degree of unsaturation (silica gel H impregnated with an aqueous solution of silver nitrate (solvent: chloroform/methanol/water 65:35:6.8; UV detection). After hydrolysis (5% HC1 anhydrous methanol), extraction with petroleum ether, and formation of FAMEs, a detailed analysis was made by GC.

As foreseen, HPLC and RP-HPLC were able to afford new possibilities in the analysis of phospholipids or other complex lipids such as sphingomyelin, cerebrosides, and ceramides [146]. However, "critical pairs" were not resolved, as complex lipids having one double bond in the side chain eluted together with the lipid, having a saturated FA with two less carbon atoms (i.e., sphyngomyelins with 24: 1 and 22:0 FA were eluted together) [147]. As Ag-HPLC separates the lipids according to the number of double bonds independently of the number of carbon atoms, by combined use of Ag-HPLC and RP-HPLC different complex lipids were separated almost completely [147]. It is the case of sphyngomyelin separated, after benzoylation, by Ag-HPLC on a commercially prepared silica-bonded silver column (Chrompack silver column) with methanol/isopropanol 8:2 as the solvent in isocratic conditions. Two well-separated peaks were obtained. By GLC it was demonstrated that peak 1 was due to 3-O benzoylated sphyngomyelin with only saturated FA (mainly $C_{18}:0$), and peak 2 contained sphyngomyelin with only monounsaturated FA (mainly $C_{24}:1$). Similar results were obtained with complex lipids containing ceramide. Combined Ag-HPLC and RP-HPLC were also applied to separate molecular species of phosphatidyl choline and other phospholipids containing more than one or two double bonds [147].

Another application of HPLC to the analysis of complex phospholipids was described by Marmer et al. [148] in the case of plasmalogen containing phosphatidyl chloline. Ag-HPLC was carried out on a cation-exchange column treated with silver nitrate under specified conditions of preparation and elution. Ag-HPLC succeeded in fractionating bovine heart phosphatidyl choline into several fractions based on overall unsaturation, not on plasmalogen content alone.

4. Steroids

The usefulness of argentation chromatography in steroid analysis has been put forward by several authors since its earliest development. Among the papers published in the 1960s covered by Morris [40], Guha and Janǎk [18], those of Copius Peereboom and Beekes [149–151] and Klein et al. [152] appear to be the most important. A large range of sterols were examined, mostly as their acetates. The great influence of the molecular environment around the double bonds on their complexation was shown as well as the considerable potentiality of Ag-TLC in conjunction with RP partition [149–151] or adsorption [152] chromatography. In fact, until now TLC remains the primary method when argentation chromatography is applied to steroid analysis, for the separation of unsaturated compounds from saturated ones as well as for the separation of unsaturated compounds according to their degree of unsaturation or for identification purposes. Many examples cited by Kirchner [68], together with others, are reported in Table 5. As far as separation of unsaturated and saturated steroids is concerned, the work of Stevens (153) provides a good example involving the synthesis of corticosteroids from sapogenins, where C-saturated and $\triangle^{9(11)}$ derivatives could be separated only by the Ag-TLC method; the order of migration is C-saturated > $\triangle^5$ > $\triangle^9$ > $\triangle^{11}$.

In the same way, in the course of an exhaustive study on forty 21-deoxy-3-oxo $\triangle^4$ steroids of the pregnane series, Lisboa [154] described the separation and characterization of several 3-oxo $\triangle^4$ C_{21} steroids by TLC using silver nitrate–treated silica gel G. The main results are the following. Steroids of the C_{21} series with $\triangle^{11}$, $\triangle^6$, $\triangle^{17(20)}$ double bonds, (but not those unsaturated at C_4 or C_{16}) form with $AgNO_3$ π-complexes, which move more slowly than the free compounds in the specified solvent systems. So separation of 6-dehydro- and 11-dehydroprogesterones from progesterone or 16-dehydroprogesterone was possible. It is assumed that these last derivatives could not form complex because of the presence of the side chain at C_{17}. The inability of double bonds linked to tertiary carbon atoms on the steroid ring to form complexes has already been cited by Morris [155] for $\triangle^5$ and $\triangle^7$ bonds in the cholestane serie. When the double bond is situated in the side chain [pregna-4,17(20)-dien-3-one] the formation of π-complex is now possible. The increased polarity of the π-complex formed by pregna-4,17(20)-dien-3-one permits its characterization in the presence of pregn-4-en-3-one, for example [154].

With the development of HPLC, the method carried out by Vonach and Schomburg [37] for olefins was applied by Tscherne and Capitano [156] to the separation of vitamins D_2 and D_3 as well as various estrogenic compounds. Using a mobile phase both with and without silver nitrate addition, and a μ-Bondapack C_{18} column

TABLE 5 Examples of AG–TLC in the Steroid Series

Separation performed	Impregnated silver nitrate adsorbents (solvents)	Refs.
Separation of human cholesterol esters according to unsaturation of the FA (from monoenoic to hexanoic)	Silica gel G (diethyl ether/hexane 1:4 or ether alone)	157
Separation and identification of sterol acetates in human serum (cholesterol, 7-dehydrocholesterol, desmosterol, 5-dehydrocholesterol, lanosterol, 24-dehydrolanosterol)	Silica gel (hexane/benzene 5:1 or benzene/ethyl acetate 5:1)	158
Separation of C saturated and $\triangle^{9(11)}$ derivative corticosteroids, produced from sapogenins and of mixtures of saturated and unsaturated androgens, progesterones, sterols, corticosteroids	Silica gel G (chloroform/ethyl acetate 1:1 or 9:1; chloroform/toluene 1:1; methyl acetate/methylene chloride 4:1)	153
Separation of a variety of chlolestanol isomers and cholesterol	Silica gel (hexane/benzene 97:3)	159
Separation of free sterols from yeast	Silica gel or neutral alumina (chloroform/acetone 95:5)	159, 160
Separation of sterol acetates	Silica gel H (benzene/hexane 3:5 or chloroform/petroleum ether (60–80°C)/acetic acid 25:75:0.5)	149, 161
Separation of sterol esters of skin surface fat into three fractions; one saturated, two unsaturated	Silica gel (1% benzene in hexane)	162
Separation of free and conjugated estrogens (as sodium sulfate)	Silica gel (isooctane/chloroform/ethanol 20:35:9)	163
Separation and characterization of forty 21-deoxy : 3-oxo-$\triangle^4$ steroids of pregnane series	Silica gel G or H; utilization of five solvent systems	164
Preparative separation of cholesterol esters (human plasma)	Silicagel–silver sulfamate (n-hexane/petroleum ether/diethyl ether/acetic acid 35:12:2:1)	78
Separation of cholestanol–cholesterol; separation of desmethyl sterols from orange vesicles	See also Ref. 151; Ag-TLC follows Ag-column chromatography	165

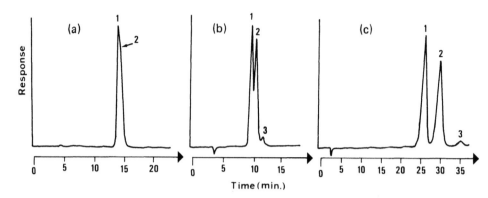

FIGURE 12   HPLC separation of vitamin D compounds using a μ-Bondapak C_{18} column 30 cm $\times$ 4 mm i.d. (a) Mobile phase, 95 mL of methanol + 5 mL of water; flow rate, 0.8 mL/min. (b) Mobile phase, 95 mL of methanol + 5 mL of water + 2 g silver nitrate; flow rate, 0.8 mL/min. 1, vitamin D_2; 2, vitamin D_3; 3, impurity. (c) Mobile phase 86.5 mL of methanol + 13.5 mL of water + 2.4 g silver nitrate; flow rate, 1.1 mL/min. (From ref. 156.)

(30 cm $\times$ 4 mm i.d.) vitamin D_2 was completely separated from vitamin D_3 in less than 35 min (Fig. 12).

It can be seen that by complexation both the retention times of vitamins D_2 and D_3 are decreased on the addition of silver nitrate in the mobile phase and that an additional double bond in vitamin D_2 provides the main basis for its separation from vitamin D_3. In the same way, better separation between estriol, estrone, estradiol, and equilin was obtained after argentation of the mobile phase than in absence of the complexing agent. The retention times are not modified, except in the case of the unsaturated equilin. An interesting point is the ability of the method to regenerate the Bondapack column completely after utilization.

In conclusion, we can say that in the field of steroid separation and characterization, argentation chromatography (TLC or RP-HPLC) appears to be a good complementary technique even if it is unable to afford alone the answer to all separation problems. But there have not been many new developments in the last decade.

V.   ARGENTATION CHROMATOGRAPHY IN THE FIELD OF HETEROATOMIC COMPOUNDS

Another development of Ag-chromatography concerns the separation of heteroatomic compounds, principally nitrogen and sulfur heterocycles. Some examples of the ability of Ag^+ ions to complex with nitriles, amines, and dinitrophenylhydrazone derivatives of carbonyl compounds were reviewed by Guha et al. (18) for papers published before 1972. Some works published since then are reported below after brief mention of the work of Tabak et al. (1970), who pioneered the use of Ag_2O instead AgNO3.

A. Nitrogen Heterocyclic Compounds

1. Separation of Pyridine and Its Homolog Derivatives

Tabak and Verzola [166] tried to extend Ag–TLC to pyridine and other nitrogen heterocycles, with both AgNO3 and AgClO4. Better results were obtained with Ag2O on SiO2 (the plates were prepared by adding a 5% AgNO3 solution to silica gel G and enough 5% NaOH solution. After drying, the plates were activated at 105 to 110°C for 1 h). About 36 solvents were tried with pyridine, picolines, lutidines, collidine, and 2- or 4-alkyl pyridines. The main conclusion was that basicity and polarity differences of pyridine and its homolog derivatives, and in some cases the stability constants of known silver ion complexes, play a major role in their separation in argentation TLC with silver oxide. It is noteworthy that Tabak et al. [167] utilized the same procedure for the separation of substituted aromatic carboxylic acids, unsaturated acids (malic and fumaric), and various aromatic amines (toluidines, chloro, and nitroanilines).

2. Separation of Polynuclear Azaheterocycles

Vivilecchia et al. [168] were the first group to apply Ag–HPLC for the separation of polynuclear azaheterocycles, using silver oxide. The silver ion–impregnated adsorbent was prepared by evaporating (30°C) a known volume of 0.1 N NaOH on the surface of Zipax by means of a rotary evaporator. The support was then treated with a saturated solution of AgNO3 in methanol/n-hexane with 1% CH3CN as the best eluting solvent. The efficiency of the separation is illustrated in Fig. 13 for the following mixture: 1 benzo[h]quinoline; 2, dibenzo[a,c]phenazine; 3, phenazine; 4, acridine; 8, benzo[c]quinoline; 9, benzo[f]quinoline.

A column of 0.5% silver nitrate on Zipax was also tried, without success; the compounds were not eluted because of much stronger complex formation with silver nitrate than with silver oxide. The elution order on silver oxide columns was rationalized by the authors in terms of steric and electronic effect in donor–acceptor complexing.

Some years later, Aigner et al. [169], studying in which form silver was present on the adsorbent (silica gel HF254) prepared under the conditions noted above [168], claimed that there was no Ag_2O on the support but an actual chemical bond between silver and the silica gel in the form of different silver silicates with complexing ability. Silver halides with well-defined composition, better stability, and good complexing properties were chosen by these authors [169], in both TLC and HPLC experiments on acridine, 5,6-denzoquinoline, pyridine, and picolines as model substances. The complexing influence is evidenced by the dominance of steric factors over basicity for the separation of azaarenes and picolines.

The separation of the same azaarenes studied by Vivilecchia was described by

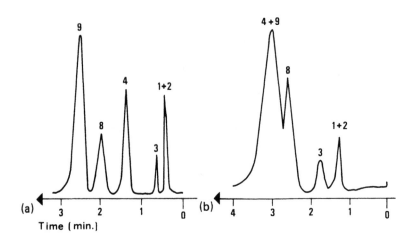

FIGURE 13 (a) Column conditions: dimensions 1 m × 2.4 mn i.d., 0.3% silver; mobile phase, 1% CH₃CN in *n*-hexane; pressure, 900 psi; linear velocity, 4.0 cm/s. (b) 1-m Zipax column; mobile phase, 1% CH₃CN in *n*-hexane; pressure, 400 psi; linear velocity, 1.2 cm/s. (From ref. 168.)

Vonach and Schomburg [37] using RP–HPLC on a 5-μm Nucleosil-5 C$_{18}$ column loaded with 10^{-2} N AgClO$_4$ and methanol/water 1:1 mixture as mobile phase. No Ag$^+$ complexation was observed for neutral nitrogen heterocycles such as carbazoles.

B. Sulfur Heterocyclic Compounds

1. Separation of Thiophenic Compounds

The isolation of thiophenic compounds by Ag–LC from aromatic shale oil fractions was described by Joyce and Uden [170] using a three-step procedure. One- and two-ring aromatic fractions were first obtained from shale oil by LC on silica, then the thiophenic compounds were separated from the aromatics by Ag–LC on a silver nitrate–coated silica column prepared as described by Heath et al. [57]. The fractions obtained were then examined by CC–gas chromatography. The one-ring thiophenes were almost completely separated from nonthiophenic compounds, and the benzothiophenes were enriched by about 96% in comparison to napthalenes.

2. Separation of S and O Polynuclear Heterocycles

Under the conditions cited above for N heterocycles (Ag–RP–HPLC), Vonach and Schomburg [37] showed that no Ag$^+$ complexation was observed for dibenzothiophene, dibenzofuran, and xanthene. The result was a surprisingly strong complexation for thiantharene with two sulfur atoms.

C. Alkyl Phenyl Sulfides

The separation of organic sulfides (acting as electron donors) was examined by Horak et al. [171] using Hg^{2+}, Ag^+, Cd^{2+}, or Pb^{2+} (acting as electron acceptors)-impregnated silica gel plates. The strength of the respective metals to complex sulfide ligand was measured by the increased sulfide retention compared to retention on untreated silica gel plates. Whereas both silica gel and Pb^{2+} and Cd^{2+} are unable to differentiate among individual sulfides R-S-Ph (with R = methyl, ethyl, n, and isopropyl, n, *tert*, *sec*, and isobutyl, benzyl), Hg^{2+} ions, and particularly Ag^+ ions, show remarkable discriminating power. The contribution to the retention of polar, molecular weight, and steric effects was discussed.

REFERENCES

1. F. G. Helfferich, *Nature, 189*: 1001 (1961).
2. F. G. Helfferich, *J. Am. Chem. Soc., 84*: 3237 (1962).
3. F. G. Helfferich, *J. Am. Chem. Soc., 84*: 3242 (1962).
4. R. A. A. Muzzarelli, A. F. Martelli, and O. Tubertini, *Analyst, 94*: 616 (1969).
5. V. A. Davankov and A. V. Semechkin, *J. Chromatogr. Chromatogr. Rev., 141*: 313 (1977).
6. H. F. Walton, in *Ion Exchange and Solvent Extraction: A Series of Advances*, Vol. 4 (J. A. Marinsky and Y. Marcus, eds.), Marcel Dekker, New York, 1973, p. 121.
7. J. D. Navratil and H. F. Walton, *Am. Lab.*, 69 (1976).
8. V. A. Davankov, S. V. Rogozhin, and A. V. Semechkin, in *Ligand-Exchange Chromatography in Chemistry and Technology of High Molecular Weight Compounds*, Vol. 4 (A. Sladkov, ed.), VINITI, Moscow, 1973, p. 5. (in Russ.). CA. *80*: 152–551 (1974).
9. M. J. S. Dewar, *Bull. Soc. Chim. Fr., 18*: C71 (1951).
10. B. W. Bradford, D. Harvey, and D. E. Chalkley, *J. Inst. Pet., 41*: 80 (1955).
11. M. E. Bednas and D. S. Russel., *Can. J. Chem., 36*: 1272 (1958).
12. J. Janák, Z. Jagaric, and M. Dressler, *J. Chromatogr., 53*: 525 (1970).
13. S. P. Wasik and W. Tsang, *Anal. Chem., 42*: 1648 (1970).
14. J. F. McKay and D. R. Latham., *Anal. Chem., 52*: 1618 (1980); *Prepr. ACS Div. Fuel Chem., 25*: 171 (1980).
15. S. Matsushita, Y. Tada, and T. Ikushige, *J. Chromatogr., 208*: 429 (1981).
16. P. Chartier, P. Gareil, M. Caude, R. Rosset, B. Neff, H. Bourgognon, and J. F. Husson, *J. Chromatogr., 357*: 381 (1986).
17. L. E. Cook and S. H. Givand, *J. Chromatogr., 57*: 315 (1971).
18. O. K. Guha and J. Janák, *J. Chromatogr., 68*: 325, (1972).
19. W. Szczepaniak and J. Nawrocki, *Chem. Anal., 20*: 3 (1975).
20. W. Szczepaniak, J. Nawrocki, and W. Wasiak, *Chromatographia, 12*: 484 (1979).
21. J. S. Shabtai, J. Herling, and E. Gil-Av, *J. Chromatogr., 2*: 406 (1959).
22. J. Herling, J. Shabtai, and E. Gil-Av, *J. Chromatogr., 8*: 349 (1962).
23. J. Shabtai, J. Herling, and E. Gil-Av, *J. Chromatogr., 11*: 32 (1963).
24. J. Shabtai, *J. Chromatogr., 18*: 302 (1965).

25. E. Gil Av, and J. Shabtai, *Chem. Ind.*, 1630 (1959).

26. E. Gil Av, and J. Herling, *J. Phys. Chem.*, *66*: 1208 (1962).

27. M. A. Muhs and F. T. Weiss, *J. Am. Chem. Soc.*, *84*: 4697 (1962).

28. A. C. Cope and E. M. Acton, *J. Am. Chem. Soc.*, *80*: 355 (1958).

29. A. C. Cope, C. L. Bumgardner, and E. E. Schweizer, *J. Am. Chem. Soc.*, *79*: 4729 (1957).

30. a. C. Cope, N. A. Level, H. H. Lee, and W. R. Moore, *J. Am. Chem. Soc.*, *79*: 4720 (1957).

31. A. C. Cope, P. T. Moore, and W. R. Moore, *J. Am. Chem. Soc.*, *82*; 1744 (1960).

32. R. F. Hirsch, H. C. Stober, M. Kowblansky, F. V. Hubner, and A. W. O'Connell, *Anal. Chem.*, *45*: 2100 (1973).

33. L. R. Chapman and D. F. Kuemmel, *Anal. Chem.*, 1598 (1965).

34. F. Mikes, V. Schurig, and E. Gil Av, *J. Chromatogr.*, *83*: 91 (1973).

35. R. R. Heath, J. H. Tumlinson, R. E. Doolittle, and J. H. Duncan, *J. Chromatogr. Sci.*, *15*: 10 (1977).

36. R. R. Heath and P. E. Sonnet, *J. Liq. Chromatogr.*, *3*: 1129 (1980).

37. B. Vonach and G. Schomburg, *J. Chromatogr.*, *149*: 417 (1978).

38. G. Schomburg and K. Zegarski, *J. Chromatogr.*, *114*: 174 (1975).

39. E. Fuggerth, *J. Chromatogr.*, *169*: 469 (1979).

40. L. J. Morris, *J. Lipid Res.*, *7*: 717 (1966).

41. R. S. Prosad, A. S. Gupta, and S. Dev, *J. Chromatogr.*, *92*: 450 (1974).

42. J. C. Kohli and K. K. Badaisha, *J. Chromatogr.*, *320*: 455 (1985).

43. B. M. Lawrence, *J. Chromatogr.*, *38*: 535 (1968).

44. G. M. Nano and A. Martelli, *J. Chromatogr.*, *21*: 349 (1966).

45. R. R. Sauers, *Chem. Ind.*, 176 (1960).

46. S. Hara, A. Ohsawa, J. Endo, Y. Sashida, and H. Itokawa, *Anal. Chem.*, *52*: 428 (1980).

47. E. Heftmann, G. A. Saunders, and W. F. Haddon, *J. Chromatogr.*, *156*: 71 (1978).

48. M. Morita, S. Mihashi, H. Itokawa, and S. Hara, *Anal. Chem.*, *55*: 412 (1983).

49. M. G. M. De Ruyter and A. P. De Leenheer, *Anal. Chem.*, *51*: 43 (1979).

50. R. P. Evershed, E. D. Morgan, and L. D. Thompson, *J. Chromatogr.*, *237*: 350 (1982).

51. L. K. T. Lam, C. Yee, A. Chung, and L. W. Wattenberg, *J. Chromatogr.*, *328*: 422 (1985).

52. L. K. T. Lam, V. L. Spernins, and L. W. Wattenberg, *Cancer Res.*, *42*: 1193 (1982).

53. S. S. Curran and D. F. Zinkel, *J. Am. Oil Chem. Soc.*, *58*: 980 (1981).

54. T. Tamaki, H. Noguehi, T. Yushima, and C. Hirano, *Appl. Entomol. Zool.*, *6*: 139 (1971).

55. N. W. H. Houx, S. Voerman, and W. M. F. Jongen, *J. Chromatogr.*, *96*: 25 (1974).

56. N. W. H. Houx and S. Voerman, *J. Chromatogr.*, *129*: 456 (1976).

57. R. R. Heath, J. H. Tumlinson, R. E. Doolittle, and A. T. Proveaux, *J. Chromatogr. Sci.*, *13*: 380 (1975).

58. R. R. Heath, A. T. Proveaux, and J. H. Tumlinson, *J. High Resolut. Chromatogr. Chromatogr. Commun.*, *6*: 317 (1978).

59. E. Heftmann and G. A. Saunders, *J. Liq. Chromatogr.*, *1*: 333 (1978).

60. W. Wasiak and W. Szczepaniak, Chromatographia, *18*: 205 (1984).

61. E. Gil Av and V. Schurig, *Anal. Chem.*, *43*: 2030 (1971).

62. V. Schurig and R. C. Chang, A. Zlatkis, E. Gil Av, and F. Mikes, *Chromatographia, 6*: 223 (1973).
63. R. L. Grob and E. J. Mc Gonigle, *J. Chromatogr., 59*: 13 (1971).
64. E. J. Mc Gonigle and R. L. Grob, *J. Chromatogr., 101*: 39 (1974).
65. V. Schurig, J. L. Bear, and A. Zlatkis, *Chromatographia, 5*: 301 (1972).
66. C. R. Scholfield, in *Geometric and Positional Fatty Acid Isomers* (H. J. Dutton and E. A. Emken, eds.), American Oil Chemists' Society, Champaign, Ill., 1979, p. 31.
67. L. J. Morris and B. W. Nichols, in *Progress in Thin-Layer Chromatography and Related Methods,*(Vol. 1. A. Niederwieser and G. Pataki, eds), Ann Arbor–Humphrey Science, Ann Arbor, Mich., 1970, p. 75.
68. J. G. Kirchner, ed., *Techniques of Chemistry,* Vol. 14, *Thin Layer Chromatography,* 2nd ed., Wiley, New York, 1978, p. 832.
69. R. G. Ackman, S. M. Barlow, I. F. Duthie, and G. L. Smith, *J. Chromatogr. Sci., 21*: 87 (1983).
70. *Off. J. Eur. Communities, L202*: 35 (1976).
71. B. Breuer, T. Stuhlfauth, and H. P. Fock, *J. Chromatogr. Sci., 25*: 302 (1987).
72. L. V. Andreev, *J. High Resolut. Chromatogr. Chromatogr. Commun., 6*: 575 (1983).
73. H. Heckers and F. W. Melcher, *J. Chromatogr., 256*: 185 (1983).
74. M. S. Huq, M. S. Khan, and S. F. Rubbi, *Bangladesh J. Sci. Ind. Res., 14*: 159 (1979); *Chem. Abstr., 92*(13): 109295a (1980).
75. M. Ozcimder and W. E. Hammers, *J. Chromatogr., 187*: 307 (1980).
76. S. Lam and E. Grushka, *J. Chromatogr. Sci., 15*: 234 (1977).
77. H. N. S. Chan and G. Levett, *Chem. Ind. (London),* 578 (1978).
78. P. P. Ilinov, *Lipids, 14*: 598 (1979).
79. M. Inomata, F. Takaku, Y. Nagai, and M. Saito, *Anal. Biochem., 125*: 197 (1982).
80. V. I. Svetashev and N. V. Zhukova, *J. Chromatogr., 330*: 396 (1985).
81. P. A. Dudley and R. E. Anderson, *Lipids, 10*: 113 (1975).
82. G. K. Bandyapadhaya and J. Dutta, *J. Chromatogr., 114*: 280 (1975).
83. V. K. S. Shukla and K. C. Srivastava, *J. High Resolut. Chromatogr. Chromatogr. Commun.,* 214 (1978).
84. C. R. Scholfield and T. L. Mounts, *J. Am. Oil Chem. Soc., 54*: 319 (1977).
85. R. Scholfield, *J. Am. Oil Chem. Soc., 57*: 331 (1980).
86. H. Rakoff and E. A. Emken, *J. Am. Oil Chem. Soc., 55*: 564 (1978).
87. C. R. Scholfield, *J. Am. Oil Chem. Soc., 56*: 510 (1979).
88. E. A. Emken, J. C. Hartman, and C. R. Turner, *J. Am. Oil Chem. Soc., 55*: 561 (1978).
89. R. O. Adlof, H. Rakoff, and E. A. Emken, *J. Am. Oil Chem. Soc., 57*: 273 (1980).
90. R. O. Adlof and E. A. Emken, *J. Am. Oil Chem. Soc., 57*: 276 (1980).
91. R. O. Adlof and E. A. Emken, *J. Am. Oil Chem. Soc., 58*: 99 (1981).
92. E. A. Emken, R. O. Adlof, and H. Rakoff, U. S. patent 132, 584 Aug. 29, 1980; *Chem. Abstr., 94*(14): 105681s (1981).
93. J. L. Sebedio, T. E. Farquharson, and R. G. Ackman, *Lipids, 17*: 469 (1982).
94. J. L. Sebedio, T. E. Farquharson, and R. G. Ackman, *Lipids, 20*: 555 (1985).
95. S. S. Radwan, *J. Chromatogr., 234*: 463 (1982).
96. M. S. F. Lie Ken Jee and C. H. Lam, *J. Chromatogr., 124*: 147 (1976).
97. M. S. F. Lie Ken Jee and C. H. Lam, *J. Chromatogr., 129*: 181 (1976).
98. R. Kleiman, G. F. Spencer, L. W. Tajarks, and F. R. Earle, *Lipids, 6*: 617 (1971).

99. A. T. Erciyes and H. Civelekoglu, *Bull. Tech. Univ. Istanbul, 36*: 381 (1983); *Chem. Abstr., 100*(19): 153268b (1984).

100. N. Bielen and S. Turina, *Prehrambeno Tehnol. Rev., 23*: 77 (1985); *Chem. Abstr., 105*(7): 59512h (1986).

101. I. Tomiyasu, S. Toriyama, I. Yano, M. Masui, and N. Akimori, *Koenshu-Iyo Masu Kenkyukai, 5*: 221 (1980) (in Japanese); *Chem. Abstr., 94*(9): 60979b (1981).

102. K. Green and B. Samuelsson, *J. Lipid Res., 5*: 117 (1964).

103. H. A. Davies, E. W. Horton, K. B. Jones, and J. P. Quilliam, *Br. J. Pharmacol., 42*: 569 (1971).

104. E. G. Daniels, in *Lipid Chromatographic Analysis,* 2nd ed. (G. V. Marinetti, ed.), Marcel Dekker, New York, 1976, p. 611.

105. J. E. Greenwald, M. S. Alexander, M. Van Rollins, L. K. Wong, and J. R. Bianchine, *Prostaglandins, 21*: 33 (1981).

106. T. K. Bills, J. B. Smith, and M. J. Silver, *Biochem. Biophys. Acta, 424*: 303 (1976).

107. N. H. Andersen and E. M. K. Leovey, *Prostaglandins, 6*: 361 (1974).

108. J. S. Bomalaski, J. C. Touschstone, A. T. Dailey, and R. B. Zurier, *J. Liq. Chromatogr., 7*: 2751 (1984).

109. W. S. Powell, *Methods Enzymol., 86*: 530 (1982).

110. M. V. Merritt and G. E. Bronson, *Anal. Biochem., 80*: 392 (1977).

111. W. S. Powell, *Anal. Biochem., 115*: 267 (1981).

112. W. S. Powell, *Anal. Biochem., 128*: 93 (1983).

113. W. S. Powell, in *Advances in Prostaglandin, Thromboxane and Leukotrienes Research,* Vol. 11 (B. Samuelsson, R Paoletti, and P. Ramvell, eds.), Raven Press, New York, 1983, p. 207.

114. L. D. Kissinger and R. H. Robins, *J. Chromatogr., 321*: 353 (1985).

115. E. W. Hammond, *J. Chromatogr., 203*: 397 (1981).

116. M. J. S. Dallas and F. B. Padley, *Lebensm. Wiss. Technol., 10*: 328 (1977).

117. F. B. Padley, *Chem. Ind. (London), 874 (1967).*

118. T. Itoh, M. Tanaka, and H. Kaneko, *J. Am. Oil Chem. Soc., 56*: 191A (1979).

119. M. H. Jee and A. S. Richtie, *J. Chromatogr., 370*: 214 (1986).

120. E. C. Smith, A. D. Jones, and E. W. Hammond, *J. Chromatogr., 205*: 188 (1980).

121. J. A. Bezard, M. A. Ouedraogo, and A. Moussa, *J. Chromatogr., 196*: 279 (1980).

122. A. Monseigny, P. Y. Vigneron, M. Levacq, and F. Zwobada, *Rev. Fr. Corps Gras, 26*: 107 (1979).

123. S. Takano and Y. Kondoh, *J. Am. Oil Chem. Soc., 64*: 380 (1987).

124. Y. Kondoh and S. Takano, *Anal. Chem., 58*: 2380 (1986).

125. P. Kalo, K. Vaara, and M. Antila, *J. Chromatogr., 368*: 145 (1986).

126. H. T. Badingo and C. De Jong, *J. Chromatogr., 279*: 493 (1983).

127. S. Chand, C. Srininvasulu, and S. N. Mahapatra, *J. Chromatogr., 106*: 475 (1975).

128. M. M. Chakrabarty, D, Bhattacharyya, A. K. Gagen, and M. K. Chakrabarty, *Sci. Cult., 46*: 336 (1980); *Chem. Abstr., 94*: (11): 82335r (1981).

129. K. Kemper, H. U. Melchert, and K. Rubach, *Lebensm. Gerichtl. chem., 42*: 105 (1988); *Chem. Abstr., 110*: 56061c (1989).

130. O. Renkonen, *Biochim. Biophys. Acta, 152*: 114 (1968).

131. N. Totani and H. K. Mangold, *Microchim. Acta, 1*: 73 (1981).

132. M. L. Blank, E. A. Cress, T. Lee, N. Stephens, C. Piantidosi, and F. Snyder, *Anal. Biochem., 133*: 430 (1983).
133. Y. Nakagawa and L. A. Horrocks, *J. Lipid Res., 133*: 430 (1983).
134. C. V. Viswanathan, *J. Chromatogr. Chromatogr. Rev., 98*: 129, (1974).
135. A. K. Das, R. Ghosh, and J. Dutta, *J. Chromatogr., 234*: 472 (1982).
136. D. A. Kennerly, *J. Chromatogr., 363*: 462 (1986).
137. C. Michalec, *J. Chromatogr., 41*: 267 (1969).
138. C. Michalec, J. Reinisova, and Z. Kolman, *J. Chromatogr., 105*: 219 (1975).
139. G. A. E. Arvidson, *J. Lipid Res., 6*: 574 (1965).
140. G. A. E. Arvidson, *J. Lipid Res., 8*: 155 (1967).
141. G. A. E. Arvidson, *Eur. J. Biochem., 4*: 478 (1968).
142. S. M. Hopkins, G. Sheehan, and R. L. Lyman, *Biochem. Biophys. Acta, 164*: 272 (1968).
143. B. Holub and A. Kuksis, *J. Lipid Res., 12*: 510 (1971).
144. S. P. Hoevet, C. V. Viswanathan, and W. O. Lundberg, *J. Chromatogr., 34*: 195 (1968).
145. N. Salem, Jr., L. G. Abood, and W. Hoss, *Anal. Biochem., 76*: 407 (1976).
146. R. H. McCluer and F. B. Jungalwala, in *Biological/Biomedical Application of Liquid Chromatography* (J. Hawk, ed.), Marcel Dekker, New York, 1979, p.7.
147. M. Smith, P. Monchamp, and F. B. Jungalwala, *J. Lipid Res., 22*: 714 (1981).
148. W. N. Marmer, T. A. Foglia, and P. D. Vail, *Lipids, 19*: 353 (1984).
149. J. W. Copius-Peereboom, in *Paper and Thin Layer Chromatography* (K. Macek and I. M. Hais, eds.), Elsevier, Amsterdam, 1965, p. 316.
150. J. W. Copius-Peereboom, *Z. Anal. Chem., 205*: 325 (1964).
151. J. W. Copius-Peereboom and H. W. Beekes, *J. Chromatogr., 17*: 99 (1965).
152. P. D. Klein, J. C. Knight, and P. A. Szczepanik, *J. Am. Oil. Chem. Soc., 43*: 275 (1966).
153. P. J. Stevens, *J. Chromatogr., 36*: 253 (1968).
154. B. P. Lisboa, *Steroids,* 319 (1967).
155. L. J. Morris, in *New Biochemical Separations* (A. T. James and L. J. Morris, eds.), Van Nostrand, Reinhold, London, 1964, p. 295.
156. R. J. Tscherne and G. Capitano, *J. Chromatogr., 136*: 337 (1977).
157. L. J. Morris, *J. Lipid Res., 4*: 357 (1963).
158. J. R. Claude and J. L. Beaumont, *Ann. Biol. Clin., 22*: 815 (1964).
159. J. D. Gilbert, W. A. Harland, G. Steel, and C. J. W. Brooks, *Biochem. Biophys. Acta, 187*: 453 (1969).
160. R. Kammereck, W. H. Lee, A. Paleokas, G. Y. Schroepfer, *J. Lipid Res., 8*: 282 (1967).
161. H. E. Vroman and C. F. Cohen, *J. Lipid Res., 8*: 150 (1967).
162. E. Haahti, T. Nikkari, and K. Juva, *Acta Chem. Scand., 17*: 538 (1963).
163. L. E. Crocker and B. A. Lodge, *J. Chromatogr., 69*: 419 (1972).
164. L. E. Crocker and B. A. Lodge, *J. Chromatogr., 62*: 150 (1971).
165. H. E. Nordby and S. Nagy, *J. Chromatogr., 79*: 147 (1973).
166. S. Tabak and M. R. M. Verzola, *J. Chromatogr., 51*: 334 (1970).
167. S. Tabak, A. E. Mauro, and A. Del'Acqua, *J. Chromatogr., 52*: 500 (1970).
168. R. Vivilecchia, M. Thiebaud, and R. W. Frei, *J. Chromatogr. Sci., 10*: 411 (1972).
169. R. Aigner, H. Spitzy, and R. W. Frei, *Anal. Chem., 48*: 2 (1976).
170. W. F. Joyce and P. C. Uden, *Anal. Chem., 55*: 540 (1983).
171. V. Horak, M. De Vallo Guzman, and G. Weeks, *Anal. Chem., 51*: 2248 (1979).

5

Ligand-Exchange Chromatography of Chiral Compounds

V. A. Davankov
USSR Academy of Sciences, Moscow, USSR

I. INTRODUCTION

Ligand -exchange chromatography (LEC) represents one of the most typical cases of complexation chromatography. Complexes dealt with in these processes comprise metal cations associated with ligands (anions or neutral molecules) that are capable of donating electron pairs onto the vacant orbitals of the metal.

Initially, metal cations in LEC were considered to be the essential part of the stationary phase. As early as 1955, Bradford et al. [1] applied silver nitrate, dissolved in polyethylene glycol, as the gas chromatography (GC) stationary phase-modifier that enhanced retention of unsaturated hydrocarbons compared to saturated ones. Obviously, Ag(I) cations have realized here their ability to form complexes via interaction with π-electrons of the carbon–carbon double bonds. Stokes and Walton have shown [2] that metal cations that are part of the stationary phase (in particular, cation-exchange resin) preserve their ability to form complexes with ammonia molecules present in an aqueous solution that stands at equilibrium with the resin. Finally, in 1961, Helfferich [3] described the substitution of organic diamine molecules for such metal-ion-coordinated ammonia molecules in the resin phase, and vice versa. He used the term *ligand exchange* for the first time, thus giving a start to ligand-exchange chromatography as a general approach to separating compounds that coordinate metal cations.

It was later recgonized that complexing metal cations, when simply added at a low concentration to the mobile phase, can also largely contribute to the chromatographic resolution of complex-forming solutes. Here, ligand exchange takes place in the mobile or both mobile and stationary phases.

The systems above correspond to the broad definition of ligand-exchange chromatography, given by Davankov and Semechkin [4], that does not specify the location of the metal cation in the chromatographic system but merely refers this type of chromatography to a process in which the formation and breaking of labile coordinate bonds to a central metal cation are responsible for the separation of complex-forming solutes.

In accordance with this definition, various kinds of ligand-exchange processes have been analyzed systemically in the book by Davankov et al. [5]. It would obviously be unwise to try presenting here a short version of the above book. We therefore prefer to discuss here only the resolution of enantiomers using chiral ligand-exchange chromatography, which is the most important and successful field of application of this technique. Herewith we concentrate on the information published after 1985 that was not yet available during the preparation of the monograph. (In addition to this, Chapter 4 deals with argentation chromatography, which was touched on only briefly in Ref. 5.) Two review chapters that have been published more recently should be mentioned here: one on column packing materials for LEC [6] and one on chiral separations using LEC [7].

II. THEORETICAL ASPECTS OF CHIRAL DISCRIMINATIONS IN CHROMATOGRAPHIC SYSTEMS

A. Chiral Recognition Mechanism: Stereochemical Postulates

Are two enantiomeric molecules, for example natural amino acid L-alanine and its mirror-image D-enantiomer, really identical from the thermodynamic point of view? Are they truly equal in all their physical and chemical properties? These questions arise now and then, particularly in connection with the crucial importance of enantiomeric purity of biomolecules to the function of living systems and with studies and speculations into the origins of chirality [8] and of life. The more recent answer to the foregoing questions is "no": D-alanine appears to be less stable by about 10^{-14} J/mol. The effect is small. At a hypothetical enantiomerization equilibrium, the enantiomeric excess of the L-isomer would amount to $10^{-6}\%$ [9,10]. Unfortunately, we do not possess a quantitation technique that could detect this small deviation from the 1:1 composition of enantiomers in a racemic mixture. Neither is there a separation technique that could make use of the small energetic nonequivalence of the enantiomers, in order to achieve their discrimination.

Therefore, the first most important stereochemical postulate still holds, stating that a special chiral selector is required in order to recognize the enantiomers and discriminate between them. The second postulate is even less evident and is called in question more frequently. It states that in order to recognize two enantiomers, the chiral selector has to enter a stereo-dependent* three-point interaction with one of the enantiomers. In this case the corresponding appropriate interaction sites of the other enantiomer would appear in an incorrect orientation to the selector, thus giving the less stable diastereomeric associate. The three-point interaction requirement unavoidably results [11] from geometric consideration of the chiral recognition of three-dimensional structures. It is generally valid for the enantioselectivity of action of chiral drugs on specific receptors [12], for the enantiospecificity of transformations of prochiral substrates by natural enzymes [13], and for the chromatographic resolution of racemic compounds on chiral stationary phases [14].

However, in many practical cases it is difficult to validate this second fundamental stereochemical postulate, because the real contact points of the chiral selector with the enantiomers are seldom known. Moreover, an interaction does not necessarily mean an immediate contact. An influence can be exerted over distances exceeding the van der Waals radius. Therefore, the calculated energies of formation of two diastereomeric associates appear to be different [15] even in cases of only two- or one-point immediate contacts between the chiral components of these associates [16]. Naturally, this is possible only if six-center forces between triplets of atoms or functionality in the two chiral species are considered, in order to evaluate, in terms of "overlap-exchange" functions, the energy differences between two possible diastereomeric associates.

B. Enantioselectivity: Influence of the Packing Surface, Mobile-Phase Additives, Sample Sizes, and Sample Concentrations

An extension of the three-point interaction rule is the model where one [17] or two [18] interactions between the components of an associate are mediated by a third, achiral structure. In LEC, these achiral species can be represented by the surface of the column packing material or solvent molecules participating in the complex formation with the central metal ion. Then the enantioselectivity of a chromatographic system may strongly depend on the surface chemistry of the packing as well as the nature of the mobile phase.

Mobile-phase components can also alter the conformation of the chiral selector, thus altering the magnitude [19,20] or inverting the sign [21] of the enantioselectivity of the column. More difficult to explain are large changes in the resolution

---

*Such interaction involves atoms or groups along three different bonds at the chiral center.

selectivity with the size of the sample [22] and also cases of concentration dependence of elution order in the resolution of enantiomers on such conformationally rigid chiral stationary phases as microcrystalline triacetylcellulose [23].

In any case, the examples above reveal manyfold factors that influence the intimate interaction mechanism between the chiral selector and the enantiomers to be recognized and separated, which can result in a dramatic change in the total enantioselectivity of chromatographic systems. It is clear that thorough studies are required if understanding of the chiral recognition mechanism is desired. Thus far, reliable achievements in this direction have been attained only in ligand-exchange chromatography and charge-transfer chromatography [5,24–26].

In chiral charge-transfer chromatography, initial attempts have been made to apply computational chemistry, calculating the absolute interaction energies of the chiral selector with two enantiomers, in order to correlate the obtained differences in the interaction energies with the experimental enantioselectivity of chiral chromatographic systems [27,28]. However, all calculations to date have been performed "in the gas phase," with only one pair of molecules involved in the association. It is not a surprising fact, therefore, that disregarding the possible influence of the sorbent matrix or the solvent makes any predictions of retention and resolution selectivity unreliable, if not impossible. The purely statistical chemometric approaches, describing the chromatographic behavior of a series of racemic compounds on a given chiral phase [29,30], seem to produce more reliable predictions.

C. Enantioselectivity and Binding Free Energies of Enantiomers to Chiral Selector

Another important theoretical issue is the relation between the overall enantioselectivity, α, of a chromatographic column, which is the ratio of capacity factors of two enantiomeric species, k'_S/k'_R, and the difference in the free energies of binding of the enantiomers to the chiral selector, $\delta \Delta G°$. In accordance with the equation

$$\delta \Delta G° \geq - RT \ln \alpha$$

the column enantioselectivity can approach the theoretical value given by the enantioselectivity of the immediate selector–enantiomer interaction only if there are no additional, nonselective interactions of the solute with the stationary phase. If the chiral selector is part of the stationary phase [chiral stationary phase (CSP)], the stronger-bonded enantiomer would reside longer in the column. Vice versa, with the chiral selector residing entirely in the mobile phase [chiral mobile phase (CMP)] and solute molecules retarded by the column packing in a different, nonspecifying manner, the enantiomer that forms stronger associates with the selector would elute first. In the practice, the chiral additive to the CMP as well as its complexes with the solute enantiomers tend to partition between two phases. In this case the chiral discrimination of the column, α, is given roughly by the ratio of complexation enantioselectivities in the stationary and mobile phase [31].

It should be pointed out that chromatography, being a unique separation technique that cumulates the small separation results of a very large number of individual solute–sorbent interaction events, makes successful use of chiral selectors that display low discriminating abilities. A minor thermodynamic enantioselectivity of $\delta \Delta G° = 0.024$ kJ/mol, corresponding to an α value of 1.01, would already suffice for many racemates to be completely resolved using chromatographic columns with plate numbers of about 1.5×10^5 (which is easily attained by capillary GC columns). No other separation technique could be based on such small selectivities.

With $\delta \Delta G°$ increasing, the column selectivity, α, rises exponentially. On doubling the value of $\delta \Delta G°$, an increase in α to the square of its initial value can be achieved, as demonstrated by Pirkle and Pochapsky [32], by making a "dimeric" solute that interacts simultaneously with two chiral sorption sites of the stationary phase.

D. Enhancing Selectivity of Enantiomeric Resolutions

Another reliable, though not really convenient approach to enhancing the enantioselectivity of chromatographic systems is the combination of chiral stationary with chiral mobile phases. To realize such a combination, one has to bind chiral selector to the sorbent matrix and add its opposite enantiomeric form to the mobile phase [33–35]. Similarly, combinations of any completely different chiral selectors in the stationary and mobile phases are feasible, though not yet reported.

If enantiomeric peaks in a complex chromatogram overlap with elution bands of other chiral or nonchiral components of the sample, a combination of the chiral column can be recommended with an achiral one that operates on completely different retention mechanisms [36]. In such a tandem system, one column would resolve the enantiomeric pairs, whereas the other would contribute to the separation of one solute from the other. The sequence of coupling the two columns is unimportant. Coupled-column chromatography can also be used to trap and compress the eluted enantiomeric peaks, which would permit their subsequent quantitation with enhanced sensitivity and precision [37]. Thus, coupled-column or, more generally, multidimensional chromatography may become a general strategy for improving the analysis of complex mixtures of chiral and nonchiral compounds.

E. Enantiomeric Purity of the Chiral Selector: Racemization

In addition to the discussion above on the principal possibility of using successfully chiral selectors of low chiral recognition power and of combining the enantioselective retention mechanisms with nonselective solute–sorbent interactions within one chromatographic system, another great advantage of chromatographic chiral resolutions should be stressed here. In all classical procedures for resolving recemates, an insufficient enantiomeric purity of the chiral selector necessarily led to an equivalent loss in the enantiomeric purity, or yield, of the resolved product. Rather, in chromatography, it is possible to obtain quanitatively both enantionmers in an enantiomerically pure state even in the case when the chiral selector has an enantiomeric purity, P, lower than 100%. As shown in Fig. 1 [38], enantiomerically impure selectors, when combined with chromatographic techniques, can still produce selectivity values α of a column that are more than sufficient for a complete resolution of solute enantiomers. The α–P relations have been found both on theoretical consideration of the problem [39] and on experimental examination of chiral GC [40] and LC [41,42] systems. The latter proved, in addition, that enantiomeric impurities in the chiral selector do not diminish column efficiency but merely result in the two sharp enantiomeric peaks approaching each other and completely coalescing into one sharp peak with $k' = 0.5$ ($k'_R + k'_S$) when the enantiomeric purity of the selector falls to zero.

The relative insensitivity of chiral resolutions toward the enantiomeric purity of the selector may be of particular importance to ligand-exchanging systems since transition metal ions are known to catalyze racemization of natural amino acids, the most common chiral selectors in LEC. It is clear that partial racemization of the selector would affect the resolution ability of the column only insignificantly. The configurational stability of chiral selectors in ligand-exchanging resins has been examined, nevertheless, with the result that N-substituted amino acid moieties, in particular N-benzylated moieties, proved to be stable under conditions of exploitation of the resins [43].

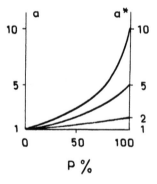

FIGURE 1   Enantioselectivity, a, of three arbitrary chromatographic systems dependent on the enantiomeric purity, P, of chiral selectors involved. Ultimate enantioselectivity, a^*, of the systems, 10, 5, and 2. (From Ref. 38.)

Certainly, the enantiomers under resolution may also undergo partial racemization (enantiomerization) during chromatography in a chiral column. This situation has been observed by Schurig et al. [44,45] on gas chromatography of configurationally labile 1-chloro-2,2-dimethylaziridine according to ligand-exchange (complexation) mechanism. Here the moderate enantiomerization process manifests itself in that the elution curve between the enantiomeric peaks does not approach the zero baseline but forms a plateau. The latter contains molecules that have inverted their configuration in the column and now travel with the speed of the antipode. The height and shape of the plateau allow calculation of kinetic parameters of the enantiomerization. Similar phenomena have been observed in LC as well. Mannschreck et al. [46–49] have shown that two enantiomeric peaks strongly overlap when the barrier to enantiomerization does not exceed 90 to 100 kJ/mol. In this case, running the column at a lower temperature slows down the inversion and considerably improves the resolution.

F. Temperature Effects

However, separating enantiomers at reduced temperatures is less applicable to ligand-exchange chromatography since ligand exchange is a relatively slow process, and therefore obtaining an acceptable plate number of a LEC column often requires its temperature to be raised to 50 to 75°C.

In this respect, one has to bear in mind that increasing the temperature up to a certain limit, given by chemical and configurational stability of all components of the system, generally improves the solute–sorbent interaction rates and the diffusion coefficients, thus generally improving the column efficiency. In contrast to this, the resolution enantioselectivity may either increase or decrease with the temperature rising. A linear relation can be expected to exist between the logarithm of selectivity, $\ln \alpha$, and the inverse of temperature, $1/T$ [38,50,51]. Herewith a certain point must exist where $\ln \alpha = 0$ and two enantiomeric peaks elute with equal k' values. On passing this point, T_{inv}, in any direction, a temperature-dependent inversion of elution orders of two isomers must take place. This phenomenon, indeed, has been observed both for GC [52,53] and LC [54] enantiomeric resolutions. Although decreasing α values with the temperature rising were reported in the majority of cases, in LEC one example was observed [55] of improved selectivities at elevated temperatures. It was shown [56] that entropic contributions to the chiral discrimination in this case predominated over the enthalpic contribution.

More detailed information on the discussion above and some additional theoretical aspects of chiral discriminations in chromatographic systems can be found in review papers [26,38,57]. Practical achievements in chiral chromatographic separations are the subject of recent books [58,59] and a series of reviews [24, 26,60–65].

III. CHIRAL LIGAND-EXCHANGING CHROMATOGRAPHIC SYSTEMS

A. Polymeric Column Packings in Chiral Ligand-Exchange Chromatography

Chromatographic column packing materials based on purely polymeric networks have played an extremely important role in the development of ion-exchange technique, ligand exchange, affinity, and size exclusion chromatographies and many other chromatographic techniques. It was only in the 1970s that these packings gave way to macroporous silica-based materials. The latter were much more pressure resistant and therefore could easily be used in the microparticulate form for highly efficient HPLC columns. However, silica-bonded phases could never reach the excellent chemical durability and high loading capacity of the polymeric prototypes. Therefore, preparative separations never dropped their preference toward the use of polymeric packings. Moreover, the most recent generation of HPLC packing comprises polymeric layers chemically bonded onto the surface of macroporous mineral matrix or totally polymeric macroporous microbeads. Thus the methods of preparation of chiral polymeric chelating resins that were developed in the early 1970s are by no means just of historic interest, but may again rise in popularity.

1. Polystyrene-Based Sorbents

The first chiral ligand-exchanging resins were synthesized by Rogozhin and Davankov [66,67] in 1968 by reacting chloromethylated cross-linked polystyrene with natural amino acids or their derivatives. Especially useful proved to be resins incorporating residues of cyclic amino acids, proline, hydroxyproline, and allohydroxyproline. The recurring units of these resins easily formed chelates with transition metal ions, in particular copper(II) ions, and when loaded with copper(II), showed a marked ability to bind additional amino acid molecules from the mobile phase. In this way diastereomeric mixed-ligand sorption complexes were formed of the following general structure:

With the chiral selector of the resin belonging to the configurational *L* (*S*) series of natural amino acids, *D* (*R*) enantiomers of the mobile ligand were found to form stronger sorption complexes. Accordingly, on elution with water, only *L*-amino acid could be found in the eluate, whereas displacement of the *D*-isomer required an ammoniac solution as the eluant (Fig. 2). These first enantiomeric resolutions demonstrated an unprecidented enantioselectivity of ligand-exchanging systems and attracted the attention of many researchers to this technique.

Numerous other α-aminocarboxylic and α-aminophosphonic acids were incorporated into the polystrene network to arrive at chiral ligand exchangers (Table 1). Davanov and coworkers applied temporary protection of all functional groups of the chiral selector, to avoid their undesired reaction with the active chloromethyl group of the polymer and thus to obtain a well-defined structure of sorption sites.

Another distinguishing feature of sorbents prepared by this researcher group is

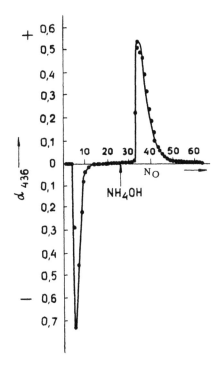

FIGURE 2 Resolution of 0.5 g of *DL*-proline on a column (475 × 9mm) containing 11 g of copper(II)-saturated polystyrene-type resin with *L*-proline (2.3 mmol/g) as the chiral selector. Particle size, dp 30–50 μm; flow rate, 7.5 mL/h; ambient temperature; polarimetric detection in a cell of 0.5 dm. Abscissa, number of fractions. (From Ref. 242.)

TABLE 1   Chiral Chelating Ligands Fixed to the Polystyrene Matrix

Fixed Ligand	Capacity (mmol/g)	Ref.	Capacity (mmol/g)	Ref.
Alanine	2.2	72		
Valine	2.3	73		
Leucine			1.5	74
Isoleucine	2.5	73		
Proline	2.3	75	1.5	76
Serine	3.0	72		
Threonine	2.9	72		
Tyrosine	2.2	72		
Hydroxyproline	3.0	72		
Allohydroxyproline	3.2	75		
Aspartic acid	2.5	77		
Pyroglutamic acid	2.2	77		
Diaminobutyric acid	2.2	78		
1-Aminobutyrolactam	2.3	78		
ε-Lysine[a]	1.7	79		
Histidine	2.0	80	0.6	81
Methionine	2.1	82		
Methionine-(*dl*)-sulfoxide	2.1	83		
Methionine (*d*)- or -(*l*)-sulfoxide	2.2	84		
S-Cysteine[a]	2.3	85		
Cysteic acid	1.2	86		
O-Hydroxyproline[a]	1.0	75		
Azetidine carboxylic acid	2.4	87		
Phenylalanine	2.1	88		
Ar-Phenylalanine[a]	1.1	88		
N-Carboxymethylvaline	1.5	89	0.5	90
N-Carboxymethylaspartic acid	1.2	89		
S-Carboxymethylcysteine	1.6	91		
S-(2,-Aminoethyl)cysteine	1.8	91		
S,*S*'-Ethylenebiscysteine	1.1	91		
N,*N*'-Ethylenebismethionine	0.9	91		
1-Aminobenzylphosphonic acid	2.0	92		
1-Amino-1-methylbenzylphosphonic acid	0.9	93		
1-Amino-1-methylpropylphosponic acid	0.6	93		
1-Amino-1-methylpentylphosphonic acid	0.8	93		
1-Aminobenzylphosphonic acid monoethyl ester	2.2	92		

[a]Bonding sites other then α-amino group.

the "macronet isoporous" structure of the polystyrene network. The latter was formed by crosslinking linear polymeric chains in solution using alkylation of their phenyl rings with bifunctional reagents, such as *p*-xylilene dichloride, in accordance with Friedel–Crafts reactions [68–71]. The most important parameters of the macronet isoporous resins are enhanced swelling ability, pressure resistance, exchange capacity, permeability, and kinetic properties. These resins displayed much higher resolving power and chromatographic performances as compared to reaction products of conventional styrene/divinylbenzene copolymers and unprotected amino acids.

Typical procedure for obtaining effective chiral ligand-exchanging resins [72] consists in reacting *p*-chloromethylated macronet isoporous polystyrene (crosslinking degree of 5 to 10%) with an amino methyl ester hydrochloride (mole ratio of 1:1.1) in a mixture of dioxane with methanol (6:1 v/v), in the presence of sodium iodide and sodium bicarbonate (0.3 and 2.5 mol, respectively) at 50 to 60°C. The subsequent hydrolysis of methyl esters is made in a neutral copper(II)-containing or a weakly alkaline aqueous media at room temperature. These conditions allow introduction of up to 2.0 to 2.8 mmol of chiral selector groups per 1 g of resin without racemization of the amino acid.

Other perspective series of polystyrene-type chiral ligand-exchanging phases represent those incorporating diamine [94] as chiral selectors as well as products of reacting chlorosulfonated polystyrene with amino acids [95,96]. In the latter case, —SO$_2$— groups link the amino function of the selector with the phenyl rings of the polymer matrix.

Of all polystyrene-based sorbents, the most selective and efficient proved to be those containing cyclic amino acids or alkylated propanediamine-1,2 as chiral selectors. As shown in Table 2, in combination with copper(II) ions, they resolve racemic mixtures of all natural amino acids. With very high selectivity values they also resolve some racemic hydroxy acids, amino amides, β-amino acids, and amino alcohols as well other complex-forming compounds (100). Due to the high enantioselectivity, preparative-scale resolutions can easily be performed. Thus up to 20 g of *DL*-proline or 6 g of threonine was quantitatively resolved with enantioselectivity values, α, of 3.95 and 1.52, respectively, on a column containing 300 g of *L*-hydroxyproline-incorporating polystrene resis [100]. Tritium-labeled amino acids were commercially produced in the optically, radiochemically, and chromatographically pure state using the same type of resins [101–103]. Certainly, rapid analytical-scale separations of enantiomers are feasible as well [99,104,105], since the macronet isoporous structure of the polymeric matrix allows a sufficiently fast mass transport and high performance of column packings.

Mechanism of chiral recognition of amino acid enantiomers during formation of diastereomeric sorption complexes in the resin phase has been the subject of thorough investigation [5,24,25,55]. Important features of the recognition model suggested by Davankov and Kurganov [55] are (a) the specific conformation of the

TABLE 2 Selectivity Values, $\alpha = k'_D/k'_L$ of Resolution of Racemic Amino Acids on the Copper(II) Forms of Polystrene Resins ($\bar{R}$) that Contain Residues of L-Proline ($\bar{R}$ Pro) (87), L-Hydroxyproline ($\bar{R}$Hyp) (97), L-Allohydroxyproline ($\bar{R}$aHyp) (98), L-Azetidine Carboxylic Acid ($\bar{R}$AzCA) (86), and N^1-Benzyl-(R)-propanediamine-1,2 ($\bar{R}$Bzpn) (99)

Racemic amino acid	$\bar{R}$Pro	$\bar{R}$Hyp	$\bar{R}$aHyp	$\bar{R}$AzCA	$\bar{R}$Bzpn
Alanine	1.08	1.04	1.04	1.06	0.70
Aminobutyric acid	1.17	1.22	1.18	1.29	0.49
Norvaline	1.34	1.65	1.42	1.24	0.48
Norleucine	1.54	2.20	1.46	1.40	0.49
Valine	1.29	1.61	1.58	1.76	0.31
Isovaline	—	1.25	—	—	—
Leucine	1.27	1.70	1.54	1.24	0.49
Isoleucine	1.50	1.89	1.74	1.68	0.62
Serine	1.09	1.29	1.24	2.15	0.62
Threonine	1.38	1.52	1.48	0.78	0.44
Allothreonine	1.55	1.45	—	—	—
Homoserine	—	1.25	—	—	—
Methionine	1.04	1.22	1.52	1.29	0.60
Asparagine	1.18	1.17	1.20	1.44	0.70
Glutamine	1.20	1.50	1.40	1.25	—
Phenylglycine	1.67	2.22	1.78	1.38	0.49
Phenylalanine	1.61	2.89	3.10	1.86	0.52
α-Phenyl-α-alanine	—	1.07	—	—	0.38
Tyrosine	2.46	2.23	2.36	1.78	—
Phenylserine	—	1.82	—	—	—
Proline	4.05	3.95	1.84	2.48	0.47
Hydroxyproline	3.85	3.17	1.63	2.25	—
Allohydroxyproline	0.43	0.61	1.48	1.46	—
Azetidine carboxylic acid	—	2.25	—	—	—
Ornitine	1.0	1.0	1.20	1.0	1.0
Lysine	1.10	1.22	1.33	1.06	1.0
Hystidine	0.37	0.36	1.32	0.56	0.85
Tryptophan	1.40	1.77	1.10	1.13	—
Aspartic acid	0.91	1.0	0.81	0.88	1.0
Glutamic acid	0.62	0.82	0.69	0.77	—

N-benzylproline ligand of the stationary phase with one of the axial positions of the copper(II) ion coordination sphere inaccessible to additional ligands because of sterical reasons, and (b) the fixation of a water or ammonia molecule in the second axial position. With the chiral stationary phase having *L*-configuration, the above axially coordinated water molecule hinders the formation of sorption complexes with *L*-amino acids. On the contrary, α side groups of *D*-amino acids would appear in the vicinity of the polystrene chain and enter hydrophobic interactions with the latter, which would additionally stabilize the *D-L* mixed ligand complex and cause the longer retention of the *D*-isomers in the chromatographic column.

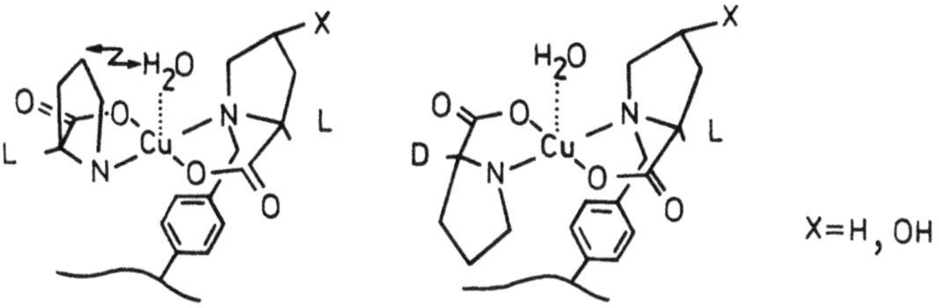

X=H, OH

Tridentate amino acid ligands, on the contrary, would form stable sorption complexes if they belong to the *L*-configurational series, which explains the experimental finding that *L*-allohydroxyproline, *L*-histidine, *L*-aspartic, and *L*-glutamic acids are second to elute from the chiral stationary phase:

X = H, OH

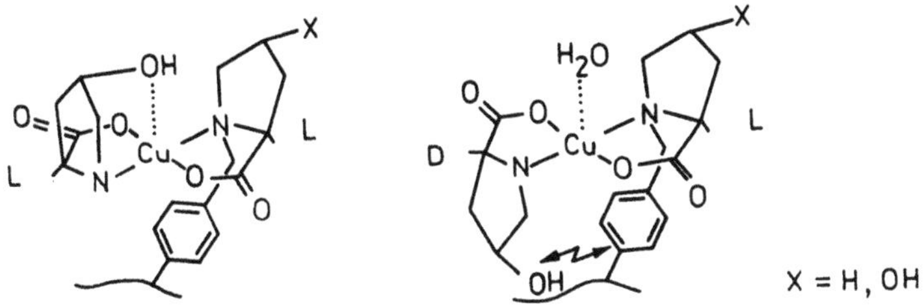

With the *L*-allohydroxyproline-incorporated resin, both bifunctional and trifunctional amino acids can act only as bidentate mobile ligands and always elute in the sequence *L* before *D*:

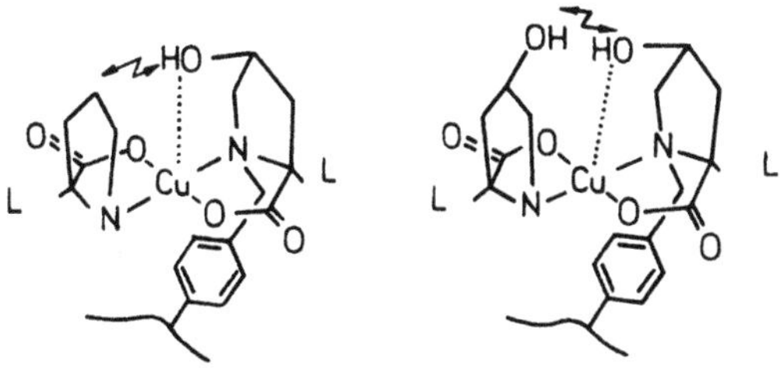

It is worth mentioning here that of many transition metal cations tested in combination with amino acid and diamine-type chiral stationary ligands, copper(II)

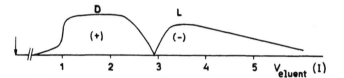

FIGURE 3 Resolution of 10 g of *DL*-threonine on a column (1200 × 60 mm) containing polyacrylamide-type resin Chirosolve *L*-proline; particle size, dp 40–80 μm; eluant, 0.005 *M* copper acetate/0.05 *M* acetic acid; flow rate, 400 mL/h; detection, UV at 254 nm. (From Ref. 109.)

ions demonstrated highest resolving power and widest application range, although, in some specific cases, octahedral Ni(II) ions [80,81,91,106] or tetrahedral Zn(II) ions [105] should be preferred.

2. Polyacrylamide-Type Resins

Cross-linked polyacrylamide, a much more hydrophilic polymeric matrix compared to that based on polystyrene, was suggested in 1977 by Lefebvre et al. [107] for binding chiral amino acid selectors. Again, a methylene —CH$_2$— group served as the link between the polymer and the grafted ligand:

$$
\begin{array}{c}
\{ \\
\text{CH}-\overset{\overset{\displaystyle O}{\|}}{\text{C}}-\text{NH}-\text{CH}_2-\text{N}-\overset{\displaystyle \begin{array}{c}\text{CH}_2-\text{CH}_2 \\ | \qquad \diagdown \text{CH}_2 \\ \overset{*}{\text{CH}} \diagup \end{array}}{\underset{\displaystyle \text{COOH}}{}} \\
| \\
\text{CH}_2 \\
\}
\end{array}
$$

When saturated with copper(II) ions, which were found to be superior to other transition metal ions, the packing successfully resolved many racemic amino acids [107–110]. To the most effective cyclic amino acid selectors—proline, hydroxy-proline, and azetidine carboxylic acid (a four-membered heterocycle), analogous six-membered compounds—pipecolic acid [108,109] and *L*-porretine (*L*-3-car-boxy-1,2,3,4-tetrahydroisoquinoline) [110] were added.

   This type of chiral packings is now commercially available from JPS Chimie (Switzerland) under the trade name Chirosolve, and it is recommended for preparative-scale resolutions of racemic amino acids and their derivatives. The loadability of the chiral column can be evaluated from Fig. 3 [109].

   The enantiomeric elution sequence of amino acids from polyacrylamide resins generally appears to be opposite to that from polystyrene sorbents. This can easily be understood from the following structural model, which involves a permanent coordination of the amide carbonyl group to the axial coordination position of the copper ion [108,111,112]. With this orientation, only *L*-amino acids would find sufficient space for free rotation of their α-group R in a mixed-ligand complex:

Accordingly, *L*-isomers are found to be retained longer on the chiral column, as compared to *D*-enantiomers. Alone proline represents an exception to this rule, probably, because its cyclic substituent R is unable to rotate around the bond to the α-carbon atom. In this case, *D*-proline would form the stronger mixed-ligand complex, leaving the upper axial coordination position for a solvent molecule.

3. Polymethacrylate and Polyvinylpyridine Matrixes

It is interesting that similar retention mechanisms for amino acid enantiomers are characteristic of chiral resins that incorporate *L*-proline or L-hydroxyproline in an epoxy-activated polymethacrylate [113] and in a matrix of polyvinylpyridine [114,115]:

(A)   (B)

Here again, *L*-amino acids, with the exception of proline, appear to be longer retained sorbates.

On the contrary, *L*-phenylalanine grafted to the polyacrylamide matrix [103] demonstrates higher affinity to *D*-isomers of all amino acids, without any exception, resolving all racemates with an enantioselectivity value of at least 1.3:

When bonded to an epichlorohydrine-activated hydrophilic polymer, TSK2000PW, L-phenylalanine, L-tryptophan, and L-histidine [116] demonstrate much lower resolving ability and an irregular elution order of the enantiomers. Information on the thermodynamics of ligand exchange in chiral resins can be found in a number of papers [112,117,118] and in a more concise form, in review publications [5,25,119].

B. Porous Silica-Bonded Chiral Ligand-Exchanging Phases

With the development of pressure-resistant microparticulate silica gel packings for HPLC, a second generation of chiral ligand-exchanging phases was introduced, comprising chiral complexing selectors that are chemically bonded onto the surface of macroporous silica. This development was based on knowledge that was already available from examining manifold polymeric chiral phases. Therefore, the most powerful chiral selectors, cyclic amino acids of the proline series, were bonded to the mineral surface. Later, other types of chiral selectors were involved in the synthesis, most important of which were chiral diamines and tartaric acid.

1. Classification of Bonding Types

According to the type of linking used to join the chiral selector to the surface, the bonded phases can be classified into the following major groups:

$$\geqslant Si - X - \underline{NH - \overset{*}{C}HR - COOH}$$

$$\geqslant Si - X - CH(OH) - CH_2 - \underline{NH - \overset{*}{C}HR - COOH}$$

$$\geqslant Si - X - NH - \underline{CO - \overset{*}{C}HR - NH} - Y$$

$$\geqslant Si - X - CO - \underline{NH - \overset{*}{C}HR - COOH}$$

The first three types of chiral bonded phases were suggested simultaneously in 1979 by three independent groups of researchers. They differ only in the manner of coupling the amino acid to the initial organomineral material. The latter possessed active chloroalkyl, 3-glycidoxypropyl, and 3-aminopropyl groups, respectively:

$$\geqslant Si - (CH_2)_n - Cl$$

$$\geqslant Si - (CH_2)_3 - O - CH_2 - \underset{O}{CH - CH_2}$$

$$\geqslant Si - (CH_2)_3 - NH_2$$

which can easily be reacted with the amino or carboxylic function of the chiral selector. Several synthesis strategies can be applied which can influence the performance of the chiral stationary phases, but their enantioselectivity is largely governed by the chemical structure of the chiral sorption sites.

An overview of chiral bonded ligand-exchanging phases is given in Table 3. One can see that the structure of bonded chiral ligands closely resembles that of the polymeric ligand exchanging phases discussed above, which gives an initial idea of structures of diastereomeric sorption complexes that are formed on interaction with mobile ligands.

These mixed-ligand complexes represent the only source of chiral recognition of mobile ligands on the CSP. Therefore, increasing the extent of the mixed-ligand complex formation would always result in the enhancing of the enantioselectivity of the column.

2. Role of pH and Mineral Ions

The most important factor influencing the complexation ability of amino acids and their derivatives is the pH of the mobile phase. Unsubstituted and N-alkyl-substituted amino acids in combination with copper(II) ions would form bis(aminoacidato)copper complexes in sufficient amounts at pH values of 4 and higher. When the amino and/or carboxy function of the stationary ligand and/or solute molecule is converted into an amide function, higher pH values are required, since the carbonyl oxygen of an amide group is a weak electron donor and consequently, weak ligand, but on releasing the amide proton in alkaline media, the amide group is converted into a strong negatively charged ligand. Therefore, increasing the pH of the eluant would always result in an increase in both resolution and retention of solute enantiomers. In practice, the pH of the mobile phase should never exceed a value of 9, where the majority of silica-bonded phases become really unstable because of partial dissolution of the mineral matrix and cleaving of the bonded ligand.

If nickel(II), zinc(II), or cadmium(II) ions are used instead of copper(II), less stable complexes are generally formed with an amino acid ligand. Therefore, higher pH values should be applied to achieve sufficient retention and resolution with these ions.

The pH of the eluant is usually stabilized by a buffer salt, generally by using sodium hydrogen phosphate in weakly acidic bufferes and ammoium acetate in alkaline buffers. In the latter case, the presence of ammonia molecules helps to prevent precipitation of copper hydroxide from the eluant due to the partial formation of copper–ammonia complexes. The ability of the buffer ions to participate in the complexation reactions and thus compete with mobile and stationary ligands for the central metal ion strongly influences the retention of solutes in ligand-exchanging chromatographic columns. Thus, retention of enantiomers can be lowered to a desired level by increasing the buffer slat concentration, which would exert but a minor influence on the enantioselectivity of the system.

The latter statement, indeed, is valid for many systems where no participation of the mineral ions in the formation of diastereomeric mixed-ligand complexes takes

TABLE 3 Chiral Silica-Bonded Ligand-Exchanging Phases

A. $\geq$ Si—X—NHCHRCOOH, Silica—Spacer—Selector

Spacer X	Selector	Racemic solutes	Refs.
—(CH₂)₂C₆H₄—CH₂—		Amino acids	120, 121
—(CH₂)—		Amino acids	121, 122
—(CH₂)₃—		Amino acids	121, 123, 124
—(CH₂)₈—	L-Proline (X = H) and L-hydroxyproline (X = OH)	Amino acids	121, 122
—(CH₂)₁₁—	L-Proline	Amino alcohols as Schiff bases	125, 126

(Selector structure: CH_2—CH with X and CH₂, —N—CH, COOH)

B. $\geq$ Si—X—NHCHRCOOH

Spacer X	Selector	Racemic solutes	Refs.
—(CH₂)₁₀—C— ‖ O	—NH—CH—COOH, CH(CH₃)₂, L-Valine	Dansyl amino acids	127

C. $\geq$ Si—(CH₂)₃—O—CH₂—CH—CH₂—NHCHRCOOH
                              |
                             OH

Selectors	Racemic solutes	Refs.
L-Proline	Amino acids	128, 129
L-Proline, L-hydroxyproline	Amino acids	130
L-Valine, L-histidine, L-proline, L-hydroxyproline, L-azetidine carboxylic acid, L-pipecolic acid, L-phenylalanine	Amino acids	131
L-Proline	Diiodothyronine, triiodothyronine, thyroxine	132
L-Proline	α-(Uracyl-N¹)-alanine	133
L-Hydroxyproline	Hydroxy acids	134
L-Proline, L-hydroxyproline, L-azetidine carboxylic acid, L-pipecolic acid; also mentioned are	Amino acids, including O-methyl-DOPA, tetra-, tri-, and diiodothyronine;	135

TABLE 3 (Continued)

Selector	Racemic solutes	Refs.
L-valine, *L*-histidine, *L*-phenyl-alanine, *L*-propylendiamide, *L*-ephedrine, and *L*-tartaric acid	histidine methyl ester; dansyl-amino acids; glycyl-amino acids; *DL*-leucyl-*DL*-leucine, -*DL*-phenylalanine, and -*DL*-thyrosine; hydroxy acids	
2-amino-1,2-diphenylethanol, 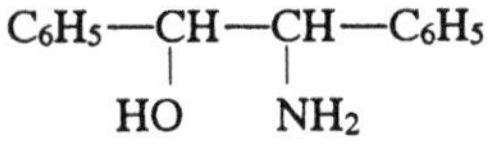 C_6H_5—CH—CH—C_6H_5 HO NH_2	Ser, Thr, Phe, Asp	136, 137
2-Carboxymethyl-amino-1,2-diphenylethanol, 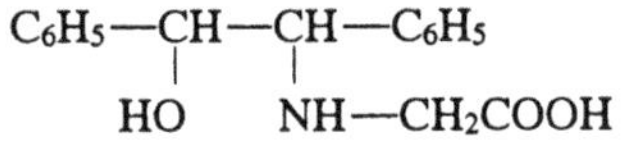 C_6H_5—CH—CH—C_6H_5 HO NH—CH_2COOH	Ala, Glu	137
2-Amino-1,2-diphenylethanol and 2-carboxymethylamino-1,2-diphenylethanol	Amino acids, benzyloxycarbonylamino acids, acetyltryptophan, phenylalanine amide, benzylhydantoin, hydroxy acids	138
trans-Cyclohexane-1,2-diamine	Amino acids, mandelic acid	139, 140

D. $\equiv$ Si—$(CH_2)_3$—NH—CCHRNH—Y
                     $\overset{\|}{O}$

Selector	Acyl group Y	Racemic solutes	Refs.
L-Proline 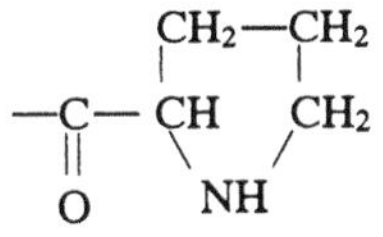 CH_2—CH_2 —C— CH CH_2 $\overset{\|}{O}$ NH	None	Trp, Phe, Tyr Dansyl-amino acids	141 142
L-Valine, 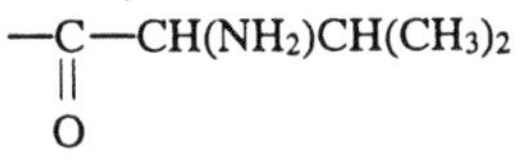 —C—CH(NH_2)CH(CH_3)₂ $\overset{\|}{O}$	None	Ser, Glu, Trp	143
L-Histidine, 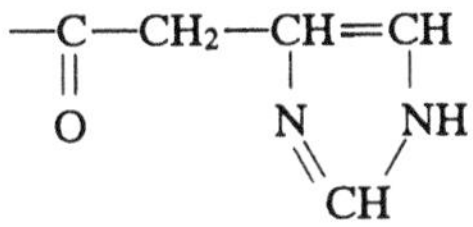 —C—CH_2—CH=CH $\overset{\|}{O}$ N NH CH	None	Amino acids, mandelic acid	144

TABLE 3   (Continued)

Selector	Acyl group Y	Racemic solutes	Refs.
L-Threonine, $-\overset{\text{O}}{\underset{\parallel}{C}}-CH(NH_2)CH(OH)CH_3$	None		145
L-Valine and L-proline	*tert*-Butyloxy-carbonyl, $-\overset{\text{O}}{\underset{\parallel}{C}}-O-C(CH_3)_3$	Dansyl amino acids	146

E. $\equiv Si-(CH_2)_3-NH-\overset{\text{O}}{\underset{\parallel}{C}}-\underset{\underset{H_3CCOO}{|}}{CH}-\underset{\underset{OCOCH_3}{|}}{CH}-\overset{\text{O}}{\underset{\parallel}{C}}-Y$

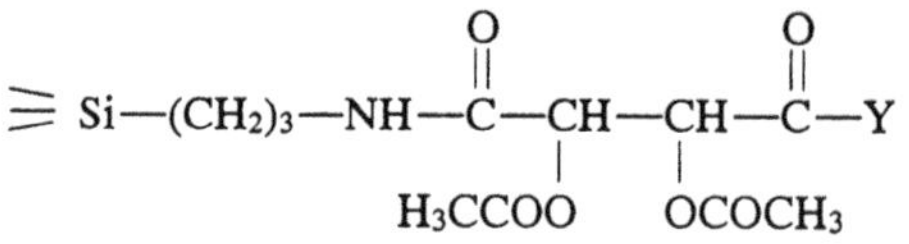

Selector	Group Y	Racemic solutes	Refs.
$L(+)$-Tartaric acid, diacetyl-	—OH	Amino acids, including DOPA, methyl-DOPA, 3-O-methyl-DOPA; mandelic acid; adrenaline, noradrenaline	147
$L(+)$-Tartaric acid, diacetyl-, -(R or S)-α-methylbenzyla-mide	$-NH-CH(CH_3)C_6H_5$	No examples	148

F.   Commercially Available Chiral Ligand-Exchanging Packings

Trade name	Racemic solutes	Refs.
Chiralpak WH	Kinurenine α-methyl-Leu, α-methyl-Met; 3-fluoro-Trp	149
Chiralpak WH	Dansyl-amino acids	150
Chiralpak WH, WE, WM	Amino acids	151
Nucleosil Chiral 1	α-Alkyl-α-amino acids	152–154
Chiral ProCu, Chiral HyproCu, and Chiral ValCu	N-Methyl-α-amino acids	154
Nucleosil Chiral 1	Hydroxy acids	155

TABLE 3   (Continued)

G.   Silica-Bonded Polystyrene

Selector	Racemic solutes	Refs.
L-Proline, *L*-hydroxyproline	Amino acids	159
N[1]-Benzyl-(*R*)-propanediamine-1,2	Amino acids	5, 7

place. However, cases are known of strong dependence of the enantioselectivity on concentration of ammonia [87,102] or pyridine [156] in the eluant, indicating involvement of these amines into the coordination sphere of the diastereomeric selector–solute complexes, which alters stereochemical contacts between their two major chiral ligands.

The second important component of mineral nature in mobile phases used in ligand-exchange chromatography is the salt of the complexing metal ion. It is evident that due to fast ligand exchange, the complexing metal ions partition between stationary and mobile phases (in accordance with the overall equilibrium complexing abilities of these two phases). Therefore, the column loses a certain amount of the complexing metal with the mobile phase. To preserve the equilibrium and stabilize the chromatographic parameters (retention and selectivity) of the system on an optimum level, one has to add some transition metal salt to the mobile phase. Generally, the chromatographic parameters would demonstrate an external dependence on copper(II) salt concentration, with the optimum value near the region of 10^{-5} to 10^{-4} M. Both higher and smaller concentrations of the salt would generally spoil the resolution. The nature of the anion of the salt is not important in buffered solutions but may have some influence on the resolution in mobile phases of very low ionic strength, for instance, through the pH of the mobile phase [157].

3.   Electrostatic and Hydrophobic Interactions

Whereas the formation of mixed-ligand sorption complexes governs the chiral resolution of the mobile ligand, two additional sorbent–solute interaction modes contribute to the retention of the latter, namely, electrostatic and hydrophobic interactions.

The electrostatic charge of the chiral stationary phase may be caused by both the selector and mineral matrix ionogenic groups. If the chiral selector is of an amino acid nature, its charge would change from positive to negative on changing the pH of the eluant from acidic to alkaline. When complexed to a bivalent transition metal cation, the sorption site of the packing would display a single positive charge. The relative retention of differently charged solute molecules may often be correlated with the charge of the sorption site, especially if the complexing metal ions are in deficiency.

When the covering of the silica surface by the chemically bonded organic layer is incomplete, surface silanol Si—OH groups may dissociate in alkaline media and strongly interact with cationic solutes. This indeed happens with small spacer groups, such as —CH_2—, separating the L-proline bonded ligand from the surface of the matrix [122]. In this case, electrostatic interactions with the accessible silanols would reveal in the fact that basic lysine happens to be the strongest retained solute of many amino acids examined. On the contrary, lysine becomes the least retained one on packings having long scaper groups, such as —$(CH_2)_8$—, which effectively protect the surface silanols.

Naturally, electrostatic interactions may be weakened by using mobile phases of high ionic strength.

With the alkyl spacer group growing in size, the stationary phase of the packing becomes more and more hydrophobic. Large solute molecules having exposed alkyl or aryl groups may now become the longest retained species. The significance of the hydrophobic contribution to the overall retention of such solutes may be visualized by adding organic modifiers, such as acetonitrile, methanol, or ethanol, to the mobile phase. These modifiers strongly diminish the hydrophobic interactions, and therefore they are widely used for adjusting the net retention of the solutes. In the case of hydrophobic stationary phases and large racemic solutes, as is the case with dansyl amino acids, one may need up to 30% of acetonitrile in the eluant.

Organic modifiers of the eluant would not affect the enantioselectivity of the resolution if hydrophobic interactions arise solely between hydrocarbonic parts of the solute and the spacer groups of the bonded phase. However, cases are known where hydrophobic interactions exist within the diastereomeric sorption complexes and where they largely contribute to the chiral recognition of the solute enantiomers. Here, resolution selectivity may vary with the varying content and type of the organic component in the mobile phase.

In addition to complexation reactions, hydrophobic interactions may be of extreme importance in the chromatography of Schiff bases of amino alcohols with salicyl aldehyde and that of dansylamino acids on bonded phases where L-proline or L-valine are "diluted" with an approximately 10-fold excess of long alkyl chains [126,127]. Such dilution entirely alters the micro environment of the chiral complexing site of the packing, even changing the enantioselectivity sign of the system compared to corresponding nondiluted chiral bonded phase. It is interesting that the latter shows higher loadibility and should be preferred in semipreparative resolutions [126], whereas the diluted phase demonstrated higher efficiency and should be used in analytical experiments. To keep the retention time of hydrophobic solutes on such packings in an acceptable region, one has to add 50 to 70% methanol to the mobile phase.

4. Column Temperature

Column temperature, the last important chromatographic parameter in ligand exchange, should be discussed briefly. It has been noted many times that increasing the column temperature to 50 to 55°C considerably improves efficiency of packings that operate in accordance with ligand-exchange mechanisms. This implies that exchange of multidentate ligands in the coordination sphere of the metal ion may be a relatively slow process. Indeed, some model copper(II) diamine complexes were found to be formed very slowly [158]. There is no other way of improving the column plate number than increasing the column temperature if all conventional optimization possibilities (by changing the type and concentration of the organic modifier and inorganic salt in the eluant as well as its pH) were exhausted. In ligand-exchange chromatography, temperature would exert only a minimum influence on the separation enantioselectivity, it would influence mainly the column efficiency. However, one example has been reported [55,56] of improving both the rate of ligand exchange and the difference in stabilities of two diastereometric mixed-ligand complexes.

5. Polymeric Bonded Phases

On the other hand, elevated temperatures may diminish the lifetime of chromatographic columns rather dramatically, especially that of packings prepared from 3-aminopropyl-activated initial material. To combat this drawback, hydrolytically stable packing was suggested, representing a new generation of chiral bonded phases—that of chiral polymers which are chemically linked to the mineral matrix in several positions of the chain [159]. The links were formed by units of methylvinyldiethoxysilane that were copolymerized with styrene. The bonded copolymer was subjected to chloromethylation and animation with L-hydroxyproline in the usual manner. The packing can be exploited at temperatures as high as 75°C for several months without losing its resolving power of efficiency.

Some typical examples of using chiral bonded phases in HPLC of racemic complex-forming compounds are presented in Figs. 4 to 6. It is worth mentioning here that ligand-exchange technique facilitates detecting many solutes with most inexpensive photometric detectors, since the solutes would elute in the form of their copper(II) complexes that strongly adsorb in the entire UV region.

C. Chiral Coating on the Surface of Porous Silica

1. Chiral Coatings in HPLC

A whole series of highly efficient chiral ligand-exchanging packings can easily be prepared by coating microparticulate silica in commercially available prepacked HPLC columns with appropriate strongly adsorbing chiral selectors. This is usually done by passing a diluted solution of a low-molecular-weight chiral ligand, or its metal complex, through the HPLC column until full saturation of the latter. Naturally, the column should then be used in combination with mobile phases that do not dissolve the coating material from the surface. Highly hydrophobic chiral selectors adsorbed on reversed-phase packings do not cause any bleeding of the column in aqueous or water-rich mixed eluants. Highly polar chiral selectors, when adsorbed on the surface of bare silica, can be used in combination with nonpolar organic solvents. Electrostatic interactions may also contribute to the permanent retention of polar selectors on changed matrixes.

Chiral polymers would adsorb especially strongly on ridged surfaces. However, they are usually deposited on the surface by stirring the packing material with a dilute solution of the polymer followed by filtration of the suspension and packing a column in a conventional manner.

2. Coated Reversed-Phase Packings

Davankov et al. dynamically coated conventional reversed-phase columns with N-alkyl derivatives of L-hydroxyproline [160] and L-histidine [161] from methanolic or aqueous methanolic solutions of these chiral selectors. Their long n-alkyl anchoring chains (C_7 to C_{16}) provided for a permanent adsorption of the coating on the hydrophobic interface layer of the packing material. In an water-rich eluant

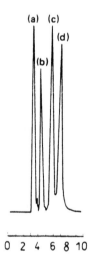

0 2 4 6 8 10

TIME, min

FIGURE 4   Resolution of two Schiff bases of RS-phenylethanolamine on a diluted with C_{18} alkyl groups C_{11}-L-proline bonded phase. Column, 150×4.6 mm; particle size, dp 5 μm; eluant, 0.005 M Cu(II)/0.2 M ammonium acetate, pH 5.0, 75% v/v methanol; flow rate, 1 mL/min; temperature, 35°C; detection UV at 350 nm. Solutes: Schiff bases of R- and S-phenylethanolamine with 2,4-dihydroxybenzaldehyde, (a) and (b), and with 2-hydroxy-4-methoxybenzaldehyde, (c) and (d). (From Ref. 243.)

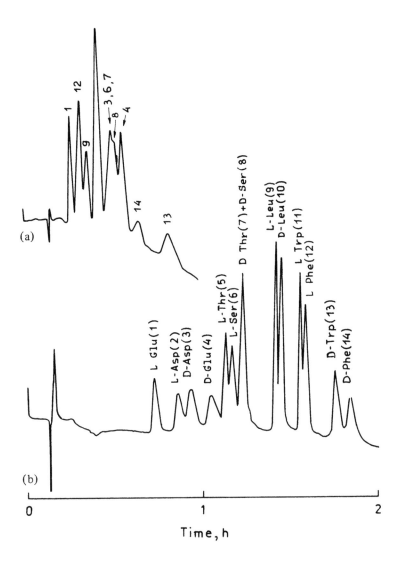

FIGURE 5   Gradient separation of seven racemic dansyl amino acids. Column: (a) Chiral-pak WH, 770 × 0.26 mm; (b) Chemcosorb ODS-UH, 150 × 0.26 mm and Chiralpak WH, 770 × 0.26 mm coupled in series; eluant, acetonitrile/aqueous solution including 0.25 M ammonium acetate and 0.1 mM copper sulfate, acetonitrile content linearly programmed from 18 to 38% in 2 h; flow rate, 5 μL/min; detection, UV at 220 nm; particle size, Chiralpak WH 10 μm, Chemcosorb ODS-UH 5 μm. (From Ref. 150.)

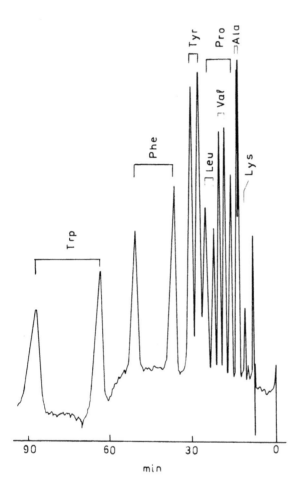

FIGURE 6 Resolution of eight racemic amino acids on polystyrene bonded phase containing residues of L-hydroxyproline. Column, 250 × 4 mm; particle size, dp 5 μm; eluant, 0.5 mM copper sulfate/0.01 M ammonium acetate, pH 4.5, 30% acetonitrile; flow rate, 0.7 mL/min; temperature, 75°C; detection, UV at 254 nm. (From Ref. 7.)

containing 10^{-4} M copper acetate and up to 15% methanol, mixtures of five to seven racemic amino acids could be resolved into individual enantiomers on a column 15 cm long, showing high analytical potentiality of the new chromatography system. This feature was also shown to be characteristic of other chiral coatings on reversed-phase columns (Table 4).

Due to the high efficiency of the coated system, even at room temperature, the detection limit of common amino acids can be as low as 10^{-10} mol, that is, about 10^{-8} g, with a simple 254-nm UV detector [160]. Subnanomole sensitivity is easily

TABLE 4 Chiral Ligand-Exchanging Coatings

A. Coated Reversed-Phase Packings

Selector	Racemic solutes	Refs.
N-Heptyl-, N-decyl-, and N-hexadecyl-L-hydroxyproline	Amino acids	160
N-Decyl-L-histidine	Amino acids	161
Palladium(II)-S-ethyl-L-cysteine	Methionine	162
N-(2-Hydroxydodecyl)-L-hydroxy-proline	Proline, thiazolidine-4-carboxylic acid	163
	5,5-Dimethyl-thiazolidine-4-car-boxylic acid	164
N,N-Dioctyl-L-alanine	Amino acids and hydroxy acids	157
	Hydroxy acids	165
	Pantothenic acid	166
	Malic acid	167
	Amino acids, acetyl-DL-Leu, amino-caprolactame, peptides, hydroxy acids	168

B. Coated Silica Gel

Selector	Racemic solutes	Refs.
Silver(I)-d-camphor-10-sulfonate	3-Methylene-7-benzylidene-bicyclo[3.3.1] nonae	169
Cobalt(III)-tris-ethylenediamine	Substituted cyclopentadienyl-rhodium(I)-norbornadiene	170
Polyacrylamide grafted with L-proline	Amino acids	171, 172
Poly-3-vinylpuridine grafted with L-proline	Amino acids	172, 173

C. Anion-Exchange Resin Dowex 1 × 2 Loaded with Chiral Complexes

Selector	Racemic solutes	Refs.
D-N-Hydroxyethyl-propylene-diaminetriacetato-ferrate(III)	N-Benzoyl- and N-acetyl amino acids	174
D-Propylenediamine-tetraacetato-copper(II) and -zince(II)	1-Phenylethyl-amine	175

achieved by postcolumn derivatization of amino acids with *o*-phthalic dialdehyde and subsequent fluorimetric detection.

The ease of the preparation of HPLC columns coated with an appropriate chiral selector offers the opportunity of regulating the elution order of the enantiomers of interest. This possibility is especially important when trace analysis of enantiomeric impurities in an optically active compound is desired. The detectability of the trace component is generally greatly enhanced by letting it elute first (i.e., before the major component of the sample) [176]. Thus the first eluting *D*-impurity in the *L*-alanine (Sigma, BRD) can easily be quantitated at the level of 270 ppm on a reversed-phase column coated with *N,N*-dioctyl-*L*-alanine [157] in combination with an aqueous eluant, 0.1 mM in copper sulfate, and a fluorimetric detector (postcolumn reaction with *o*-phthalaldehyde).

Reversed-phase columns coated with *N,N*-dioctyl-*L*-alanine were also shown to be efficient in resolving chiral hydroxy acids (Table 5). Copper(II) ions were used in the eluant to organize the complexation interactions of the chiral selector with the hydroxy acid enantiomers. However, to enhance a selective detection of the latter in their mixture with amino acids, which were all detected at 254 nm, an additional postcolumn reaction with 2.5 mM iron(III) perchlorate in 0.04 N perchloric acid was carried out followed by a selective photometric detection of hydroxy acids at 420 nm [165], as shown in Fig. 7.

Chiral-coated systems have been used successfully for obtaining important information on the mechanism of chiral recognition of bidentate ligands by the stationary chiral selector. Thus, from scheme (B) on page 211, it can be easily understood that for steric reasons, *L*-amino acids should be preferred to *D*-isomers in mixed-ligand copper(II) complexes formed by a chiral polymeric coating that contains *L*-proline in poly-3-vinylpyridine [173].

On the contrary, *D*-amino acids should be retained longer on a reversed-phase packing containing *N*-alkyl-*L*-hydroxyproline or *N*-alkyl-*L*-histidine as chiral complexing sorption sites [17,18,160,161]:

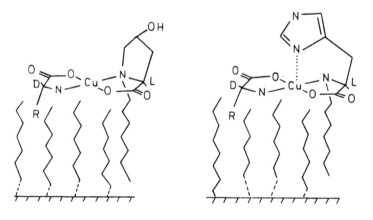

TABLE 5   Resolution of Racemic 2-Hydroxy Acids Using Ligand-Exchange Chromatography on a Reversed-Phase Column Coated with N,N-dioctyl-L-alanine[a].

| Racemic compound | Concentration | | Capacity factors | | Resolution, R_s |
	CuSO$_4$ (mM)	Acetonitrile (%)	k'_D	k'_L	
50-mm column					
Lactic acid	2	10	4.9	5.8	1.20
2-Hydroxy-n-butyric acid	2	10	10.6	15.1	3.05
2-Hydroxyvaleric acid	2	10	29.3	42.0	2.89
2-Hydroxycaproic acid	2	10	93.9	133.8	3.42
Leucic acid	2	10	77.4	99.0	2.38
Glyceric acid	1		20.8	15.0	2.23
Malic acid	0.5	10	36.8	31.2	0.39
Tartaric acid	1	10	88.2	55.4	0.73
4-Hydroxy-3-methoxymandelic acid	2	10	18.5	27.1	3.08
4-Hydroxymandelic acid	2	10	20.9	31.6	4.02
3-Hydroxy-4-methoxymandelic acid	2	10	24.8	36.8	4.84
3-Hydroxymandelic acid	2	10	33.8	48.8	3.93
Mandelic acid	2	10	65.0	89.4	2.88
10-mm column					
β-Phenyllactic acid	2	10	49.5	76.0	1.34
α-Phenyllactic acid	2	10	56.0	73.5	0.90
p-Chloromandelic acid	2	10	167.0	233.0	1.39

Source: Ref. 165.
[a] Conditions: room temperature; mobile-phase flow rate, 1.0 mL/min; packing, 3-μm ODS-silica of 100 Å mean pore diameter.

Here the important role of an achiral component of the chromatographic system, the hydrophobic interface layer, was recognized in separation of the solute enantiomers. The difference in the total binding energy of the L- and D-amino acids originates from the fact that only the latter can enter additional hydrophobic interactions with the support, which stabilize the D-L sorption complex.

3.   Coated Silica

Finally, it should be emphasized here, that the application field of chiral ligand-exchanging coatings is far from being exhausted, as indicated by initial examples of combining bare silica packings with polar chiral selectors. Thus silver(I)-d-camphor-10-sulfonate, when deposited on silica, forms π-complexes with aliphatic un-

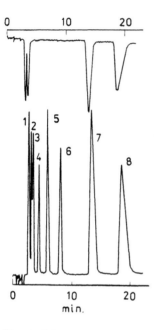

FIGURE 7 Resolution of two hydroxy acids and two amino acids on a column (50 × 4.6 mm) with reversed-phase material (3 μm) coated with *N, N*-dioctyl-*L*-alanine, MC1 GEL CRSIOW. Eluant, 2mM copper sulfate in aqueous acetonitrile (10%); flow rate, 1.0 mL/ min; room temperature. Detection: upper, UV at 420 nm after reaction with a solution of 2.5 mM $Fe(ClO_4)_3$ in 0.04 *N* $HClO_4$, 0.1 mL/min; lower, UV at 254 nm. Solutes: 1 and 2, *DL*-lactic acid; 3 and 4, *DL*-tyrosine; 5 and 6, DL-phenylalanine; 7 and 8, *DL*-2-hydroxybutyric acid. (From Ref. 165.)

saturated hydrocarbons and can be used for preparative resolutions of racemic diene [169]. Chiral cations of tris-ethylenediaminecobalt(III), when retained by electrostatic forces on deprotonized surface silanols, efficiently resolve (Fig. 8) racemic metalloorganic compounds, such as cyclopentadienylrhodium(I)-norbornadiene derivatives [170]. The latter example is especially interesting in that the chiral selector represents a kinetically inert, stable cobalt(III) complex with a totally saturated octahedral coordination sphere. Therefore, only the outer coordination sphere (i.e., the solvation shell of the complex) is involved in the formation of adducts with solute enantiomers to be discriminated. The field of chiral ligand-exchange chromatography using the outer coordination sphere of stable complexes is barely touched thus far [5,6,177].

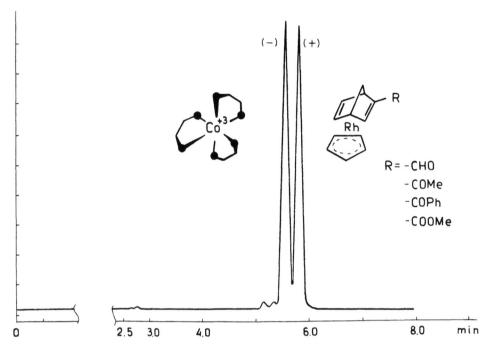

FIGURE 8   Resolution of racemic Rh(I) complex, R = -CHO, on a column (250 × 4mm) with Silasorb Si 600, dp 5 μm, coated with (+) -[Co(en)₃]³⁺. Eluant, dichloroethane/cyclohexane/isopropanol 30:20:1; flow rate, 1 mL/min; detection, UV at 250 nm. (From Ref. 170.)

4. Thin-Layer Chromatography Using Chiral Ligand-Exchanging Coatings

Our last comment in this section relates to the thin-layer chromatography of racemic compounds using chiral ligand-exchanging coatings. Due to simplicity of preparation of chiral plates by dipping reversed-phase plates into diluted solutions of N-(2-hydroxydodecyl)-L-hydroxyproline and a copper(II) salt, ready-to-use plates have been developed by Macherey-Nagel in cooperation with Degussa (BRD). Chiralplate proved to be extremely versatile in the resolution of racemic amino acids, N-methylamino acids, N-formylamino acids, α-alkyl amino acids, dipeptides, lactones, α-hydroxy acids, thiazolidine derivatives, halogenated amino acids, and other important compounds. As shown in the review paper by Guenther [178] and references therein, this technique allows determination of trace levels of one enantiomer in an excess of the other down to 0.1%. Chiralplate was also used in forced-flow planar chromatographic separations [179]. Interestingly, a definite correlation was found to exist between the resolution of the enatiomers of substituted thiazolidine carboxylic acids and dipole moments of the substituents [180].

D. Chiral Complexes as Additives to the Mobile Phase

1. Combination of Chiral Eluants with Achiral HPLC Columns

In addition to the chemically bonded chiral stationary phases and chiral coatings discussed above, a third general approach has been developed for realizing ligand exchange in a HPLC mode. It uses "chiral eluants" in combination with an achiral, conventional HPLC column. The essential component of the eluant is a complex of a chiral ligand with a transition metal cation. The complex partition between the stationary and mobile phases depends on its adsorption capacity and its solubility in the eluant. If a reversed-phase packing is being used in combination with a chiral eluant, which is extremely popular in ligand-exchange chromatography (see Table 6), hydrophobic additives would tend to concentrate on the sorbent surface, whereas small and hydrophilic chiral selectors, such as bis(L-prolinato)copper(II), would stay predominantly in the aqueous-organic mobile phase. Certainly, it is possible to adjust the distribution coefficient of the chiral selector to a desired level by changing the content and type of the organic additive (acetonithile or methanol) to the mobile phase. But in practice, one would pay much more attention to the retention time of solutes under resolution, which is also influenced by the solvent strength. Anyway, one has to keep in mind that the system is labile and requires equilibration on any change in chromatography conditions.

Another drawback of a system operating with a chiral eluant is loss of the chiral selector. On the other hand, with many rather inexpensive chiral compounds commercially available, this technique offers innumerable variants for optimization of high-performance analytical resolutions.

Indeed, from Fig. 9 [206] one can judge that at the efficiency attainable, such difficult analytical problems can be tackled as are the determination of D-amino acid traces in fermented foodstuffs [232] or in urine and cerebrospinal fluid [198]. In these cases, dansyl derivatives of amino acids were analyzed in eluants containing 2 mM copper(II) and 4 mM L-phenylalnine amide (pH 7.3) or 5 mM L-histidine methyl ester (pH 5.5), respectively, with an acetontrile gradient up to 50%.

Three general procedures were suggested to accomplish a total enantiomeric analysis of a mixture of all common amino acids. Weinstein et al. [199,200] suggested that the initial amino acid mixture must be separated first into three fractions using a conventional cation-exchanging resin and volatile aqueous pyridine buffers. The amino acids within the groups are then separated into the individual enantiomers on an RP column with N,N-dipropyl-L-alanine and copper acetate in the aqueous mobile phase. Subnanomole sensitivity can be achieved by postcolumn derivativation of the enantiomers with o-phthalaldehyde and subsequent fluorimetric detection.

TABLE 6 Resolution of Racemates Using Chiral Eluant Technique with Reversed-Phase Packings

Chiral selector	Racemic solutes	Refs.
2-(L)-alkyl-4-octyldiethylenetriamine (where alkyl = ethyl, isopropyl, isobutyl) in combination with Zn(II), Cd(II), Ni(II), Hg(II), Ni(II), Zn(II)	Dansyl amino acids	181, 182
	Dansylglycylamino acids	182
L-Proline + Cu(II)	Amino acids	183
	Dansylamino acids	184–190
	3-Fluorinated alanines	191
	α-Halogenomethyl-substituted alanines	191
	α-Hydroxy acids	192
N-Methyl-L-proline + Cu(II)	Amino acids	193
N-Benzyl-L-proline + Cu(II)	Amino acids	17
L-Prolyl-octylamide + Ni(II)	Dansyl amino acids	182, 194
+ Ni(II), Zn(II)	Phenoxypropionic acids	195
L-Prolyl-dodecylamide + Ni(II)	Dansyl amino acids	194
L-Prolyl-glycine, L-prolyl-L-valine and L-prolyl-L-tyrosine in combination with Cu(II), Zn(II), Co(II), Hg(II)	Amino acids	183
L-Pipecolic acid + Cu(II)	Amino acids	193
	Dansyl amino acids	190
L-Azetidine carboxylic acid + Cu(II)	Dansyl amino acids	190
L-Arginine + Cu(II)	Dansyl amino acids	185, 189
L-Histidine + Cu(II)	Dansyl amino acids	185, 189, 196
L-Histidine methyl ester + Cu(II)	Dansyl amino acids	196–198, 187–189
N,N-Dimethyl-L-methionine sulfoxide + Cu(II)	Amino acids	193

TABLE 6 (Continued)

Chiral selector	Racemic solutes	Refs.
N,N-Dipropyl-*L*-alanine + Cu(II)	Amino acids	199–202
	Amino acid amides	202
	α-Methyl amino acids	202–203
	Dansylamino acids	204
	α-Hydroxy acids	205
	Amino acids	201
N,N-Dialkyl-*L*-alanine + Cu(II)		
(where alkyl = methyl, ethyl, propyl, butyl)		
L-Alanine amide + Cu(II)	Dansyl amino acids	206
L-Valine + Cu(II)	α-Hydroxy acids	192
N,N-Dimethyl-*L*-valine	Amino acids	201
+ Cu(II)	α-Methylamino acids	203
	α-Hydroxy acids	205, 207
+ Cu(II), Zn(II), Co(II), Hg(II)	Amino acids	183
N,N-Dialkyl-*L*-valine + Cu(II)	Amino acids	201
(where alkyl = methyl, ethyl, propyl, butyl)		
L-Valine amide + Cu(II)	Dansyl amino acids	206
L-Isoleucine + Cu(II)	α-Hydroxy acids	192
	Pyridone carboxylic acids	208
N,N-Dimethyl-*L*-leucine + Cu(II)	Amino acids	17, 193
L-Phenylalanine + Cu(II)	Amino acids	209–212
	α-Hydroxy acids	192, 209, 213
	2-Pentafluoroethyl-alanine	214
	2-Trifluoromethyl-alanine	214
	9-(3,4-Dihydroxybutyl)guanaine	215
	Pyridone carboxylic acids	208

N-Methyl-L-phenylalanine + Cu(II)	Amino acids	209
	α-Hydroxy acids	209
N,N-Dimethyl-L-phenylalanine + Cu(II)	Amino acids	209
	α-Hydroxy acids	209
L-Phenylalanine amide + Cu(II)	Dansyl amino acids	206, 231, 232
L-Tyrosine amide + Cu(II)	Dansyl amino acids	206
N-(p-Toluenesulfonyl)-L-phenylalanine + Cu(II)	Amino acids	216–218
N,N-Dimethyl-D-phenylglycine + Cu(II)	Amino acids	193
N-(p-Tolueneasulfonyl)-D-phenylglycine + Cu(II)	Amino acids	218, 219
L-Aspartyl-alkylamide + Cu(II) (where alkyl = ethyl, butyl, hexyl, octyl)	Amino acids	220
L-Aspartyl-cyclohexylamide + Cu(II)	Amino acids	221, 222
L-Aspartyl-L-phenylalanine methyl ester (Aspartame) + Cu(II), Zn(II)	Amino acids	223–225
(R,R)-(+)-Tartaric acid mono-n-octylamide + Cu(II)	Homoserine-dehydrochalcone conjugate	226
	Dansyl amino acids	189
N,N,N',N'-Tetramethyl-(R)-propanediamine-1,2 + Cu(II)	Amino acids	227
	Substituted phenylethanolamines	227
Diamino-diamido-type ligands + Cu(II),	Amino acids	228
	Mandelic acid	228
	Dansyl amino acids	229, 230

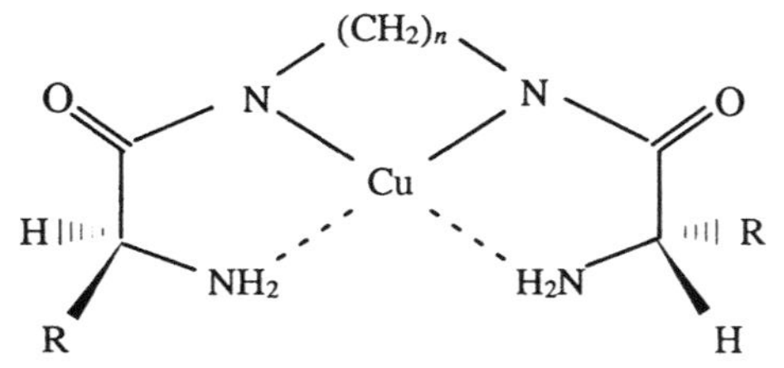

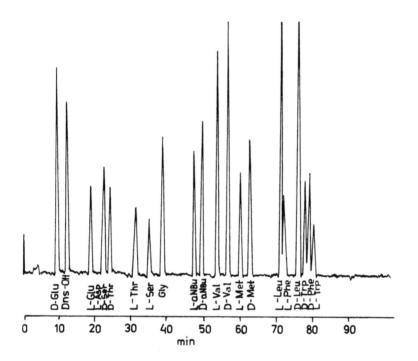

FIGURE 9 Enantiomeric resolution of a synthetic mixture of dansylamino acids. Column: Novapak C18, 4μm, 150 × 4mm. Chiral eluant: 4 mM L-valine amide, 2 mM copper acetate, 0.3 M sodium acetate, pH7. Acetonitrile gradient form 20 to 28%. (From Ref. 206.)

Nimura and coworkers [219] prefer a two-column system, 5 and 20 cm long, which allows resolution of polar amino acids on the longer column and that of longer retained nonpolar amino acids on the shorter one. In both cases copper complex with N-(p-toluenesulfonyl)-D-phenylglycine is used as the chiral additive to the eluant and an acetontrile gradient is applied in the last step, the chiral resolution of heavy amino acids on the short ODS column.

Finally, Karger and co-workers [194] separate dansyl derivatives of amino acids using conventional reversed-phase conditions and a linear methanol gradient and then subject some of the eluted individual racemic peaks to the enantiomeric resolution on a subsequent reversed-phase column percolated with a L-prolyl-n-dodecylamide/nickel complex solution.

To confirm assignment of individual amino acid enantiomers in complex matrixes, treatment of the mixture with enantioselective enzymes (e.g., D-amino acid oxidase) was suggested [188], which gradually alters the peak heights of all enantiomers belonging to the D-configurational series. Such manipulations with aqueous solutions are easy to combine with the subsequent enantiomeric analysis under reversed-phase conditions with a *L*-proline containing chiral eluant.

Another elegant example of validating the peak assignment is the resolution of mixtures of α-amino acids and α-hydroxy acids using bis(*N,N*-dimethyl-*L*-valinato)copper on an RP column and selective detection of hydroxy acids after their postcolumn reaction with Fe(III) ions at 436 nm. In this manner *D*-malic acid was determined [207] in apple juice suspected of being adulterated with synthetic *DL*-malic acid.

Besides the amino acids and hydroxy acids, many other types of complexing compounds can be successfully resolved into enantiomeric pairs using ligand-exchange technique. Thus 9-(3,4-dihydroxybutyl)guanine (I), which has antiherpes activity, and fluazifop (II), a major metabolite of a selective herbicide, were resolved using *L*-phenylalanine-Cu(II) [215] and *L*-prolyl-*n*-octylamide-Ni(II) [195], respectively. Resolution of a quinolinone carboxylic acid [208] is presented in Fig. 10.

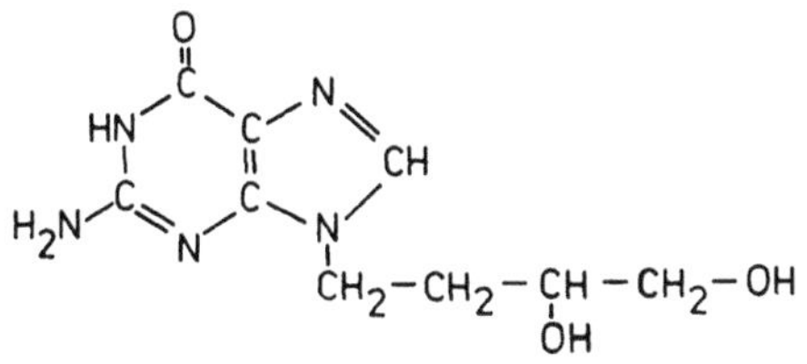

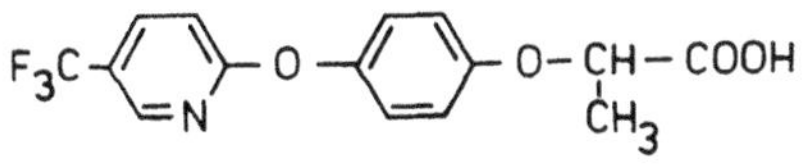

Milligram-scale resolutions of expensive racemic compounds using chiral eluant technique are feasible (Fig. 11) [214]. However, in this case, one is bound to face the problem of removing the chiral eluant additive from collected fractions of interest. This should not always be as easy as is the removing of copper(II) and phenylalnine, on an ion exchanger Dowex 50, from fractions containing enantiomers of strongly acidic polyfluoroalanine [214].

Concerning the mechanism of chiral discriminations in chromatographic systems that operate with chiral eluants, it should be emphasized that the complexation reactions taking place in the mobile phase and on the packing surface exert an opposite influence on the total residence time of two enantiomers in the chromatographic column [31]. However, due to the fact that the achiral surface of the sorbent actively interferes with the between-ligand interactions within the adsorbed mixed-ligand diastereomeric complexes, the enantioselectivity of complex

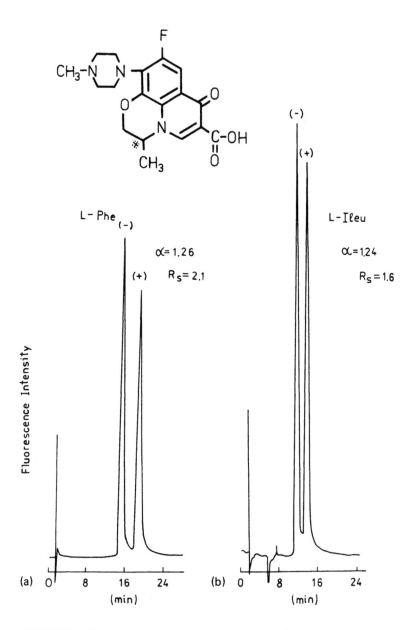

FIGURE 10 Typical chromatograms of ofloxacin enantiomers on an ODS column (150 ×
6mm, dp 5μm) using 3 mM CuSO₄ and 6 mM L-phenylalanine (a) or 6 mM L-isoleucine (b)
as the chiral eluant containing 15% methanol. (From Ref. 208.)

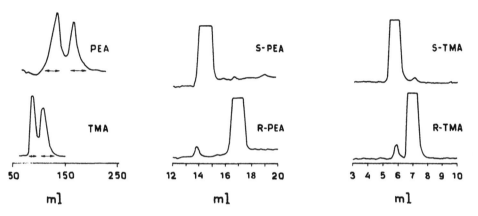

FIGURE 11 Preparative resolution of 4 mg of 2-pentafluoroethylalanine (PEA) and 4 mg of 2-trifluoromethylalanine (TMA) dissolved in 1 mL of mobile phase. Mobile phase: 2 mM L-phenylalanine, 1 mM copper acetate, 3 mM potassium acetate, pH 4.4. Column: 250 × 15 mm, ODS, dp 40 μm. Right: Analytical HPLC of collected fractions with the same mobile phase but 12% acetonitrile (PEA) or 6% acetonitrile (TMA). (From Ref. 214.)

formation on the stationary phase usually predominates over that in the bulk solution, thus determining the overall selectivity of the chromatographic column. From the schema below, one can easily understand that additional hydrophobic interactions between the substituent R of an L-amino acid in mixed complexes with L-proline of L-leucine and the alkyl chains of the reversed-phase packing contribute to the stabilization of the L-L diastereomers and cause the longer retention of L-amino acids compared to D-isomers [17].

On the contrary, N-benzyl-L-proline would adsorb on the hydrophobic surface with its bulky benzyl group, thus causing longer retention of D-isomers of amino acids:

TABLE 7 Resolution of Racemates Using Chiral Eluant Technique
in Normal-Phase Systems

Chiral selector	Metal ion	Stationary phase	Racemic solutes	Refs.
L-Proline (in 0.05–1.0 M NaOAc, pH 5.5)	Cu(II)	Sulfonated poly-styrene-type cation exchanger	Amino acids	236
L-Proline (in hexane/ n-propanol/ water 60:37.5:2.5)	Cu(II)	LiChrosorb Si 60	Amino acids: triiodothy-ronine and thyroxine	210
N,N,N′,N′-Tetra-methyl-(R)-propane-diamine (in aceto-nitrile/water 90:10, 5 mM Me₄pn)	Cu(II)	LiChrosorb Si 100	Amino acids and mandelic acid	228

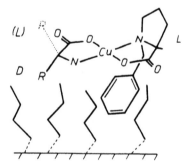

In the absence of a hydrophobic surface (i.e., in bulk solutions), enantioselectivity of formation of complexes with chiral selectors is usually found to be much less expressed than it is in heterogeneous reversed-phase systems [233–235].

Bare silica or polymeric cation-exchange resins can also be used in combination with chiral complex-doped eluants (Table 7). Remarkably, the elution order of amino acid enantiomers may reverse entirely on changing bare silica for a reversed-phase packing, which demonstrates once again the importance of interactions between the sorbent surface and diastereomeric sorption complexes for the chiral recognition of enantiomeric solutes [228].

2. Electrokinetic Methods

True enantioselectivity of the ternary complex formation in the mobile phase (i.e., the ratio of thermodynamic stability constants of two diastereomeric complexes in

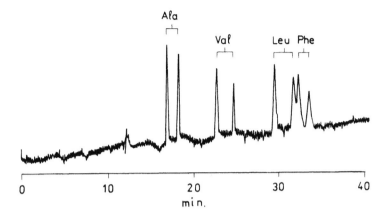

FIGURE 12   Resolution of a mixture of four racemic amino acids as N-(3,5-dinitroben-zoyl) O-isopropyl ester derivatives by electrokinetic capillary chromatography. Column: fused silica tubing 50 cm × 50 μm i.d. Micellar solution: 25 mM sodium N-dodecanoyl-L-valinate in 25 mM borate/50 mM phosphate buffer, pH 7.0. Total applied voltage, ca. 10 kV; current, 26 μA; detection, UV at 230 nm; temperature, ambient (ca. 20°C). (From Ref. 241.)

bulk solution) is the sole driving force for enantiomeric resolutions in high-performance electromigration methods. In a typical experiment, a chiral support electrolyte containing 2.5 mM Cu(L-His)$_2$ and 10 mM ammonium acetate of pH 7 to 8 was observed to move through a fused silica capillary column (75 cm × 75 μm I.D.) due to electro-osmotic forces toward the cathode. Therefore, cations, neutral species, and anions injected at the anode end of the column could be detected at the cathode end in a single run. Amino acids bonded to Cu(II)/L-His were found to migrate faster than free neutral amino acid species because the Cu(II)/L-His carries a positive charge under neutral pH conditions and exhibits an additional acceleration in the electric field of about 30 kV applied to the column. In the electrophoregram, D-dansyl amino acids appeared to migrate faster than the L-isomers, which implies that the latter were less involved into complexation with the chiral selector. However, the difference in the migration rates of the L- and D-enantiomers (corresponding to the enantioselectivity of the mixed-ligand complex formation) was really small [237]. It was only due to the extremely high efficiency of the electrokinetic technique (a plate number of more than 100,000 for the column above) that the enantiomeric peaks could be resolved at all.

The high sensitivity of the electrokinetic methods should be pointed out. Using a laser fluorimetric detector, as little as 50 mol of dansyl amino acids could be determined [238] with a satisfactory signal-to-noise ratio in a chiral electrolyte comprising 5.0 mM aspartame, 2.5 mM copper sulfate, and 10 mM ammonium acetate at pH 7.2.

The greatest drawback of electrokinetic resolutions is the poor selectivity of separations of solutes having equal electrostatic charge. This drawback can be overcome by using chiral selectors that are able to form micelles in the support electrolyte, as is the case with copper(II) complexes of N,N-didecyl-L-alanine [239]. Indeed, the partition of solutes between the micelles and the bulk solution offers an additional dimension for discriminating the solutes [240,241]. It is extremely important that the enantioselectivity of complexation may also rise significantly on passing from homogeneous support electrolytes to quasi-heterogeneous micellar solutions. Figure 12 shows an example of highly selective separation of four identically charged solutes combined with highly enantioselective resolution of each of them into individual isomeric pairs.

REFERENCES

1. B. W. Bradford, D. Harvey, and D. E. Chalkley, *J. Inst. Pet.*, *41*: 80 (1955).
2. R. H. Stokes and H. F. Walton, *J. Am. Chem. Soc.*, *76*: 3327 (1954).
3. F. G. Helfferich, *Nature*, *189*: 1001 (1961).
4. V. A. Davankov and A. V. Semechkin, *J. Chromatogr.*, *141*: 313 (1977).
5. V. A. Davankov, J. D. Navratil, and H. F. Walton, *Lingand Exchange Chromatography*, CRC Press, Boca Raton, Fla., 1988.
6. V. A. Davankov, in *Packings and Stationary Phases in Chromatographic Techniques* (K. K. Unger, ed.), Marcel Dekker, New York, 1990, p. 541.
7. V. A. Davankov, in *Chiral Separations by HPLC* (A. M. Krstulovic, ed.), Ellis Horwood, Chichester, West Sussex, England, 1989.
8. M. de Min, G. Levy, and J. C. Micheau, *J. Chim. Phys.*, *85*: 60 (1988).
9. S. F. Mason and G. E. Tranter, *Mol. Phys.*, *53*: 1091 (1984).
10. G. E. Tranter, *Mol. Phys.*, *56*: 825 (1985).
11. V. R. Meyer and M. Rais, *Chirality*, *1*: 167 (1989).
12. L. H. Easson and E. Stedman, *Biochem. J.*, *27*: 1257 (1933).
13. A. G. Ogston, *Nature*, *162*: 963 (1948); ibid., *167*: 693 (1951).
14. C. E. Dalgliesh, *J. Chem. Soc.*, 1952: 3940.
15. S. Topiol and M. Sabio, *J. Chromatogr.*, *461*: 129 (1989).
16. S. Topiol, *Chirality*, *1*: 69 (1989).
17. V. A. Davankov and A. A. Kurganov, *Chromatographia*, *17*: 686 (1983).
18. V. A. Davankov, V. R. Meyer, and M. Rais, *Chirality*, *2*: 208 (1990).
19. I. W. Wainer and Ya-Qin Chu, *J. Chromatogr.*, *455*: 316 (1988).
20. I. Fitos and M. Simonyi, *J. Chromatogr.*, *450*: 217 (1988).
21. P. Macaudiere, M. Lienne, M. Caude, R. Rosset, and A. Tambute, *J. Chromatogr.*, *467*: 357 (1989).
22. J. Vindevogel, J. Van Dijck, and M. Verzele, *J. Chromatogr.*, *447*: 297 (1988).
23. C. Roussel, J. Louisstein, F. Beauvais, and A. Chemlal, *J. Chromatogr.*, *462*: 95 (1989).
24. V. A. Davankov, A. A. Kurganov, and A. S. Bochkov, *Adv. Chromatogr.*, *22*: 71 (1983).
25. V. A. Davankov, *Adv. Chromatogr.*, *18*: 139 (1980).

26. W. H. Pirkle and T. C. Pochapsky, *Adv. Chromatogr.*, *27*: 73 (1987); *Chem. Rev., 89*: 347 (1989).
27. U. Norinder and E. G. Sundholm, *J. Liq. Chromatogr.*, *10*: 2825 (1987).
28. K. B. Lipkowitz, D. A. Demeter, R. Zegarra, R. Larter, and T. Darden, *J. Am. Chem. Soc., 110*: 3446 (1988).
29. P. Erlandsson, R. Issaksson, I. Nilsson, and S. Wold, *J. Chromatogr., 466*: 364 (1989).
30. R. Daeppen, V. R. Meyer, and H. Arm, *J. Chromatogr., 464:* 39 (1989).
31. V. A. Davankov, A. A. Kurganov, and T. M. Ponomareva, *J. Chromatogr., 452:* 309 (1988).
32. w. H. Pirkle and T. C. Pochapsky, *J. Chromatogr., 369*: 175 (1986).
33. M. Fujita, Y. Yoshikawa, and H. Yamatera, *Chem. Lett., 11*: 473 (1975).
34. C. Petersson and G. Gioeli, *J. Chromatogr., 435*: 225 (1988).
35. G. Schill, *Swiss Chem., 10*: 34 (1988).
36. T. Takeuchi, H. Asai, and D. Ishii, *J. Chromatogr., 407*: 151 (1987).
37. A. Walhagen and L. -E. Edholm, *J. Chromatogr., 473*: 371 (1989).
38. V. A. Davankov, *Chromatographia, 27*: 475 (1989).
39. S. V. Rogozhin, V. A. Davankov, V. V. Korshak, V. Vesa, and L. A. Belchich, *Izv. Akad. Nauk SSSR Ser. Khim.*, 502 (1971).
40. U. Beitler and B. Feibush, *J. Chromatogr., 123*: 149 (1976).
41. C. Pettersson, A. Karlsson, and C. Gioeli, *J. Chromatogr., 407*: 217 (1987).
42. W. H. Pirkle, R. Daeppen, and D. S. Reno, *J. Chromatogr., 407*: 211 (1987).
43. I. I. Piesliakas, S. V. Rogozhin, and V. A. Davankov, *Zh. Obshch. Khim., 44*: 468 (1974).
44. V. Schurig and W. Buerkle, *J. Am. Chem. Soc., 104*: 7573 (1982).
45. V. Schurig, *J. Chromatogr., 441*: 135 (1988).
46. M. A. Guyegkeng and A. Mannschreck, *Chem. Ber., 120*: 803 (1987).
47. A. Mannschreck, E. Gmahl, T. Burgemeister, F. Kastner, and V. Sinnwell, *Angew. Chem. Int. Ed. Engl., 27*: 270 (1988).
48. A. Mannschreck and L. Kiessl, *Chromatographia, 28*: 263 (1989).
49. A. Mannschreck, D. Andert, A. Eiglsperger, E. Gmahl, and H. Buchner, *Chromatographia, 25*: 182 (1988).
50. U. Beitler and B. Feibush, *J. Chromatogr., 123*: 149 (1976).
51. M. Koppenhoefer and E. Bayer, *Chromatographia, 19*: 123 (1984).
52. K. Watanabe, R. Charles, and E. Gil-Av, *Proceedings, Chromatography '86*, Chiba, Japan, 1986, p. 83.
53. V. Schurig, J. Ossig, and R. Link, *Angew. Chem., 101*: 197 (1989).
54. H. Hess, G. Burger, and H. Musso, *Angew. Chem., 90*: 645 (1978).
55. V. A. Davankov, Y. A. Zolotarev, and A. A. Kurganov, *J. Liq. Chromatogr., 2*: 1191 (1979).
56. A. A. Kurganov, L. Y. Zhuchkova, and V. A. Davankov, *J. Inorg. Nucl. Chem., 40*: 1081 (1978).
57. V. A. Davankov, in *Chiral Separations by HPLC (A. M. Krstulovic, ed.)*, Ellis Horwood, Chichester, West Sussex, England, 1989.
58. S. G. Allenmark, *Chromatographic Enantioseparation: Methods and Applications*, Ellis Horwood, Chichester, West Sussex, England, 1988.

59. A. M. Krstulovich, ed., *Chiral Separations by HPLC: Applications to Pharmaceutical Compounds,* Ellis Horwood, Chichester, West Sussex, England, 1989.
60. R. Daeppen, H. Arm, and V. R. Meyer, *J. Chromatogr., 373*: 1 (1986).
61. A. M. Krstulovic, *J. Chromatogr., 488*; 53 (1989).
62. I. W. Wainer, *Trends Anal. Chem., 6*: 125 (1987).
63. M. de Min, G. Levy, and J. C. Micheau, *J. Chim. Phys., 85*: 603 (1988).
64. D. W. Armstrong, S. M. Han, *CRC Crit. Rev. Anal. Chem., 19*: 175 (1988).
65. M. Mack and H. E. Hauck, *Chromatographia, 26*, 197 (1988).
66. S. V. Rogozhin and V. A. Davankov, Ger. Offen. 1932190 (1970); *Chem. Abstr., 72*: 90875c (1970).
67. S. V. Rogozhin and V. A. Davankov, *Dokl. Akad. Nauk SSSR, 192*: 1288 (1970); *Chem. Commun.,* 490 (1971).
68. V. A. Davankov, S. V. Rogozhin, and M. P. Tsyurupa, U. S. Patent 3729457 (1970); *Chem. Abstr., 75*: 6841v (1971).
69. V. A. Davankov and M. P. Tsyurupa, *Angew. Macromol. Chem., 91*: 127 (1980).
70. G. I. Rosenberg, A. S. Shabaeva, V. S. Moryakov, T. G. Musin, M. P. Tsyurupa, and V. A. Davankov, *React. Polym., 1*: 175 (1983).
71. V. A. Davankov and M. P. Tsyurupa, *React. Polym.,13*: 27 (1990).
72. V. A. Davankov, S. V. Rogozhin, and I. I. Piesliakas, *Vysokomol. Soed., 14B*: 276 (1972).
73. V. A. Davankov, S. V. Rogozhin, I. I. Piesliakas, and V. S. Vesa, *Vysokomol. Soed., 15B*: 115 (1973).
74. E. Tsuchida, H. Nishikawa, and E. Terada, *Eur. Polym. J., 12*: 611 (1976).
75. V. A. Davankov, *Habilitationsschrift,* INEOS, Moscow, 1975.
76. J. Jozefonvicz, M. A. Petit, and A. Szubarga, *J. Chromatogr., 147*: 177 (1978).
77. S. V. Rogozhin, I. A. Yamskov, V. A. Davankov, T. F. Kolesova, and V. M. Voevodin, *Vysokomol. Soedin., 17A*: 564 (1975).
78. S. V. Rogozhin, I. A. Yamskov, and V. A. Davankov, *Vysokomol. Soedin., 17B*: 107 (1975).
79. S. V. Rogozhin, I. A. Yamskov, and V. A. Davankov, *Vysokomol. Soedin., 16B*: 849 (1974).
80. S. V. Rogozhin, V. A. Davankov, and I. A. Yamskov, *Isv. Akad. Nauk SSSR Ser. Khim.,* 2325 (1971).
81. N. Spassky, M. Riex, J. Guette, M. Guette, M. Sepulchre, and J. Blanchard, *J. Compt. Rend., C287*: 589 (1978).
82. V. A. Davankov, S. V. Rogozhin, I. A. Yamskov, and V. P. Kabanov, Izv. Akad. Nauk SSSR Ser. Khim., 2327 (1971).
83. S. V. Rogozhin, V. A. Davankov, I. A. Yamskov, and V. P. Kabanov, *Zh. Obshch. Khim., 42*: 1614 (1972).
84. B. B. Berezin, I. A. Yamskov, and V. A. Davankov, *J. Chromatogr., 261*: 301 (1983).
85. S. V. Rogozhin, I. A. Yamskov, and V. A. Davankov, *Vysokomol. Soedin., 15B*: 216 (1973).
86. S. V. Rogozhin, V. A. Davankov, I. A. Yamskov, and V. P. Kabanov, *Vysokomol. Soedin., 14B*: 472 (1972).
87. V. A. Davankov and Y. A. Zolotarev, *J. Chromatogr., 155*: 295 (1978).

88. I. A. Yamskov, B. B. Berezin, V. E. Tikhonov, L. A. Belchich, and V. A. Davankov, *Bioorgan. Khim., 4*: 1170 (1978).
89. S. V. Rogozhin, I. A. Yamskov, A. S. Pushkin, L. A. Belchich, L. Y. Zhuchkova, and V. A. Davankov, *Izv. Akad. Nauk SSSR Ser. Khim.,* 2378 (1976).
90. R. V. Snyder, R. J. Angelichi, and R. B. Meck, *J. Am. Chem. Soc., 94*: 2660 (1972).
91. I. A. Yamskov, B. B. Berezin, and V. A. Davankov, *Makromol. Chem., 179*: 2121 (1978).
92. Y. P. Belov, S. V. Rogozhin, and V. A. Davankov, *Izv. Akad. Nauk SSSR Ser. Khim.,* 2320 (1973).
93. Y. P. Belov, V. A. Davankov, and S. V. Rogozhin, *Izv. Akad. Nauk SSSR Ser. Khim.,* 1856 (1977).
94. A. A. Kurganov, L. Y. Zhuchkova, and V. A. Davankov, *Makromol. Chem., 180*: 2102 (1979).
95. V. S. Vesa, *Zh. Obshch. Khim., 42*: 2780 (1972).
96. D. Muller, J. Jozefonvicz, and M. -A. Petit, *J. Inorg. Nucl. Chem., 42*: 1665 (1980).
97. V. A. Davankov and Y. A. Zolotarev, *J. Chromatogr., 155*: 285 (1978).
98. V. A. Davankov and Y. A. Zolotarev, *J. Chromatogr., 155*: 303 (1978).
99. V. A. Davankov and A. A. Kurganov, *Chromatographia, 13*: 339 (1980).
100. V. A. Davankov, Y. A. Zolotarev, and A. A. Kurganov, *J. Liq. Chromatogr., 2*: 1191 (1979).
101. N. F. Myasoedov, O. B. Kusnetsova, O. V. Petrenik, V. A. Davankov, and Y. A. Zolotarev, *J. Labelled Compd. Radiopharmaceut., 17*: 439 (1979).
102. Y. A. Zolotarev, N. F. Myasoedov, V. I. Penkina, O. V. Petrenik, and V. A. Davankov, *J. Chromatogr., 207*: 63 (1981).
103. Y. A. Zolotarev, N. F. Myasoedov, V. I. Penkina, I. N. Dostovalov, O. V. Petrenik, and V. A. Davankov, *J. Chromatogr., 207*: 231 (1981).
104. V. A. Davankov, Y. A. Zolotarev, and A. V. Tevlin, *Bioorgan. Khim., 4*: 1164 (1978).
105. V. A. Shirokov, V. A. Tsyryapkin, L. V. Nedospasova, A. A. Kurganov, and V. A. Davankov, *Bioorgan. Khim., 9*: 878 (1983).
106. I. A. Yamskov, S. V. Rogozhin, and V. A. Davankov, *Bioorgan. Khim., 3*: 200 (1977).
107. B. Lefebvre, R. Audebert, and C. Quivoron, *Isr. J. Chem., 15*: 69 (1977).
108. B. Lefebvre, R. Audebert, and C. Quivoron, *J. Liq. Chromatogr., 1*: 761 (1978).
109. G. Jeanneret-Gris, C. Soerensen, H. Su, and J. Porret, *Chromatographia, 28*: 337 (1989).
110. J. Jeanneret-Gris, J. Porret, and K. Bernauer, *Chromatographia* (in press).
111. D. Muller, J. Jozefonvicz, and M. A. Petit, *J. Inorg. Nucl. Chem., 42*: 1083 (1980).
112. F. Lafuma, J. Boue, R. Audebert, and C. Quivoron, *Inorg. Chim. Acta, 66*: 167 (1982).
113. I. A. Yamskov, B. B. Berezin, V. A. Davankov, Y. A. Zolotarev, I. N. Dostovalov, and N. F. Myasoedov, *J. Chromatogr., 217*: 539 (1981).
114. D. Charmot, R. Audebert, and C. Quivoron, *J. Liq. Chromatogr., 8*: 1753 (1985).
115. D. Charmot, R. Audebert, and C. Quivoron, *J. Liq. Chromatogr., 8*: 1769 (1985).
116. N. Watanabe, H. Ohzeki, and E. Niki, *J. Chromatogr., 216*: 406 (1981).
117. Y. A. Zolotarev, A. A. Kurganov, and V. A. Davankov, *Talanta, 25*: 493 (1978).
118. Y. A. Zolotarev, A. A. Kurganov, A. V. Semechkin, and V. A. Davankov, *Talanta, 25*: 499 (9178).
119. V. A. Davankov, *Pure Appl. Chem., 54*: 2159 (1982).

120. A. S. Bochkov, Y. A. Zolotarev, Y. P. Belov and V. A. Davankov, *Progress in Chromatography*, Proc. 2nd Danube Symposium, Carlsbad, Czechoslovakia, 1979, Paper B3.23.
121. P. Roumeliotis, K. K. Unger, A. A. Kurganov, and V. A. Davankov, *Angew. Chem. Int. Ed. Engl.*, *21*: 930 (9182).
122. P. Roumeliotis, A. A. Kurganov, and V. A. Davankov, *J. Chromatogr.*, *266*: 439 (1983).
123. P. Roumeliotis, K. K. Unger, A. A. Kurganov, and V. A. Davankov, *J. Chromatogr.*, *255*: 51 (1983).
124. K. Sugden, C. Hunter, and G. Lloyd-Jones, *J. Chromatogr.*, *192*: 228 (1980).
125. L. R. Gelber, B. L. Karger, J. L. Neumeyer, and B. Feibush, *J. Am. Chem. Soc.*, *106*: 7729 (1984).
126. C. H. Shieh, B. L. Karger, L. R. Gelber, and B. Feibush, *J. Chromatogr.*, *406*: 343 (1987).
127. B. Feibush, M. J. Cohen, and B. L. Karger, *J. Chromatogr.*, *282*: 3 (1983).
128. G. Guebitz, W. Jellenz, G. Loefler, and W. Santi, *J. High Resolut. Chromatogr. Chromatogr. Commun.*, *2*: 145 (1979).
129. G. Guebitz, W. Jellenz, and W. Santi, *J. Liq. Chromatogr.*, *4*: 701 (1981).
130. G. Guebitz, W. Jellenz, and W. Santi, *J. Chromatogr.*, *203*: 377 (1981).
131. G. Guebitz, F. Juffmann, and W. Santi, *Chromatographia*, *16*:103 (1982).
132. G. Guebitz and F. Juffmann, *J. Chromatogr.*, *404*: 391 (1987).
133. S. V. Galushko, I. P. Shishkina, A. T. Pilipenko, and L. P. Prikaschikova, *Zh. Analt. Khim.*, *43*: 1719 (1988).
134. G. Guebitz and S. Mihellyes, *Chromatographia*, *19*: 257 (1984).
135. G. Guebitz, *J. Liq. Chromatogr.*, *9*: 519 (1986).
136. Y. Yuki, K. Saigo, K. Tachibana, and M. Hasegawa, *Chromatography '86*, Chiba, Japan, 1986, p. 155.
137. Y. Yuki, K. Saigo, K. Tachibana, and M. Hasegawa, *Chem. Lett.*, 1347, (1986).
138. Y. Yuki, K. Saigo, H. Kimoto, K. Tachibana, and M. Hasegawa, *J. Chromatogr.*, *400*: 65 (1987).
139. M. Sinibaldi, V. Carunchio, C. Corradini, and A. M. Girelli, *Chromatographia*, *18*: 459 (1984).
140. C. Corradini, F. Federici, M. Sinibaldi, and A. Messina, *Chromatographia*, *23*: 118 (1987).
141. A. Foucault and M. Caude, *J. Chromatogr.*, *185*: 345 (1979).
142. W. Lindner, *Naturwissenschaften, 67*: 354 (1980).
143. H. Engelhardt and G. Ahr, *Chromatographia, 14*: 227 (1981).
144. N. Watanabe, *J. Chromatogr.*, *260*: 75 (1983).
145. J. Castells, J. M. Daga, F. Lopez-Calahorra, and D. Velasco, *React. Polym.*, *6*: 207 (1987).
146. H. Engelhardt and S. Kromidas, *Naturwissenschaften, 67*: 353 (1980).
147. H. G. Kicinski and A. Kettrup, *Fresenius Z. Anal. Chem.*, *320*: 51 (1985).
148. W. Lindner and I. Hirschboeck, *J. Pharmaceut. Biomed. Anal.*, *2*: 183 (1984).
149. J. Florance, A. Galdes, Z. Konteatis, Z. Kosarych, K. Langer, and C. Martucci, *J. Chromatogr.*, *414*: 313 (1987).
150. T. Takeuchi, H. Asai, and D. Ishii, *J. Chromatogr.*, *407*: 151 (1987).

151. A. Ichida, *Int. Lab.*, 26 (March/April 1989).
152. H. Brueckner, J. Bosch, T. Graser, and P. Fuerst, *J. Chromatogr., 395*: 569 (1987).
153. H. Brueckner, *Fresenius Z. Anal. Chem., 327*: 32 (1987).
154. H. Brueckner, *Chromatographia, 24*: 725 (1987).
155. H. Scopan, H. Guenther, and H. Simon, *Angew. Chem., 99*: 139 (1987).
156. V. A. Davankov, S. V. Rogozhin, A. V. Semechkin, V. A. Baranov, and G. S. Sannikova, *J. Chromatogr., 93*: 363 (1974).
157. H. Kiniwa, Y. Baba, T. Ishida, and H. Katoh, *J. Chromatogr., 461*: 397 (1989).
158. A. A. Kurganov, T. M. Ponomareva, and V. A. Davankov, *Dokl. Akad. Nauk SSSR, 293*: 623 (1987).
159. A. A. Kurganov, A. V. Tevlin, and V. A. Davankov, *J. Chromatogr., 261*: 223 (1983).
160. V. A. Davankov, A. S. Bochkov, A. A. Kurganov, P. Roumeliotis, and K. K. Unger, *Chromatographia, 13*: 677 (1980).
161. V. A. Davankov, A. S. Bochkov, and Y. P. Belov, *J. Chromatogr., 218*: 547 (1981).
162. H. Lam-Thanh, S. Fermandjian, and P. Fromageot, *J. Chromatogr., 235*: 139 (1982).
163. E. Busker and J. Martens, *Fresenius Z. Anal. Chem., 319*: 907 (1984).
164. E. Busker, K. Guenther, and J. Martens, *J. Chromatogr., 350*: 179 (1985).
165. H. Kathoh, T. Ishida, S. Kuwata, and H. Kiniwa, *Chromtographia, 28*: 48I (1989).
166. T. Arai, H. Matsuda, and H. Oizumi, *J. Chromatogr., 474*: 405 (1989).
167. H. Katoh, T. Ishida, Y. Baba, and H. Kiniwa, *Bunseki Kagaku, 38*: 249 (1989).
168. H. Katoh, T. Ishida, Y. Baba, and H. Kiniwa, *J. Chromatogr., 473*: 241 (1989).
169. P. A. Krasutskii, V. N. Rodionov, V. P. Tichonov, and A. G. Yurchenko, *Teor. Eksp. Khim., 20*: 58 (1984).
170. A. A. Kurganov, I. T. Chizhewski, V. A. Davankov, A. I. Yanovski, and Y. T. Struchkov, *Metalloogr., Khim., 1*: 913 (1988).
171. J. Boue, R. Audebert, and C. Quivoron, *J. Chromatogr., 204*: 185 (1981).
172. D. Charmot, R. Audebert, and C. Quivoron, *J. Liq. Chromatogr., 8*: 1769 (1985).
173. D. Charmot, R. Audebert, and C. Quivoron, *J. Liq. Chromatogr., 8*: 1743 (1985).
174. F. Humbel, D. Vonderschmitt, and K. Bernauer, *Helv. Chim. Acta, 53*: 1983 (1970).
175. K. Bernauer, M. -J. Jeanneret, and D. Vonderschmitt, *Helv. Chim. Acta, 54*: 297 (1971).
176. J. A. Perry, J. D. Rateike, and T. J. Szczerba, *J. Chromatogr., 389*: 57 (1987).
177. H. Yoneda, *J. Chromatogr., 313*: 59 (1985).
178. K. Guenther, *J. Chromatogr., 448*: 11 (1988).
179. S. Nyiredy, K. Dallenbach-Toelke, and O. Sticher, *J. Chromatogr., 450*: 241 (1988).
180. K. Kovacs-Hadady and I. T. Kiss, *Chromatographia, 24*: 677 (1987).
181. J. N. LePage, W. Lindner, G. Davies, D. E. Seitz, and B. L. Karger, *Anal. Chem., 51*: 433 (1979).
182. W. Lindner, J. N. LePage, G. Davis, D. E. Seitz, and B. L. Karger, *J. Chromatogr., 185*: 323 (1979).
183. E. Gil-Av, A. Tishbee, and P. E. Hare, *J. Am. Chem. Soc., 102*: 5115 (1980).
184. S. K. Lam and F. K. Chow, *J. Liq. Chromatogr., 3*: 1579 (1980).
185. S. Lam, F. Chow, and A. Karmen, *J. Chromatogr., 199*: 295 (1980).
186. S. K. Lam, *J. Chromatogr., 234*: 485 (1982).
187. S. Lam, *J. Chromatogr. Sci., 22*: 416 (1984).

188. S. Lam, G. Malikin, M. Murphy, L. Freundlich, and A. Karmen, *J. Chromatogr., 468:* 359 (1989).
189. S. Lam and A. Karmen, *J. Liq, Chromtogr., 9:* 291 (1986).
190. J. van der Haar, J. Kip, and J. C. Kraak, *J. Chromatogr., 445:* 219 (1988).
191. J. Wagner, E. Wolf, B. Heintzelmann, and C. Gaget, *J. Chromatogr., 392:* 211 (1987).
192. R. Horikawa, H. Sakamoto, and T. Tanimura, *J. Liq. Chromatogr., 9:* 537 (1986).
193. E. Gil-Av and S. Weinstein, in *Handbook of HPLC for the Separation of Amino Acids, Peptides and Proteins,* Vol. 1 (W. S. Hancock, ed.), CRC Press, Boca Raton., Fla., 1984, p. 429.
194. Y. Tapuchi, N. Miller, and B. L. Karger, *J. Chromatogr., 205:* 325 (1981).
195. G. S. Davy and P. D. Francis, *J. Chromatogr., 394:* 323 (1987).
196. S. Lam and A. Karmen, *J. Chromatogr., 239:* 451 (1982).
197. S. Lam and A. Karmen, *J. Chromatogr., 289:* 339 (1984).
198. S. Lam, H. Azumaya, and A. Karmen, *J. Chromatogr., 302:* 21 (1984).
199. S. Weinstein, M. H. Engel, and P. E. Hare, *Anal. Biochem., 121:* 370 (1982).
200. S. Weinstein, *TRAC, 3:* 16 (1984).
201. S. Weinstein, *Angew. Chem. Suppl.,* 425 (1982).
202. A. Duchateau, M. Crombach, M. Aussems, and J. Bongers., *J. Chromatogr., 461:* 419 (1989).
203. S. Weinstein and N. Grinberg, *J. Chromatogr., 318:* 117 (1985).
204. S. Weinstein and S. Weiner, *J. Chromatogr., 303:* 244 (1984).
205. J. Benecke, *J. Chromatogr., 291:* 155 (1984).
206. E. Armani, L. Barazzoni, A. Dossena, and R. Marchelli, *J. Chromatogr., 441:* 287 (1988).
207. L. W. Doner and P. J. Cavender, *J. Food Sci., 53:* 1898 (1988).
208. T. Arai, H. Koike, K. Hirata, and H. Oizumi, *J. Chromatogr., 448:* 439 (1988).
209. R. Wernicke, *J. Chromatogr. Sci., 23:* 39 (1985).
210. E. Oelrich, K. Preusch, and E. Wilhelm, *J. High Resolut. Chromatogr. Chromatogr. Commun., 3:* 269 (1980).
211. M. Lebl, P. Hrbas, J. Skopkova, J. Slaninova, A. Machova, T. Barth, and K. Jost, *Collect. Czech. Chem. Commun., 47:* 2540 (1982).
212. L. R. Gelber and J. L. Neumeyer, *J. Chromatogr., 257:* 317 (1983).
213. W. Klemisch, A. von Hodenberg, and K. -O. Vollmer, *J. High Resolut. Chromatogr. Chromatogr. Commun., 4:* 535 (1981).
214. J. W. Keller and K. Niwa, *J. Chromatogr., 469:* 434 (1989).
215. U. Forsman, *J. Chromatogr., 303:* 217 (1984).
216. N. Nimura, T. Suzuki, Y. Kasahara, and T. Kinoshita, *Anal. Chem., 53:* 1380 (1981).
217. N. Nimura, A. Toyama, and T. Kinoshita, *J. Chromatogr., 234:* 482 (1982).
218. N. Nimura, A. Toyama, Y. Kasahara, and T. Kinoshita, *J. Chromatogr., 239:* 671 (1982).
219. N. Nimura, A. Toyama, and T. Kinoshita, *J. Chromatogr., 316:* 547 (1984).
220. C. Gilon, R. Leshem, and E. Grushka, *J. Chromatogr., 203:* 365 (1981).
221. C. Gilon, R. Leshem, and E. Grushka, *Anal. Chem., 52:* 1206 (1980).
222. E. Grushka, R. Leshem, and C. Gilon, *J. Chromatogr., 255:* 41 (1983).
223. C. Gilon, R. Leshem, Y. Tapuhi, and E. Grushka, *J. Am. Chem. Soc., 101:* 7612 (1979).

224. G. Gundlach, E. L. Sattler, and U. Wagenbach, *Fresenius Z. Anal. Chem., 311*: 684 (1982).
225. S. Kato, H. Kamakura, and S. Ariizumi, *Shokuhin Eiseigaku Zasshi, 27*: 272 (1986); *Chem. Abstr., 105*: 202486k (1986).
226. G. E. DuBois and R. A. Stephenson, *J. Agric. Food Chem., 30*: 676 (1982).
227. W. F. Lindner and I. Hirschboeck, *J. Liq. Chromatogr., 9*: 551 (1986).
228. A. A. Kurganov and V. A. Davankov, *J. Chromatogr., 218*: 559 (1981).
229. R. Marchelli, A. Dossena, G. Castani, F. Dallavalle, and S. Weinstein, *Angew. Chem., 24*: 336 (1985).
230. E. Armani, A. Dossena, R. Marchelli, and R. Virgili, *J. Chromatogr., 441*: 275 (1988).
231. R. Marchelli, A. Dossena, G. Castani, G. G. Fava, and M. F. Belicchi, *Chem. Commun.,* 1985, 1672.
232. G. Palla, R. Marchelli, A. Dossena, and G. Castani, *J. Chromatogr., 475*: 45 (1989).
233. T. Takeuchi, R. Horikawa, and T. Tanimura, *Anal. Chem., 56*: 1152 (1984).
234. T. Takeuchi, R. Horikawa, and T. Tanimura, *J. Chromatogr., 284*: 285 (1984).
235. J. M. Brode and D. L. Leussing, *Anal. Chem., 58*: 2237 (1986).
236. P. E. Hare and E. Gil-Av, *Science, 204*: 1226 (1979).
237. E. Gassmann, J. E. Kuo, and R. N. Zare, *Science, 230*: 813 (1985).
238. P. Gozel, E. Gassmann, H. Michelsen, and R. N. Zare, *Anal. Chem., 59*: 44 (1987).
239. A. S. Cohen, A. Paulus, and B. L. Karger, *Chromatographia, 24*: 15 (1987).
240. J. Snopek, I. Jelinek, and E. Smolkova-Keulemansova, *J. Chromatogr., 452*: 571 (1988).
241. A. Dobashi, T. Ono, S. Hara, and J. Yamaguchi, *J. Chromatogr., 480*: 413 (1989).
242. V. A. Davankov and S. V. Rogozhin, *Dokl. Akad. Nauk SSSR, 193*: 94 (1970).
243. C. H. Shich, B. L. Karger, L. R. Gelber, and B. Feibush, *J. Chromatogr., 406*: 343 (1987).

6

Special Topics: Applications of Complexation Chromatography to the Analysis of Coal and Petroleum Products

D. Cagniant
University of Metz, Metz, France

I. INTRODUCTION

Petroleum distillates and heavy residues and coal liquid products are all very complex mixtures of hydrocarbons (saturates, S; olefins, O; and more or less hydrogenated polycyclic aromatics, PAHs), heterocycles (principally nitrogen and sulfur heterocycles), polar compounds (principally hydroxylated aromatics), and oligomeric systems. Obviously, these mixtures represent a real challenge to analysts. Nevertheless, knowledge of their composition, expressed in terms of principal chemical groups (such as saturates, olefins, and aromatics), or in some cases in terms of main chemical families (such as mono-, di-, triaromatics) or even in terms of individual structures, is important for many reasons (e.g., fundamental or applied research, laboratory or industrial scales, upgrading processes, environmental problems, etc.). Generally speaking, according to the objectives, more or less tedious procedures are carried out involving fractionation steps, followed by structural identification by means of several complementary analytical methods, sometimes coupled with spectroscopic methods.

The aim of this chapter is to show how complexation chromatography (argentation chromatography and charge-transfer chromatography) can participate in the analysis of petroleum and coal liquid products. In fact, many studies cited in Chapters 3 and 4 for the separation of model compounds were carried out because of their application to petroleum or coal products. As the analytical problems are almost the same whatever the sample origin may be (petroleum, shale oil, or coal products), the subject will be treated on the basis of the methods used.

II. HYDROCARBON GROUP TYPE ANALYSIS

In the case of petroleum products, hydrocarbon group type analysis (HGTA) [i.e., the quantitative estimation of the three main groups: saturates (S), olefins (O), and polycyclic aromatic hydrocarbons (PAHs)] has the greatest interest for the evaluation of feedstocks and for refinery process control. Indeed, many problems arise from the presence of olefinic compounds in greater or lesser amounts. For instance, analysis of the fractions produced by heavy petroleum distillate conversion is essential to the evaluation of process efficiency. Nevertheless, this is a difficult task, as it requires the separation and quantification of olefins in the presence of other groups [1]. Furthermore, as these olefins play an important role in product stability, as in visbreaking distillates, their amounts allow a distinction to be made between conversion or straight-run residues [1]. In the same way, the determination of octane number for gasolines requires accurate evaluation of the respective amounts of saturates, olefins, and PAHs. Then the gasoline and jet fuel specifications are checked by normalized tests such as the well-known FIA-ASTM D 1319 [2]. Despite its widespread use for about 40 years, the fluorescent indicator adsorption (FIA) method presents numerous limitations inherent in the method itself and is not useful for highly colored products. Several HPLC methods have been developed leading to more accurate HGTA, applicable to various distillate products [3].

HGTA is also of special interest in the case of high-boiling distillates and residues obtained from shale oil because of their greater content of olefinic compounds than in petroleum products. Their instability causes more problems in shale oil processing than those encountered in petroleum processing [4]. It is certainly in the field of HGTA that argentation chromatography has been most successful because of its ability to separate olefinic compounds, as shown in Chapter 4, particularly using HPLC methods.

In case of coal products (tars and pitches) issued from hydroliquefaction or carbonization processes, the analytical objectives are somewhat different, as these products are constituted primarily of PAHs and nitrogen heterocycles, although in addition to saturates and occasionally olefins in amounts depending on the processes (e.g. low- or high-temperature carbonization processes). The largest part of the analytical method concerns PAH separation according to the number of aromatic nuclei. Furthermore, owing to their carcinogenic and (or) mutagenic activities, many studies have been undertaken to characterize individual structures among PAHs and polycondensed nitrogen heterocycles [5]. It is in the field of PAH analysis that charge-transfer chromatography has found useful applications, as have many other chromatographic methods (glass capillary column chromatography, HPLC on amino silane, RP–HPLC, etc.).

A. Application of Argentation Chromatography

As early as 1955, Bradford et al. [6], in reviewing the chromatographic analysis of hydrocarbon mixtures, described the successful identification by gas chromatography (GC) of traces of ethane in ethylene, using a column containing 30% of a saturated solution of silver nitrate in glycol. Ethane was eluted before ethylene, and a complete separation was obtained.

Following this initial and limited example, the primary results were obtained by application of HPLC to HGTA, even in the case of particular heavy and colored products. Although amino-bonded phases were often used, several papers have reported the utilization of Ag–HPLC with either coated or bonded silica gel columns. However, as Hayes and Anderson [7,8] and Norris, Campbell et al. [9] have pointed out, nearly all HPLC approaches to HGTA suffer from a common drawback: deriving accurate response factors that are at once applicable to a wide range of different distillate products containing, simultaneously, saturates, olefins, and PAHs. The ideal detector for HGTA would be sensitive to hydrocarbons and would give a response independent of carbon number. Obviously, UV detectors are excluded, as they do not respond to saturates, are not useful in the short-wavelength region where olefins adsorb, and give widely varying responses for PAHs. Refractive index detectors require calibration for typical saturates, olefins, and PAHs. In the case of IR detectors, a suitable wavelength having a nearly constant response for many hydrocarbons would be selected, as described in the work of Matsushita et al. [10]. It was suggested [9a] that this method be limited to a given sample type (i.e., gasoline). Recent improvements were brought about by the utilization of a dielectric constant detector [7,8] and of supercritical fluid chromatography (SCF) with a flame ionization detector (FID) [9]. As far as the columns are concerned, two main techniques were developed, using either the dual-column technique [4,10] or one column with backflushing [11].

1. HGTA by Ag–HPLC on Silver-Coated Silica Gel Columns

McKay and Latham [4] described a dual-column HPLC method for the separation and determination of saturates (S), olefins (O), and PAHs in high-boiling distillates and residues of shale oils, as well as in whole shale oils. This dual-column technique uses silica gel to separate PAHs from (O + S) and silica gel coated with silver nitrate to separate O from S. Columns (stainless steel 61 × 0.78 cm i.d.) contained 15 g of gel wetted with cyclohexane. The silver nitrate–coated gel was prepared dissolving the nitrate in water, mixing the solution with silica gel, and removing the water on a rotary evaporator. The gel was activated at 110° C for 12 h. The HPLC apparatus is shown in Fig. 1.

For example, in a routine separation, after removing acid, basic, and neutral ni-

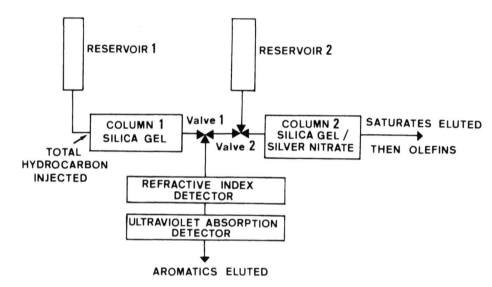

FIGURE 1   Dual-column HPLC apparatus. (From Ref.. 4.)

trogen compounds by preliminary chromatographies on ion-exchange resins and attapulgus clay, the shale oil sample (200 mg), dissolved in cyclohexane (1 mL) was placed on column 1. Cyclohexane (reservoir 1) was pumped through the two columns (40 min at 1 mL/min under about 350 psi). Under these conditions PAHs were retained on column 1, while saturates and olefins pass through column 2 and were separated by pumping cyclohexane from reservoir 2. Saturates were eluted within 30 min at 1 mL/min, and olefins were then eluted by pumping benzene/cyclohexane 20:80 for 30 min at 1 mL/min. Simultaneously, benzene/methanol 40:60 was pumped through column 1 for 60 min to elute PAHs. The total separation required about 2 h.

This dual-column method was used to separate hydrocarbons from high-boiling distillates (210 to 370° C) and residues (>535° C) as well as from the whole shale oils. The quantitative data were obtained gravimetrically, after solvent elimination from the fractions collected. The recovery of separated hydrocarbon types was generally greater than 90%, except in the case of whole shale oils analysis, where unknown amounts of light hydrocarbons were lost during the solvent removal step. The separation efficiency was proved by infrared (IR) spectrometry at two wavelenghts: 1640 cm⁻¹ (monoolefin stretching band) and 1600 cm⁻¹ (aromatic carbon–carbon stretching band).

A similar procedure was described independently by Matsushita et al. [10]. The silver nitrate–coated gel was prepared as before [4], using acetonitrile instead of water. The gel was activated at 110° C for 12 h. First column, 50 × 0.75 cm i. d.; second column, 4 × 0.75 cm i. d. Peak areas were measured with a computing integrator after IR detection at 1450 cm⁻¹. The order of elution (carbon tetrachloride as mobile phase) was saturates < olefins < PAHs. The time required for the separation of gasoline samples is 25 min (12 to 15 min for S, 15 or 18 min for O, and from 18 to 25 min for PAHs), without backflushing. Accurate and reproducible quantitations were claimed. An example is shown in Fig. 2.

With practical samples, response factors of individual groups were determined with standard gasoline by the FIA method and were found to be 1.000 for saturates (S), 0.605 for olefins (O), and 0.941 for PAHs. The basic calculations for each group in volume percent are as follows:

T = area (S) + area × 0.604 (O) + area × 0.94 (PAHs)
$\%S$ = area (S) × 100/T
$\%O$ = area (O) × 60.4/T
$\%\,PAHs$ = area (PAHs) × 94.1/T

The detection limits for O was 0.1 vol %.

Recent developments in supercritical fluid chromatography (SCF) prompted Norris and Rawdon to evaluate SCF as a routine technique [11]. One of the early

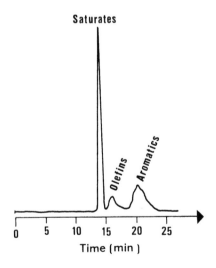

FIGURE 2 Chromatogram of a gasoline without back-flushing. Columns, LS-320 (50 cm × 7.5 mm i.d.) and LS-320 coated with silver nitrate (4 cm × 7.5 mm i.d.); mobile phase, carbon tetrachloride: flow rate, 1.2 mL/min; sample size, 4 μL; detector, infrared. (From Ref. 10.)

problems they investigated [9a] was the HGTA of fuels, and their main objective was to determine if SCF with a FID detector could provide a satisfactory alternative to the FIA technique.

The supercritical fluid chromatograph used is a modified HPLC chromatograph (Hewlett-Packard 1082 B) [12] with a modified gas chromatographic FID detector [13]. Two columns in series provided separation of the three groups: the first column (24 × 0.46 cm i. d., 5 μm silica) separates (S + O) from PAHs; the second column (25 × 0.46 cm i.d.), with 20% silver nitrate–coated R Sil 5, permits S/O separation.

Some chromatograms are reported in Fig. 3 for gasoline (3a), shale oil (3b) and creosote oil (3c) with CO_2 as the supercritical fluid. Then SCF offers a feasible alternative to the FIA technique for HGTA in petroleum distillates and can be extended to shale- and coal-derived liquids.

An improved hydrocarbon group resolution of olefinic gasolines was investigated by Chartier et al. to separate saturates, olefins, and PAHs on the same column using backflushing [14]. Silver-coated silica gel was made by percolating a 3% w/w solution of silver nitrate in acetonitrile through a silica gel column (Lichrosorb Si60, "in situ procedure"):

$$\equiv Si-\bar{O}-H + Ag(CH_3CN)_x^+, NO_3^- \rightleftharpoons \equiv Si-\overset{\displaystyle H}{\underset{\displaystyle Ag(CH_3CN)_y^+, NO_3^-}{O}}$$
$$+ (x - y)CH_3CN$$

The column (stainless, 15 × 0.48 cm i. d.) was washed with 5 mL of isopropanol and equilibrated with pentane (50 mL). It is assumed that about 15% of the silanol sites are covered with silver ions. An example of separation of a heavy gasoline sample prepared from fluid catalytic cracking (FCC) (bp < 280° C) is reported in Fig. 4.

Another one-column procedure was used for HGTA by Hayes and Anderson [8]. An "olefin-selective" HPLC column (15 × 0.46 cm i. d.) was prepared by in situ flushing silica (5-μm particles) with aqueous silver nitrate. A silica precolumn adsorbs polar heteroatomic compounds present in the samples. Freon 123 (2,2-dichloro-1,1,1-trifluorethane) was used for the mobile phase. The most significant feature of the procedure is the use of a dielectric constant (DC) detector to measure small changes in the dielectric constant of the liquid steam eluting from the column.

The predetermined time for backflushing was set at 200 s (flow rate, 1.0 mL/min), a sufficient time to obtain the elution of any three ring PAHs prior to the backflush. Under these conditions, the chromatographic run was complete in 8

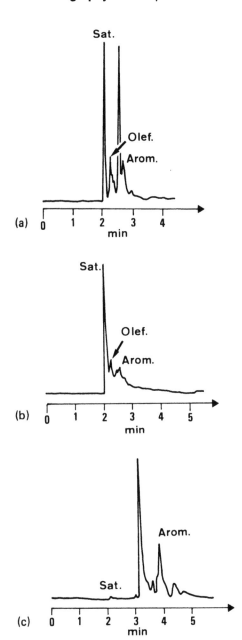

FIGURE 3   SCF chromatograms of gasoline (3a), shale oil (3b), and creosote oil (3c). Saturates are attenuated 8×. Conditions: columns (see the text); column temperature, 35°C; inlet pressure, 3600 psi; outlet pressure, 2600 psi; mobile phase, 3 mL/min of dry carbon dioxide; FID flows, 120 mL/min hydrogen, 600 mL/min air. (From Ref. 9a.)

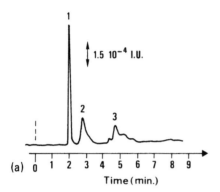

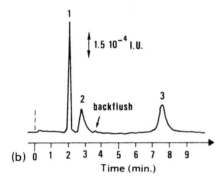

FIGURE 4 Separation of a heavy FCC gasoline sample on a silver(I)-coated silica gel. Column, 15 cm × 0.48 cm i.d.; stationary phase, Lichrosorb Si60 modified by percolating a 3% (w/w) solution of silver nitrate in acetonitrile; mobile phase, pentane (10 ppm water): (a) Without backflush; (b) with backflush; detector, differential refractometer (Water R 401) peaks (1) = S; (2) = 0; (3) = PAHs. (From Ref. 14.)

min, and only 2 min of requilibration is necessary before another sample injection. As shown in Fig. 5, the order of elution in this case was S < PAHs < O.

The method was tested on a complex "simulated" gasoline mix (about 400 compounds) and on a wide range of distillate products evaluated both by HPLC-DC and FIA (including motor gasolines, jet fuels, catalyzed cracked naphthas, and diesel fuels). The great majority of model polar compounds tested are either irreversibly adsorbed on the precolumn or eluted outside the retention windows of the hydrocarbon group types.

The authors concluded that this new HGTA procedure is ideal for the simple, accurate, and rapid determination of hydrocarbon groups in a wide range of fuel

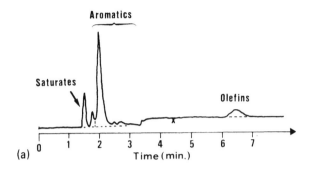

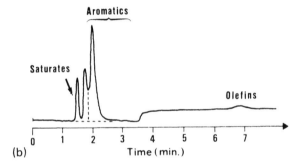

FIGURE 5 HPLC-DC profile of hydrogenation sequence of a light pyrolysis fuel oil: (a) pyrolysis fuel oil charge; (b) hydropyrolized fuel oil. It can be seen that after hydrostabilization there is a significant increase in the alkylbenzene section of the aromatics. This gain is matched by the loss in the olefinic content. (From Ref. 8.)

distillates without operator intervention. Use of unit response factors gave analytical results within 1% absolute for a complex standard solution. Furthermore, the method seems particularly suitable to on-line process stream analysis.

2. HGTA by Ag–HPLC on Silver-Bonded Silica Gel Columns

We previously reported the HGTA by Chartier et al. [14] using Ag–HPLC on silver-coated silica gel column. This group described in the same way [14] the utilization of column with silver chemically bonded to silica. In this case, a stainless steel column (25 × 0.48 cm i. d.) packed with Lichrosorb Si60 was equilibrated in pure water. A 3% w/w (0.176 M) solution of silver nitrate in 0.5 M aqueous ammonia was pumped at a flow rate of 1 mL/min. The column was washed successively with water (15 mL), isopropanol (10 mL), and chloroform (300 mL), then dried under nitrogen at 160° C for 8 h to remove ammonia. After this treatment, the column was equilibrated with 50 mL of pentane at a flow rate of 1 mL/min.

The role of the aqueous ammonia solution is to make easier the exchange between hydrogen and silver ions on the silica gel surface, probably as follows:

$$\equiv SiOH + Ag(NH_3)_2 \qquad\qquad \equiv SiOAg(NH_3) + NH_4 \qquad\qquad (2)$$

and

$$\equiv SiOH + NH_3 \qquad\qquad SiONH_4 \qquad\qquad (3)$$

It is assumed that about 30% of the silanol groups are bonded to silver. Used on the same FCC gasoline sample, this procedure gave a chromatogram similar to the one reported in Fig. 4.

Another type of silver-bonded silica (Lichrosorb Si60) was described by Felix et al. [15], who reported the successful synthesis and use of a new bonded acceptor stationary phase which ligands a silver salt. Starting with 4-aminophtalonitrile, the following ligand was synthesized:

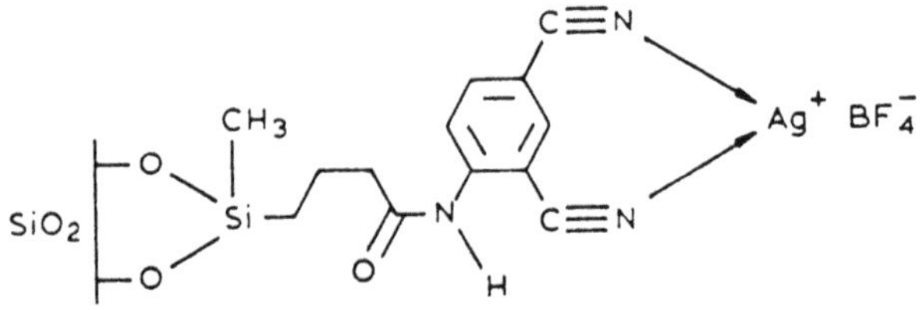

After packing, the column was washed and conditioned [15]. It was tested on 24 model PAHs, with hexane as mobile phase (flow rate, 1 mL/min). As foreseen, with a charge-transfer reaction mechanism, the retention time increases with the number of aromatic nuclei. The organometallic silver complex is very stable toward water, even if the eluants are saturated with water, giving a lifetime at the column of several months. Nevertheless, this new phase is incompatible with usual HPLC solvents such as dichloromethane.

In conclusion, this silver-bonded silica gel phase permits a complete HGTA, as can be seen from the chromatograms reported in Fig. 6, except when fuel samples contain PAHs with three or more rings, because they coelute with the olefins.

B. Application of Charge-Transfer Chromatography

During the last 40 years many studies have been undertaken to characterize polycyclic aromatic compounds (PAC) in complex mixtures of various origins. Bartle et al. [5] reviewed the modern analytical methods developed until 1981 for environmental PAHs, principally because of the carcinogenic and /or mutagenic properties of some PACs (hydrocarbons or heterocycles). In addition, the renewal of interest in research and development in the coal chemistry since the 1970s, and the increasing interest in heavy petroleum distillates and residues, gave rise to many analytical procedures suitable to the characterization of heavy coal and pe-

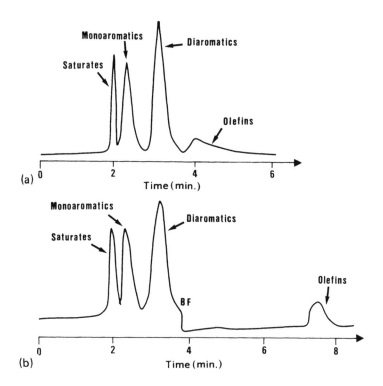

FIGURE 6 Separation of a catalytic gasoline (6a); of a catalytic gas oil (6b) with back-flushing. Mobile phase, hexane; flow rate, 1 mL/min; detection, differential refractometer. (From Ref. 15.)

troleum products. Many examples can be cited, such as the evaluation of catalytic hydroliquefaction runs, the determination of the influence of carbonization conditions on coal tar and pitch compositions, and the evaluation of the role played by the pitch constituents in mesophase formation. Furthermore, all upgrading processes require knowledge of the feedstock chemical composition.

As already pointed out, the PAH structures are extremely complex because they contain condensed structures from two to seven or more nuclei, often partially hydrogenated, more or less alkylated, in addition to heterocyclic compounds, hydroxylated PAC, and more complicated oligomeric structures. At last, the samples are often only partially soluble in the solvents used in chromatography and are generally of low volatility.

According to the complexity of the samples, several complementary methods are used in procedures of varying difficulty, and some of them are based on charge-transfer complexation (CTC). Zander et al. [16–20] developed the *charge-transfer fractionation* (CTF) of pitches by complexation with picric acid or with iodine. Although this procedure does not involve charge-transfer chromatography, it is an interesting way to separate the constituents of complex mixtures of PAHs, such as coal tar or pitches, according to their first ionization potential and have proved a useful means of studying the thermal reactivity of coal tar materials [18]. The same procedure has been carried on by Yokono et al. [21,22].

The methods reported in Chapter 3 showed that three groups of electron attractors were used most often in CTC, with PAC acting as electron donors: nitroaromatics, tetrachlorophtalimido groups, and caffeine (or related derivatives). Most of these methods have applications in petroleum and coal product analysis.

1. Separation of PAC with Nitroaromatics Acceptors

The first example of CTC applied to petroleum product analysis is perhaps the method proposed as early as 1949 by Godlewicz [23], reported here only from a historic point of view. Trinitrobenzene is proposed as an useful indicator in the chromatographic separation (in four fractions) on silica gel of gasolines, lubrificating oils, and other petroleum distillates and residues. A relationship was established between the fraction constitutions (saturates, partially hydrogenated aromatic rings, aromatics, and resinous compounds) and the color developed on the column. The author claimed that his method permits a rapid and quantitative estimation of both the most valuable and undesirable constituents contained in various lubricating oils.

The following two examples illustrate old studies using CTC combined with chromatographic separation but are not really considered as CTC chromatography. Extractable (Soxhlet-methanol/benzene) environmental PAHs were isolated [24] by means of successive chromatographies, the last of them being carried out on alumina/silica gel. The PAHs were separated from the methylene dichloride eluate by addition of 2,4,7-trinitroflurenone (TNF). This tedious procedure was followed by percolation of the complex (in CH_2Cl_2) through a silica gel bed achieved by other fractionations of the PAH concentrates. Nevertheless, this method was the first that resolved any fossil-fuel contribution to the PAH fraction of sedimental samples, with PAH concentrates in the microgram to milligram range. A somewhat similar procedure was described by Jewell [25], performing preparative TNF charge-transfer chromatography of aromatics, applied to petroleum residues, shale oils, and coal liquids. Another approach to CTC chromatography applied to mixtures originating from petroleum, developed by Tye and Bell [26], was reported in Section II.A.3 or Chapter 3 for model compounds. The results of all of these studies [23–26] permit an evaluation of the progress realized in this particular field of analytical chemistry during the last decade as well by liquid chromatography and high-performance liquid chromatography.

Liquid chromatography. The separation of hydroaromatic compounds from aromatics by picric acid on dual alumina columns [27–30] was introduced in Chapter 3, where the principle and the experimental procedure were described for model compounds. We shall consider now some applications to coal industrial liquid synfuels [27,28] and hydroliquefaction run products carried out in batch experiments at a laboratory scale [29,30].

The evaluation of hydroaromatics is interesting from several points of view. As Wozniak et al. [28] suggested, one solution to the problem caused by the mutagenic properties of coal-derived synfuels could be their catalytic hydrogenation, which reduces the level of PAHs mutagens but increases the level of hydroaromatics (up to 80% of the fuel). Therefore, knowledge of the composition of hydroaromatics is important for upgrading development and for the production of safe environmentally synfuels. On the other hand, these hydroaromatics have well-known hydrogen-donor properties and are important constituents of recycling solvents in hydroliquefaction processes. A better understanding of the composition of hydroaromatics according to the experimental conditions (autoclaves, temperature, catalysts, etc.) could be used to improve the yield of hydroliquefaction processes. The major hydroaromatic components in fuels from a solvent-refined coal process (SRC II), from hydrotreatment of an H-coal raw distillate, and from an integrated two-step liquefaction (ITSL) process were characterized by Wozniak and Hites [27,28]. They used the picric acid–coated alumina method for the separation, and high-resolution gas chromatography (CC) and gas chromatography/mass spectrometry (GC–MS) for the identification and quantification, with reference to data derived from 139 hydroaromatic standards [31].

Several important conclusions concerning the liquefaction process chemistry can be drawn from these studies. Nevertheless, we focus our attention here on the characteristics of the chromatographic procedure for hydrotreated H-coal products.

1. Aliphatic hydrocarbons and alkybenzenes elute immediately from the column with hexane in fraction T_1 (see Chapter 3, section II.A.1), as picrate formation occurs only with hydrocarbons more condensed than naphthalene.
2. The hydroaromatic fraction, T_2, is relatively free of PAHs except for fluorene, a major component of the synfuel that elutes in both the T_2 and T_3 fractions.
3. The PAH fraction, T_3, contains fluorene, phenanthrene, pyrene, and their alkylated analogs. The major constituents have four or fewer nuclei. The only hydroaromatic compound to elute in this fraction was 4,5-dihydropyrene. The most important fact is that the less aromatic among the hydroaromatic compounds elute as expected in the hydroaromatic (T_2) fraction, while the more aromatic ones are retained by picric acid like PAHs, and elute in the T_3 fraction.
4. The remaining PAHs, with four or more nuclei, and the polar compounds were eluted from the column with benzene (the T_4 fraction).

Diack et al. [29,30] applied the procedure described in Chapter 3 for model compounds to a hydrogenated anthracenic oil (H-AO) and to several catalytic hydroliquefaction products. For the H-AO sample, Table 1 summaries the structures identified in the hydroaromatic fraction T2 (75% of the initial H-AO) by GC–MS, their retention indexes, and their concentrations established by GC analysis. In addition to 26% of unidentified structures, about 34% of PAHs were eluted with hydroaromatic compounds (40%), primarily anthracene/phenanthrene, fluorene, and alkylated naphthalenes. On the other hand, the aromatic fraction, T3 is constituted primarily of three- and four-ring PAHs and of some hydroaromatic structures with a high aromatic nuclei content, such as dihydropyrene. Regardless of these partial separations between aromatics and hydroaromatics in complex mixtures, some interesting results were obtained for hydroliquefaction products performed at a laboratory scale [32,33]. Starting from a bituminous coal (Freyming, France), the picric acid method was applied to the aromatic fraction of the hydroliquefaction product which was first separated from saturated and polar compounds by chromatography on alumina. The T2 fraction (26% of the aromatic fraction) and the T3 fraction (18% of the aromatic fraction) were submitted to GS–MS analyses. The chromatograms are shown in Figs. 7 and 8. Regardless of the presence in the T2 fraction of aromatic polycyclic hydrocarbons (retention times between 14 and 38 min), the efficiency of the picric acid method is well demonstrated by the useful separation of compounds eluting between 40 and 50 min. Indeed, the main structures identified by mass spectroscopy are principally issued from the -methylnaphthalene used as solvent in the hydroliquefaction experiments.

These structures are of two types: derivatives of dihydro benzo[f]- or benzo[k]fluoranthenes and derivatives of binaphtyle [30].

High-performance liquid chromatography on chemically bonded NO2 phases. The characterization of tars and pitches from coal pyrolysis was studied exhaustively by Zander [34] and his collaborators [35-37]. They showed [34,35] that HPLC can be applied to the analysis of PAHs with molecular weights up to 600 (about 70% of classical high-temperature pyrolysis pitches). As it is advantageous to have a specific group separation between PAHs and nitrogen-containing compounds (PANH), they showed [36] that this can easily be achieved by using a charge-transfer type of stationary phase, such as chemically bonded NO2 phases (Nucleosil-5 NO2 5 μm from Macherey-Nagel). Different linear relationships between the log retention (log k') and carbon number were observed [34,36] for three groups of coal tar components: biaryls, PAHs, and PANHs (see Fig. 9) (more than 100 components of coal tars). It can be seen that hydrocarbons (biaryls and PAHs) can be group-specifically separated from PANHs up to C numbers of approximately 30. An application is shown in Fig. 10 [34,35]. A coal tar pitch (softening point 70° C) was analyzed by HPLC using the NO2 phase (mobile phase n-hexane/chloroform). The main peaks were identified by their retention indices and by

TABLE 1   Structures Identified in the T_2 Fraction of Hydrogenated Anthracene Oil

Peak	Compound name	Molecular weight	Retention index	Concentration (%)
1	Tetralin	132	196.62	0.84
2	Napthalene	128	200	1.13
3	2-Me-tetralin	146	207.22	0.18
4	1-Me-tetralin	146	208.66	0.08
5	5-Me-tetralin	146	216.48	0.30
6	2-Me-naphthalene	142	220.97	0.93
7	1-Me-naphthalene	142	222.98	0.52
8	Bicyclohexane	166	223.53	0.05
9	C_2-tetralin	160	225.73	0.30
10	C_2-tetralin	160	227.76	<0.04
11	C_2-tetralin	160	227.06	0.13
12	C_2-tetralin	160	227.76	<0.04
13	C_2-tetralin	160	228.53	0.05
14	C_2-tetralin	160	230.54	0.05
15	C_2-tetralin	160	230.94	<0.04
16	C_2-tetralin	160	233.48	0.10
17	Tetrahydro acenaphthene	158	236.09	1.10
	+ biphenyl	154	236.09	2.20
18	Me-bicyclohexane	180	238.98	0.65
19	C_2-naphthalene	156	241.44	0.29
20	C_2-tetralin	160	243.33	0.40
21	C_2-naphthalene	156	244.15	0.20
22	C_2-naphthalene	156	244.52	0.15
23	C_2-naphthalene	156	245.33	0.62
24	Me-cyclohexyl benzene	174	246.36	0.12
25	Me-tetrahydroacenaphthene	172	248.85	0.09
26	Me-tetrahydroacenaphthene	172	249.51	0.12
27	Acenaphthene	154	253.13	1.54
28	di-Me-biphenyl	182	253.84	0.51
29	Me-biphenyl	168	254.48	0.49
30	di-Me-biphenyl	182	255.42	<0.04
31	Tri-Me-naphthalene	170	255.92	<0.04
32			258.04	<0.04
33	di-Benzofuran	168	259.20	3.63
34	Tri-Me-naphthalene	170	261.40	<0.04
35	Tri-Me-naphthalene	170	262.72	0.07
36	Tri-Me-naphthalene	170	264.37	0.05
				0.36
41	Fluorene	166	270.04	4.36

TABLE 1   (continued)

Peak	Compound name	Molecular weight	Retention index	Concentration (%)
42	Me-dibenzofuran	182	272.20	0.68
43	Octahydrophenanthrene	186	273.31	0.94
44	Octahydrophenanthrene	186	274.09	0.74
				2.23
50	Dihydrophenanthrene	180	286.67	5.72
51	Dihydroanthracene	180	287.58	1.17
52	Dihydro-Me-phenanthrene	194	288.85	0.80
53	Tetrahydro-Me-phenanthrene	196	291.39	0.28
54	Octahydrophenanthrene	186	293.07	1.57
55	Tetrahydrophenanthrene	182	296.13	3.88
56	Decahydrofluoranthene	212	297.18	0.37
56	Anthracene/phenanthrene	178	300	11.80
58	Dihydro-Me-phenanthrene	194	304.16	0.32
				2.39
62	Me-phenanth/anthracene	192	316.03	0.37
63	Me-phenanth/anthracene	192	318.97	1.83
64	Me-phenanth/anthracene	192	321.20	0.77
65	Me-phenanth/anthracene	192	322.35	0.82
66	Tetrahydrofluoranthrene	206	328.89	8.57
67	Phenylnaphthalene	204	330.62	0.73
				1.26
71	Dihydropyrene	204	342.63	2.84
72	Fluoranthene	202	343.97	0.53
73	Dihydrofluoranthene	204	346.01	
74	Me-phenylnaphthalene	218	348.12	
75	Pyrene	202	351.26	
76	Dihydropyrene	204	353.93	
79	Tetrahydrobenzofluorene	220	362.59	
80	Tetrahydrobenzofluorene	220	363.95	
83	Hexahydrobenzanthracene	234	376.59	
84	Hexahydrobenzanthracene	234	380.47	
85	Dihydrobenzanthracene	230	383.23	
86	Dihydrobenzanthracene	230	385.05	
88	Octahydrochrysene	236	391.28	
89	Octahydrotriphenylene	236	393.65	
90	Tetrahydrochrysene	232	296.63	

$$\sum\nolimits_{73-90} = 7.60$$

$$\sum = 80.14$$

Source: Ref. 29 and 30.

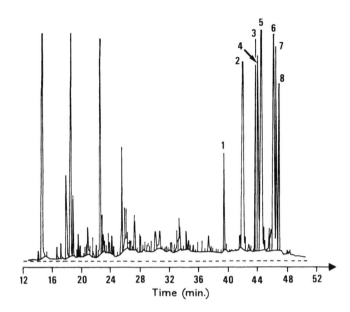

FIGURE 7 Hydroliquefaction sample (batch autoclave catalyst Fe_2O_3: solvent, 1-Me-naphthalene; pH_2 = 14 MPa; t = 450°C (1 hr); heating rate, 3°C/min). Gas chromatogram of the T_2 fraction. Peak 1 (m/z = 254), DH-BF*; peaks 2, 4, 5 (m/z = 268), Me-DH-BF; peak 3 (m/z = 268), Me-binaphthyle; peaks 6–8 (m/z = 282), di-Me-DH-BF.
*DHBF, dihydrobenzofluoranthene. (From Ref. 29 and 30.)

measuring the UV adsorption spectra of the separated substances [37]. PANHs appeared at high retention times.

Lankmayer and Müller [38] compared Nucleosil-5 NO2 (from Macherey-Nagel), Lichrosorb RP-18, and Lichrosorb NH2 for the separation of PAHs commonly found in dust sample extracts. They proved that the nitrophenyl phase provided the best separation and that nanogram amounts of PAHs can easily be detected by using a UV detector. Dust samples from the ventilation system of the Katschberg tunnel (Tauernautobahn, Salzburg, Austria) that had been exposed to automobile exhaust gases were taken as typical specimens of PAHs containing materials.

Several commercially available chemically bonded phases (Nucleosil 5μm NO2, NH2, CN, and sulfonic acid) were also compared by Matsunaga [39]. Nucleosil-5 NO2 gave the largest capacity factors k' and the largest selectivity for PAHs. Two examples are shown in Figs. 11 and 12. Figure 11 is a chromatogram of a coal tar, obtained on a Nucleosil-NO2 column. A clear separation was achieved in the aromatic region. A relatively good separation of PAHs was also obtained in a

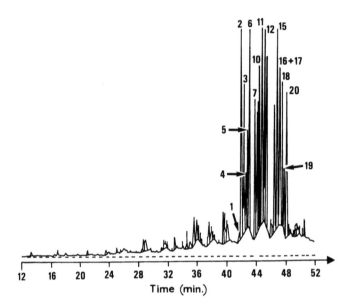

FIGURE 8 Hydroliquefaction sample (conditions as in Fig. 7). Gas chromatogram of the T₃ fraction. Peaks 1, 3 (m/z 268), Me-DH-BF*; peaks 2, 4, 12, 13, 20 (m/z = 268), Me-binapthyle; peak 5 (m/z = 254, 2, 2′-binaphtyle; peaks 6, 14–19 (m/z 282), di-Me-binaphtyle; peak 7–11 (m/z = 282), di-Me-DHBF.
*DH-BF dihydrobenzofluoranthene. (From Ref. 29 and 30.)

heavy petroleum distillate (Fig. 12). Nevertheless, overlapping of aromatics and polar compounds was observed in shale oils and petroleum fractions. Generally, the resolution was poorer for heavier fractions.

Besides Nucleosil-5 NO₂, the outstanding suitability of some of the charge-transfer bonded phases (listed in Tables 7 and 8 of Chapter 3) in the separation of higher-boiling aromatic compounds of petroleum distillates was established by Eppert and Schinke [40]. From 3-(2,4,5,7-tetranitrofluorenimino), 3-(2,4-dinitrophenyl amino) (DNAP), and 3-(2,4,6-trinitrophenyl amino) (TNAP) propylsilyl phases bonded to spheric silica gel, the picramyl derivatives (TNAP) was the preferred acceptor applied to the analysis of diesel fuels prepared in several technological conditions. The conditions of the separation of PAHs in subclasses (according to the number of aromatic nuclei) as well as the influence of the eluant on the selectivity and the behavior of heterocompounds were discussed.

The DNAP and TNAP phases, also studied by Thompson et al. [41,42], in comparison with their homologs DNAO and TNAO (see Tables 8 and 9 of Chapter 3) were applied to the separation, according to aromatic ring number, of liquid fossil fuels (petroleum distillates, coal liquids) and shale oils. The cases of a SRC II proc-

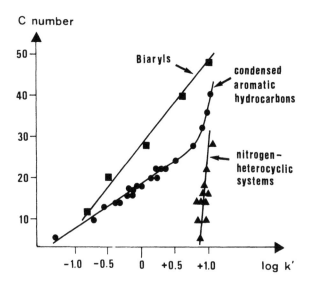

FIGURE 9   Diagram showing the relationship between HPLC retention behavior and molecular size of different classes of compounds (NO_2 phase). (From Ref. 34.)

ess coal liquid and of a petroleum distillate are shown in Figs. 13 and 14. The basic compositional differences between the aromatic compounds contained in petroleum and coal liquids account for the efficiency of the separation: the coal liquid sample is composed essentially of one- to three-ringed derivatives with short alkyl groups. On the other hand, the alkyl groups of petroleum products are of significantly longer length than those in coal liquids. Naphthenic structures are also in larger amounts. The percentages of each compounds classes were calculated for each column after measuring their adsorbtivity using an UV spectrophotometer at 254 nm. The results were compared with gravimetric data obtained from a preparative open-column method (alumina). The DNAP column results are closest to those obtained with preparative gravimetric techniques.

A solvent-refined coal sample was analyzed by Welch and Hoffman [43] by using their 2,4-dinitrophenylmercaptopropylsilica phase. Four main peaks were separated: A, monoaromatics such as tetralines, indanes, and alkylbenzenes; B, diaromatics such as naphthalenes and biphenyles; C, diaromatics such as fluorenes and dibenzofurans; and D, : triaromatics, such as anthracenes.

By using a dielectric constant detector (see Section II.A.2), two Partisil PAC columns, and one tetranitrofluorenimino (TENF) column, Hayes and Anderson [7] carried out an HGTA of hydrocarbon distillate samples. The PAC columns sepa-

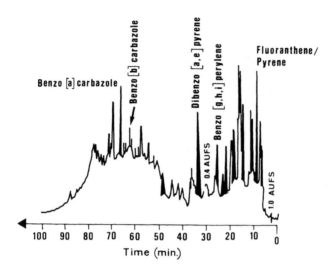

FIGURE 10 HPLC chromatogram of a coal tar pitch (Nucleosil-5 NO₂ phase). Columns 200 mm × 4mm i.d.; 20 mm × 4mm i.d.; mobile phase, *n*-hexane (chloroform gradient elution); UV detection, 300 nm. (From Ref. 34.)

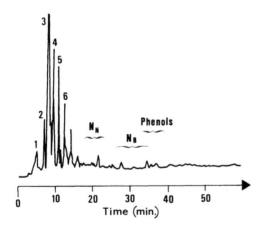

FIGURE 11 HPLC of a coal tar (Nucleosil-NO₂). Gradient elution from dried hexane to dried methylene chloride at 2%/min, 1 mL/min. "N_N," "N_B," and "Phenols" indicate the regions where nonbasic nitrogen, basic nitrogen compounds, and phenols should appear. Peak identification by UV spectrum: (3) phenanthrenes, (4) pyrenes, (5) chrysenes, (6) benzo [a] pyrenes. (Form Ref. 39.)

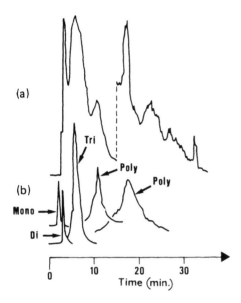

FIGURE 12   HPLC separation of heavy petroleum distillate (Kuwait 350–500°C distillate) (Nucleosil-NO$_2$). Gradient elution, hexane (10 min) and then to methylene chloride, at 2.5% min. (a) Original sample; (b) fractions obtained by preparative alumina chromatography of (a). (From Ref. 39.)

rated the saturates from aromatics, and the addition of the TENF column improved the resolution between the alkybenzene and the alkyl naphthalene groups. *N*-Butyl chloride was selected for the mobile phase. The authors point out the adaptability of the method to on-line analysis, which makes it very attractive to facilities involved in the hydroprocessing of petroleum and synthetic feedstocks.

2. Determination of Polycyclic Aromatic Compounds in Petroleum and Coal Products on a Tetrachlorophtalimido-propyl–Bonded Silica

The characteristics of the tetrachlorophtalimidopropyl–bonded silica phase, used for the separation of PAHs, were discussed in Chapter 3. Several applications to the analysis of petroleum and coal liquids are discussed below.

Petroleum products.   The separation of PAHs according to their aromatic nuclei number was described by Holstein for German crude oils [44,45], gas oils, and hydrogenated gas oils [45]. More recently, the levels of PAHs in oils were determined by Jadaud et al. [46] using a three-step process: extraction and concentration by CTC on an improved TCI phase [47], oxidation of thiophenes, followed by the

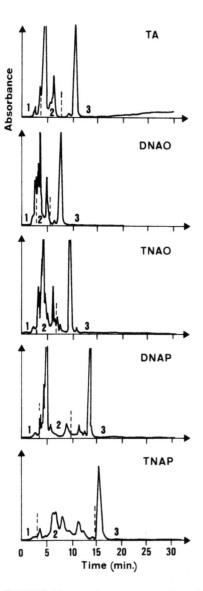

FIGURE 13 Comparative separation of an SRC II process coal liquid 200 to 325°C hydrocarbon sample according to aromatic ring number. TA, triamine silica–DNAO = (8-(2,4-dinitroanilino)octyl)silica; TNAO, (8-(2,4,6-trinitroanilino)octyl)silica: DNAP (3-(2,4-dinitroanilino)propyl)silica; TNAP: (3-(2,4,6-trinitroanilino)propyl)silica (From Ref. 42.)

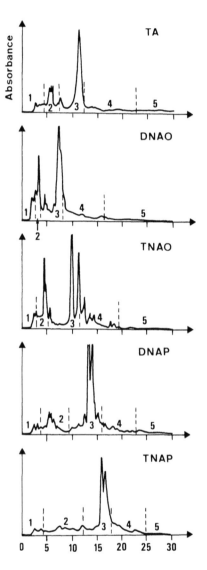

FIGURE 14 Comparative separation of a Tosco process shale oil hydrocarbon sample according to aromatic ring number. For the legend of the phases, see Fig. 13. (From Ref. 42.)

separation of the resulting sulfones by adsorption chromatography on bare silica. Indeed, the determination of the concentration of PAHs in oils on TCI requires the separation of PAHs from heterocycles, especially thiophenes, which are strongly retained by CTC. The separation of PAHs was achieved by CTC on TCI with iso-octane/methylene chloride as the mobile phase. The method described [46] has been applied to the measurement of the PAH concentration in a series of new, used, and regenerated lubrificating oils.

Although the selectivity was higher in CTC, the resolutions were comparable and the duration of the analysis was shorter using reversed-phase liquid chromatography. However, as the solubility of PAHs is very low in an aqueous medium, particularly in the case of real samples, CTC was preferred by the authors [47] for efficient and reproducible separation of PAHs. An example of the separation of PAHs in an used lubrificating oil, before and after the oxidation of thiophenes, is shown in Fig. 15.

Coal products. Several examples of the application of TCI phase on coal product analysis were provided by Holstein et al. in case of coal-derived recycle oil [48-50] anthracenic oil, shale oil, pitches (softening points 35°C and 90;C) [44], and bituminous coal liquefaction products [50]. Some typical chromatograms are shown in Figs. 16 and 17.

3. Separation of Polycyclic Aromatic Compounds in Petroleum and Coal Products by CTC with Caffeine and Its Derivatives

As early as 1965, the TLC separation of model PAHs on silica gel plates impregnated with caffeine, described by Lam and Berg [51], was applied to an untreated coal tar sample. From the positions and fluorescence colors of the spots, after five runs, compared with reference compounds, and from the adsorption spectra of the extracted spots, the following PAHs were identified and quantified (in weights percent of the tar): anthracene 0.21; pyrene, 0.58; 3,4-benzopyrene, 0.19; perylene, 0.05; and 1,12-benzoperylene, 0.12. Interestingly, about 25 years later, the same procedure, using the advantages of modern technology, was applied by Funk et al. [52] to the detection of selected carcinogen PAHs in drinking water.

The HPLC separation of petroleum asphaltenes on grafted silica (Lichrosorb Si60 10 μm) with caffeine-type ligands (theobromine and theophylline) was performed by Felix and Bertrand [53]. Using a new phase, 7-(n-propyl)theophylline silica [54], the same group [55,56] studied petroleum residues (a Safanya vacuum residue, asphalts, and asphaltenes) from different crude oils [56]. A comparison of the results obtained from several samples by using UV detection at different wavelengths allowed the authors [56] to propose a structural model for asphaltenes: "a cobweb structure" composed with condensed aromatic nuclei (no more than six to seven) and alkyl chains. The molecules, with a low content of naphthenes and a great amount of condensed aromatic nuclei, give a strong charge-transfer complex with caffeine and have the greatest retention time, and vice versa.

As in the case of CTC-TCI, the advantage of the caffeine-bonded phase over reversed-phase packing (C18), with better resolving power, is the utilization of nonaqueous mobile phase (hexane/CH2Cl2). Then more than 90% of the sample can be detected.

Another case of complexation chromatography, used for HPLC separation of PAHs in shale oils, was cited by Mourey et al. [57]. A chemically bonded pyrrolidone phase [chlorodimethyl-3-(N-pyrrolidone)propylsilane, prepared from chlorodimethylsilane and N-allylpyrrolidone] interacts electronically with the PAHs in a fashion not yet elucidated. The separation is effected according to the number of aromatic rings and the type of ring annelation, in both normal- and re-

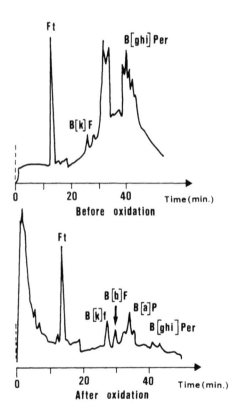

FIGURE 15 Separation of PAHs in a used lubricating oil before and after oxidation of thiophenes. Column 25 cm × 4.6 mm i.d.; stationary phase, tetrachlorophtalimidopropyl-bonded Lichrosorb Si100 (10 μm); ligand density, 2.7 μmol/m²; mobile phase: 0–13 min, isooctane/methylene chloride (70:30 v/v); 13–23 min, 70:30 to 50:50 (v/v); 23–37 min, 50:50 to 0:100 (v/v). Solutes: Ft, fluoranthene; B(k)F, benzo [k] fluoranthene; B(b)F, benzo [b] fluoranthene; b(a)P, benzo [a] pyrene; B(ghi)Per, benzo [ghi] perylene. (From Ref. 46.)

versed-phase modes. The phase can be used to give a profile of PAHs in shale oil samples.

III. SEPARATION OF HETEROATOMIC COMPOUNDS

A. Sulfur Compounds: Sulfides and Sulfur Heterocycles

Sulfur compounds are present to some extent in all fuels, especially in petroleum crude oils. The main classes of sulfur compounds are polycyclic aromatic sulfur heterocycles (PASHs) and aliphatic sulfides or alkyl aryl sulfides. Since the

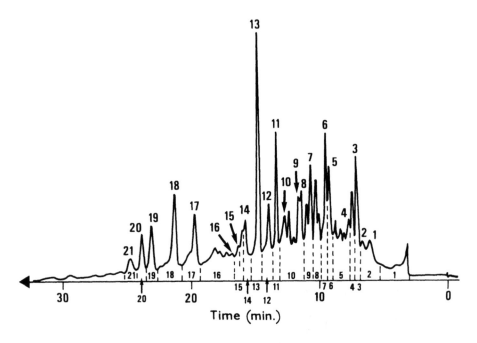

FIGURE 16   HPLC chromatogram of recycle oil. Numbers on the baseline denote fractions separated; numbers on the peaks as follows: Peaks: *1*, 3,3'-di-Me-B and (or) 3,4-di-Me-B; *2*, 4,4'-di-Me-B *3*, 9,10-dihydro-P; *4*, 1,2,3,6,7,8-hexahydro-Py; *5*, 1,2,3,10*b*-tetrahydro-Ft; *6*, dibenzofuran; *7*, F; *8*, 1,2,3,4-tetrahydro-P; *9*, 1,2,3,4-tetrahydro-A; *10*, dibenzothiophene; *11*, 9,10-dihydro-Py; *12*, 9,10-dihydro-Me-Py; *13*, P; *14*, 2 and (or) 3-Me-P; *15*, 1-Me-P; *16*, 3,6-di-Me-P; *17*, Ft; *18*, Py; *19*, 4-Me-Py; *20*, carbazol; *21*, 1-Me-Py. Me, methyl; A, anthracene; F, fluorene; P, phenanthrene; B, biphenyl; Ft, fluoranthene; Py, pyrene. Chromatographic conditions: column, 3 ×TCI, 25 cm ×4.1 mm i.d., 3 cm^3/min; linear gradient from 100%/*n*-hexane to 100% methylene chloride in 30 min; UV detector, 280 nm. (From Ref. 49.)

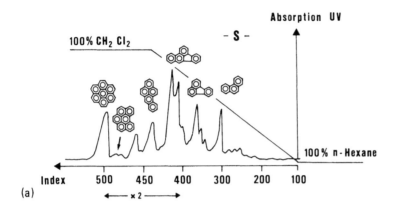

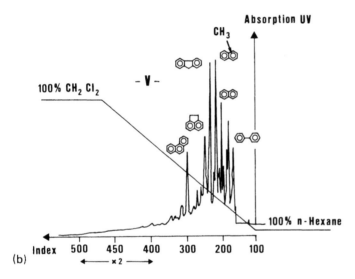

FIGURE 17   HPLC chromatograms on a TCI phase of two recycle oils, previously fractionated by silica–alumina chromatography. (a) Sample S (from hot separator of a coal hydroliquefaction pilot plant) (Saarbergwerke, Furstenhausen, RFA). (b) Sample V (from vacuum distillation of the same process). (From Ref. 50.)

PASHs are more thermally stable than the aliphatic sulfur compounds [58], PASHs are generally present in higher amounts than aliphatic sulfides in thermally treated heavy petroleum oils (and coal liquids).

Knowledge of the sulfur components in petroleum is of both theoretical interest and practical value to the petroleum industry, taking into account their role in catalytic processing (they are deleterious in oil refining because they poison catalysts and cause corrosion) and environmental problems (they give noxious sulfur dioxide emission if used directly in fuel and could themselves be mutagens and/or carcinogens). Then the problem of separating sulfur compounds from the hydrocarbons and other materials in petroleum received much attention for many years. Many separation methods were proposed, often tedious and not giving satisfactory results for the higher-boiling fractions of crude oils. Some of these methods involve complexation chromatography, by complexation of the sulfur lone pair of electrons with metal salts. They are classified hereafter according to the type of separation carried out.

1. Separation of Mercaptans, Organic Sulfides, and Disulfides from PAHs

In 1966, the separation of organic sulfides occurring petroleum, petroleum fractions, and extracts of bituminous rocks, even at low concentrations, was described by Orr. Using liquid–liquid chromatography on mercuric acetate [59] or aqueous zinc chloride [60], alkyl and cycloalkyl sulfides were completely separated from hydrocarbons, thiophenes, thiols, and aromatic sulfides. At the same time, Snyder [61] used a mercuric ion-impregnated cation-exchanged resin to the removal of sulfides during the analysis of nitrogen and oxygen compounds in petroleum distillates.

An extensive review of the separation methods for sulfur compounds from PAHs has been presented by Rall et al. [62], but for the higher-boiling fractions of crude oils, even the best method [62] based on the formation of sulfonium halides was not suitable for the separation of alkyl aryl and diaryl sulfides from PAHs.

The "soft acid" character of organic sulfides (see Chapter 1) has been used [63] for their separation from PAHs by liquid chromatography on a copper-loaded carboxylic cation-exchange resin (Bio-Rex 70, 200 to 400 mesh from Bio-Rad, slurried with a solution of copper sulfate). The sulfides strongly adsorbed from n-pentane solution were recovered from the column, without loss, after rapid elution of PAHs, by backwashing with a mixture of pentane and ethyl ether. Sulfides were isolated by this procedure from petroleum aromatic concentrates that had been prepared by classical alumina–silica gel chromatography [i.e., aromatic concentrates from high-boiling (370 to 335°C) distillate of Wilmington CA crude oil].

Another example is the separation of mercaptans, sulfides, disulfides, and dibenzothiophene from PAHs by thin-layer chromatography on mercury- or silver-loaded silica gel plates, with n-hexane as developing solvent. Mercuric acetate and silver nitrate were found effective for class separation of sulfur compounds [64]. Two-dimensional development (n-hexane, benzene) was carried out to attain a clear separation in three groups: mercaptans (strongly retained), sulfides, and dibenzothiophene (developed with anthracene). The method was applied to the analysis of transformer oil and aromatic concentrates of high-boiling distillates

(280 to 440°C) of Kuwait crude oil. The coloration of the spots sprayed with palladous chloride solution are characteristic of the sulfur compounds groups.

2. Separation of Sulfur Heterocycles from PAHs

As already mentioned in Chapter 4, the isolation of thiophenic compounds from aromatic shale oil fractions by argentation liquid chromatography on a silver nitrate–coated silica column was described by Joyce and Uden [65]. However, this method is limited to the separation of only one- and two-ring thiophens.

More recently, the isolation of sulfur heterocycles (PASHs) from petroleum and coal materials was carried out by Gundermann et al. [66] and Nishioka et al. [67]. Generally speaking, the identification of the PASHs in such complex mixtures necessitates their separation from PAHs. This is a difficult task, due to the similar properties of these two classes of polycyclic compounds and the much greater concentrations of the PAH group. Thus the procedures applied to real samples (coal or petroleum products) are somewhat tedious.

For example, the separation scheme for "organic coal sulfur" proposed by Gundermann et al. [66] involved the following sequential steps, starting with the coal sample: cyclohexane extraction, size exclusion chromatography (SEC), silica gel column chromatography applied to a mixture of selected SEC fractions, and finally, preparative thin-layer complexation chromatography on $PdCl_2$ (or $CuCl_2$)-impregnated silica gel. For example, dibenzothiophen can be separated from phenanthrene.

By using a column of silica gel impregnated with $PdCl_2$ (n-hexane 1:1 as eluting solvent), Nishioka et al. [67] analyzed the PAH/PASH mixtures, separated at first into chemical classes by adsorption chromatography on neutral alumina from two industrial samples: a solvent-refined coal heavy distillate (SRC II HD 260 to 450°C boiling-point range) and a catalytically cracked petroleum residue. The PAHs separated in the second fraction (P2) of the complexation chromatography were analyzed by GC and GC–MS, Some or all of the PASHs were eluted as Pd complexes. These complexes can be desulfurized during the GC analysis if the sulfur atom is not in an interior ring of the molecule [i.e., naphtho(2,3-b)thiophen gives 2-ethylnaphthalene]. This drawback can be avoided by the addition of diethylamine before GC analysis, as the amine, being a stronger Lewis base than the PASHs, forms a strong ligand with Pd and replaces PASHs in the complex. According to the examples shown in Fig. 18, the method is effective for the isolation of two- or six-ring PASHs from complex coal and petroleum liquids, and the fractionation time is only 30 min for the complexation step.

B. Nitrogen Compounds: Azaarenes

As in the case of sulfur compounds, the petroleum industry is concerned with the problem of the analysis (isolation, identification, and quantitation) of nitrogen compounds, although they represent only a small percentage of the heteroatomic

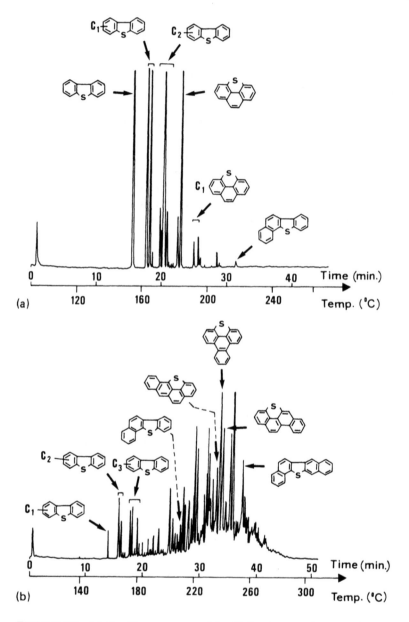

FIGURE 18 (a) Gas Chromatogram of the P2 fraction (see the text) of an SRC II HD. Conditions: 70–120°C at 10°C min⁻¹ and 120–265°C at 4°C min⁻¹ after an initial 2-min isothermal period; helium carrier gas. Flame photometric detector. (b) Gas chromatogram of the P2 fraction (see the text) of a catalytically cracked petroleum bottoms. Conditions: 70–140°C at 10°C min⁻¹ and 140–280°C at 4°C min⁻¹ after an initial 2-min isothermal period; helium carrier gas. Flame photometric detector. (From Ref. 67.)

derivatives and are in much smaller quantities than their sulfur analogs. Nitrogen compounds are more concentrated than sulfur compounds in coal liquid products. On the other hand, the drawbacks are the same for nitrogen and sulfur compounds as to their deleterious effects on catalysts and environmental pollution.

Nevertheless, as the utilization of petroleum distillates is concerned principally with the presence of sulfur compounds, and as the coal liquids are not yet competitive, few analytical methods were applied to nitrogen derivative detection in fossil fuels. The oldest methods are based on the well-known complexation with ferric chloride, and the most recent ones use the developments of CTC on selected bonded phases.

1. Complexation with Ferric Chloride

The selective complexation of nitrogen compounds by ferric chloride supported on a clay mineral kaolin was used by Jewell and Snyder [68] for the separation of neutral and nonbasic nitrogen compounds from petroleum. Since the nonbasic as well as the basic nitrogen compounds are capable of forming isolable complexes with ferric chloride, the method can be carried out in two ways: either total removal of all nitrogen compounds or selective separation between basic and neutral/nonbasic ones. In the latter case, the basic compounds must be eliminated first with mineral acids. Quantitative recovery of the nitrogen compounds from the complexes is best achieved using a strong anion-exchange resin of the quaternary ammonium hydroxide type. Nevertheless, no structural identification was undertaken in this study.

2. Application of Charge-Transfer Complexation on Bonded Phases

Some examples were cited in Chapter 3 but only for model compounds. The behavior of biaryls, condensed aromatics, and azaaromatics (PANHs) in HPLC using silica modified with nitrophenyl groups was investigated by Zander [34]. As already mentioned (Fig. 10) nitrogen-containing heterocycles of relatively low molecular weight (cf. benzocarbazoles) appeared at high retention times and are clearly separated from PAHs in a coal tar sample.

On-line preconcentration followed by liquid chromatography of azaarenes using 3-(2,4-dinitrobenzenesulfonamido) (DNSP) propylsilica and 3-(2,4-dinitroanilino) (DNAP) propylsilica was carried out by Nondek and Chvalovsky [69].

As PAHs with up three or four rings and weak nitrogen bases of the pyrrole and aniline classes are retained less than azaarenes, the use of these phases makes it possible to determine trace amounts of azaarenes in gasoline, kerosene, and diesel fuels. PANHs are retained reversibly in mobile phases of moderate polarity (dichloromethane) and their quantitative recovery can be achieved. On the other hand, the relatively strong but reversible complexation between the nitrogen bases and the nitroaromatic stationary phases makes it possible to preconcentrate these compounds from hydrocarbon mixtures such as gasoline or diesel fuel. Then on-line preconcentration is carried out with a small precolumn packed with DNAP or DNSP silica placed upstream of the analytical column. For example, the detection limit of pyridine, added to a diesel fuel virtually free from PANHs, is about 50 and 100 ppb using DNAP and DNSP silica, respectively. Aniline, quinoline, pyridine, and isoquinoline were thus identified in a kerosene sample. These impurities are present in ppm amounts.

Pharr et al. [70] investigated two chloro-bonded phases and their application to the separation, by charge-transfer interactions, of N-heterocyclic compounds. Indeed, chemically bonded 3-(2,2,2-trichloroethoxy)propylsilica and 3-(2,2,2-tri-chloroacetamido)propylsilica were used for the HPLC separation of PAHs and PANHs. The selectivity of the bonded phases for different compounds depends on solubility, adsorption effects, and CT interactions. A balance between these parameters must be maintained for the selective retention of the compounds by a judicious choice of the mobile phase (hexane, hexane-dichloromethane, or hexane-methyl *t*-butyl ether mixture). Retention data for thiophenes and nitrogen heterocycles were obtained. The procedure was applied to some shale oil distillates, before and after treatment with a few drops of mercury, to remove elemental sulfur and sulfur-containing compounds. Two main regions can be distinguished in the chromatograms: the first one (peak A) corresponds to one- and two-aromatic-ring compounds; the second one (a large mass of peaks, "peaks B") was attributed to polar compounds such as phenols, pyrroles, and furans.

Remark. Another application of strong electron donor–acceptor interaction in the field of PANHs was used by Zander [71] in fluorescence analysis. Indeed, silver nitrate selectively enhances the phosphorescence/fluorescence quantum yield ratio of azaaromatics PANHs relative to that of carbazole and PAHs, because of the strong electron-donor properties of PANHs.

REFERENCES

1. J. M. Colin and G. Vion, *Symposium International*, Lyon, June 25–27 1984, Technip, Paris, p. 292.
2. *Annual Book of ASTM Standards,* Part 23, American Society for Testing and Materials, Philadelphia, 1980.
3. G. Felix, E. Thoumazeau, J. M. Colin, and G. Vion, *J. Liq. Chromatogr. 10*: 2115 (1987).
4. J. F. McKay and D. R. Latham, *Anal. Chem., 52*: 1618 (1980).
5. K. D. Bartle, M. L. Lee, and S. A. Wise, *Chem. Soc. Rev., 10*: 113 (1981).
6. B. W. Bradford, D. Harvey, and D. E. Chalkley, *J. Inst. Pet., 41*: 80 (1955).
7. P. C. Hayes Jr. and S. D. Anderson, *Anal. Chem., 57*: 2094 (1985).
8. P. C. Hayes Jr. and S. D. Anderson, *Anal. Chem., 58*: 2384 (1986).
9a. A. Norris and M. C. Rawdon, *Anal. Chem., 56*: 1767 (1984).
9b. R. M. Campbell, N. M. Djordjevic, K. E. Markides, and M. L. Lee, *Anal. Chem., 60*: 356 (1988).

10. S. Matsushita, Y. Tada, and T. Ikushige, *J. Chromatogr.*, *208*: 429 (1981).
11. M. G. Rawdon and T. A. Norris, *Am. Lab. (Fairfield Commun.)*, *16*: 17 (1984).
12. D. R. Gere, R. Board, and D. McManigill, *Anal. Chem.*, *54*: 736 (1982).
13. M. G. Rawdon, *Anal. Chem.*, *56*: 831 (1984).
14. P. Chartier, P. Gareil, M. Caude, R. Rosset, B. Neff, H. Bourgognon, and J. F. Husson, *J. Chromatogr.*, *357*: 381 (1986).
15. G. Felix, A. Thienpont, M. Emmelin, and A. Faure, *J. Chromatogr.*, *461*: 347 (1989).
16. G. P. Blumer, H. W. Kleffner, J. Palm, and M. Zander, *Erdoel Kohle Erdgas Petrochem. Brennst. Chem.*, *35*: 259 (1982).
17. J. W. Stadelhofer, W. Gemmeke, and M. Zander, *Light Met.*, 1211 (1983).
18. K. D. Bartle and M. Zander, *Erdoel Kohle Erdgas Petrochem. Brennst. Chem.*, *36*: 15 (1983).
19. M. Zander, *Erdoel Kohle Erdgas Petrochem. Brennst. Chem.*, *38*: 496 (1985).
20. M. Zander, *Fuel*, *66*: 1459 (1987).
21. T. Yokono, T. Imamura, and Y. Sanada, *Sekiyu GakkaiShi*, *30*: 59 (1987); *Chem. Abstr.*, *106*: 87379j (1987).
22. T. Yokono, N. Takahashi, and Y. Sanada, *Energy Fuels*, *1*: 227 (1987).
23. M. Godlewicz, *Nature*, *164*: 1132 (1949).
24. W. Giger and M. Blumer, *Anal. Chem.*, *46*: 1663 (1974).
25. D. M. Jewell, *Anal. Chem.*, *47*: 2048 (1975).
26. T. Tye and Z. Bell, *Anal. Chem.*, *36*: 1612 (1964).
27. T. J. Wozniak and R. A. Hites, *Anal. Chem.*, *55*: 1791 (1983).
28. T. J. Wozniak and R. A. Hites, *Anal. Chem.*, *57*: 1320 (1985).
29. M. Diack, R. Gruber, D. Cagniant, H.. Charcosset, and R. Bacaud, 2nd International Rolduc Symposium on Coal Science, 1989, Rolduc, The Netherlands, *Fuel Process. Techol.*, *24*: 151 (1990).
30. M. Diack, Thesis, University of Metz, France, 1990.
31. T. J. Wozniak, Ph.D. dissertation, Indiana University, Bloomington, Ind., 1984.
32. M. Besson, R. Bacaud, H. Charcosset, V. Cebolla-Burillo, and M. Oberson, *Fuel Process. Technol. 12*: 91 (1986).
33. R. Bacaud, H. Charcosset, and M. Jamond, 2nd International Rolduc Symposium on Coal Science, 1989, Rolduc, The Netherlands.
34. M. Zander, *Fuel Process. Technol.*, *20*: 69 (1988).
35. G. P. Blümer, R. Thoms, and M. Zander, *Erdoel Kohle Erdgas Petrochem. Brennst. Chem.*, *31*: 197 (1978).
36. G. P. Blümer and M. Zander, *Fresenius Z. Anal. Chem.*, *288*: 277 (1977).
37. R. Thoms and M. Zander, *Z. Anal. Chem.*, *282*: 443 (1976).
38. E. P. Lankmayr and K. Müller, *J. Chromatogr.*, *170*: 139 (1979).
39. A. Matsunaga, *Anal. Chem.*, *55*: 1375 (1983).
40. G. Eppert and I. Schinke, *J. Chromatogr.*, *260*: 305 (1983).
41. P. L. Grizzle and J. S. Thomson, *Anal. Chem.*, *54*: 1071 (1982).
42. J. S. Thomson and J. N. Reynolds, *Anal. Chem.*, *56*: 2434 (1984).
43. K. J. Welch and N. E. Hoffman, *J. High Resolut. Chromatogr. Chromatogr. Commun.*, *9*: 417 (1986).
44. W. Holstein, *Chromatographia*, *14*: 468 (1981).
45. W. Holstein, *Erdoel Kohle Erdgas Petrochem. Brennst. Chem.*, *40*: 175 (1987).

46. P. Jadaud, M. Caude, R. Rosset, X. Duteurtre, and J. Henoux, *J. Chromatogr. 464*: 333 (1989).
47. P. Jadaud, M. Caude, and R. Rosset, *J. Chromatogr. 439*: 195 (1988).
48. H. Deymann and W. Holstein, *Erdoel Kohle Erdgas Petrochem. Brennst. Chem., 34*: 353 (1981).
49. W. Holstein and D. Severin, *Chromatographia, 15*: 231 (1982).
50. Ph. Cléon, M. C. Fouchères, D. Cagniant, D. Severin, and W. Holstein, *Chromatographia, 20*: 543 (1985).
51. J. Lam and A. Berg, *J. Chromatogr., 20*: 168 (1965).
52. W. Funk, V. Glück, B. Schuch, and G. Donnevert, *J. Planar Chromatogr. 2*: 28 (1989).
53. G. Felix and C. Bertrand, *Symposium International*, Lyon, June 25–27, 1984, Technip, Paris, p. 368.
54. G. Felix and C. Bertrand, *J. Chromatogr., 319*: 432 (1985).
55. G. Felix, C. Bertrand, and Van Gastel, *Chromatographia, 20*: 155 (1985).
56. G. Felix and C. Bertrand, *Analusis, 15*: 28 (1987).
57. T. H. Mourey, S. Siggia, P. C. Uden, and R. J. Crowley, *Anal. Chem., 52*: 885 (1980).
58. C. D. Czogalla and F. Boberg, *Sulfur Rep., 3*: 121, 1983.
59. W. L. Orr, *Anal. Chem., 38*: 1558 (1966).
60. W. L. Orr, *Anal. Chem., 39*: 1163 (1967).
61. L. R. Snyder, *Anal. Chem., 41*: 314 (1969).
62. H. T. Rall, C. J. Thompson, H. Y. Coleman, and R. L. Hopkins, *U. S. Bur. Mines Bull., 659* (1972).
63. J. W. Vogh and J. E. Dooley, *Anal. Chem., 47*: 816 (1975).
64. T. Kaimai and A. Matsunaga, *Anal. Chem., 50*: 268 (1978).
65. W. F. Joyce and P. C. Uden, *Anal. Chem., 55*: 540 (1983).
66. K. D. Gundermann, H. P. Ansteeg, and A. Glitsch, *Proceedings of the International Conference on Coal Science*, Pittsburgh, 1983, p. 631.
67. M. Nishioka, R. M. Campbell, M. L. Lee, and R. N. Castle, *Fuel, 65*: 270 (1986).
68. D. M. Jewell and R. E. Snyder, *J. Chromatogr., 38*: 351 (1968).
69. L. Nondek and V. Chvalovsky, *J. Chromatogr. 312*: 303 (1984).
70. D. Y. Pharr, P. C. Uden, and S. Siggia, *J. Chromatogr. Sci., 26*: 432, 1988.
71. M. Zander, *Z. Naturforsch., 33a*: 998 (1978).

Abbreviations

CHROMATOGRAPHIC METHODS

Ag	argentation
CT	charge transfer
CTC	charge-transfer chromatography
DA–CC	donor–acceptor complex chromatography
EDA	electron donor–acceptor
GC	gas chromatography
GC–MS	gas chromatography coupled to mass spectrometry
GC–CC	gas chromatography on glass capillary column
GLC	gas–liquid chromatography
GSC	gas–solid chromatography
HPLC	high-performance (pressure) liquid chromatography
HPTLC	high-performance thin-layer chromatography
HPTFC	high-performance thin-film chromatography
LC	liquid chromatography
LEC	ligand-exchange chromatography
MPLC	medium-pressure liquid chromatography
NARP	nonaqueous reversed-phase chromatography
PARC	partial argentation resin chromatography

RPLC	reversed-phase liquid chromatography
RP-HPLC	reversed-phase high-pressure liquid chromatography
SCF	supercritical fluid chromatography
TLC	thin-layer chromatography

THEORIES, METHODS, AND DETECTORS

DC	dielectric constant detector
FIA	fluorescent indicator adsorption
HGTA	hydrocarbon group type analysis
HSAB	hard-soft acid-base
LFER	linear free-energy relationship
PCRD	postcolumn reactor detector

LIGANDS FOR BONDED PHASES

CSPs	chiral stationary phases
DNA	2,4-dinitroanilino
DNAO	8-(2,4-dinitroanilino)octyl
DNAP	3-(2,4-dinitroanilino)propyl
DNBP	3,5-dinitrobenzamidopropyl
DNF	2,6-dinitrofluorenimino
DNFA	dinitrofluoroanilinopropyl
DNPS	bis(3-nitrophenyl)sulfone
DNSP	2,4-dinitrophenylsulfamidopropyl
PCA	pentachloroanilinopropyl
PFA	pentafluoroanilinopropyl
PFBP	pentafluorobenzamidopropyl
TCI	tetrachlorophtalimido
TENF	2,4,5,7-tetranitrofluorenimino
TNAO	8-(2,4,6-trinitroanilino)octyl
TNAP	3-(2,4,6-trinitroanilino)propyl or picramidopropyl
TNFI, TRNF	2,4,7-trinitrofluorenimino

DERIVATIVES AND RADICALS

AA	arachidonic acid
ACAC	acetylacetonate
ANA	aminophtalamide
BEHP	bis(2-ethylhexyl) phosphate
CHB	cyclohexylbutyrate
CHD	bis(1,2-cyclohexanedione dioximato)

Cl–silica gel	3-chloroproylsilyl bonded silica gel
DHP	di-*n*-hexyl phosphinate
DMF	dimethylformamide
DMG	bis(dimethylglyoximato)
DNP	2,4-dinitrophenyl
DDQ	2,3-dichloro-5,6-dicyanoquinone
m–DNB	*m*-dinitrobenzene
DCNB	2,4-dinitrochlorobenzene
DNPh	2,4-dinitrophenylhydrazone
DOPC	dioleolylphosphatidyl choline
EN	ethylene diamine
FA	fatty acid
FAME	fatty acid methyl ester
G–gel 60	poly(2,3-epoxipropylmethacrylate)
HETE	12-hydroxyeicosatetraenoic acid
HHT	12-hydroxyheptadecatrienoic acid
HNDPS	hexanitrodiphenylsulfide
Ndc	*n*-nonyl-β-diketone
N–DTP	2-thienyl-1-naphthyldithiophosphinic acid
nPrS	n-propylsulfur
OPTA	*N*-(2-hydroxyethyl)propylene diaminotriacetate
PAs	phosphatidic acids
PC	phosphatidyl choline
PCSM	polychloromethylstyrene
PI	phosphatidyl inositol
PY	pyridine
PGs	prostaglandines
PAC	polycyclic aromatic compound
PAHs	polycyclic aromatic hydrocarbons
PANHs	nitrogen heterocycles
PASHs	sulfur heterocycles
PCP	pentachlorophenyl
Phen	1,10-phenanthroline
PMDA	pyromellitic dianhydride
Quin	quinoleine
R–DTP	*p*-phenylene bis(methoxyphenyl)dithiophosphinic acid
SAL	bis(salicylaldimino)
SDVB-0.3	polystyrene-0.3% vinylbenzene copolymer
TETRYL	2,4,6-trinitrophenyl *N*-methyl nitramine
TFAC	3-trifluoroacetylcamphorate
TFAC-1R	3-trifluoroacetyl-(1*R*)-camphorate

THF	tetrahydrofuran
THP	tetrahydropyran
TCNE	tetracyanoethylene
TMUA	tetramethylureic acid
TNB	1,3,5-trinitrobenzene
TNF	2,4,7-trinitrofluorenone
TNT	2,4,6-trinitrotoluene
TNDPS	tetranitrodiphenyl sulfide
TPhB	tetraphenylborate

Index

285

[Olefins]
  polyolefins, 157–159
Orbitals
  bonding, 2
  frontier, 2, 4
    energy of, 14
    selectivity and, 14
  molecular
    theory, 1, 2
  nonbonding, 2

Packings (*see also* Polymers, Silica)
  in chiral LEC, 204–228
  in EDA chromatography, 33–95
  in GC, 33–38
  in LC, 45–48
  in LEC, 41–45, 198
  in TLC, 38–40
PAHs
  by Ag/LC, 152
  by CTC/GC, 9, 126
  by CTC/HPLC
    on CSPs, 127–130
    with nitro groups, 109–116, 258, 260–269
    with other acceptors, 118–123, 267, 270–273
  by CTC/LC, 10, 38, 45, 46, 69, 104–108
  by CTC/RP-LC, 22
  by CTC/TLC, 67–69, 98–103, 270–274
  in coal products
    anthracenic oil, 270
    carcinogenic, 248, 256
    mutagenic, 248, 256
    pitches, 263, 270
    recycle oil, 270
    tars, 260, 270
  as electron acceptors, 6
  as electron donors, 6
  in HGTA, 248–273
  nitro derivatives, 117, 118
  in petroleum products, 46, 258, 264–272

[PAHs]
  asphaltenes, 273
  crude oils, 267
  gas oils, 267
  heavy distillates, 264
  lubrificating oils, 45, 267, 270
  retention indexes, 15, 16, 122, 124, 125
  Rf values, 98, 99, 102, 103
  in shale oils, 273
  structural parameters, 16
Peptides, 27, 42
  di, 27, 42
  enkephalin, 42
  oligo, 133, 135
  poly, 42
Pesticides
  sulfur containing
    in CTC/TLC, 143
Petroleum products (*see also* HGTA, PAHs)
  olefins in, 248
  PAHs in, 248
  PANHs, 275, 276, 278
  PASHs, 275, 276
  saturated in, 248
Pharmaceuticals (*see* Drugs)
Phenols, 58–60, 66, 76
  amino, 77
Pheromones, 69, 70, 72, 152, 166
Polymers, (*see also* Dextrans)
  natural, 11, 133–137
  packings with, 204–212, 219
    chiral ligands for, 206
    polyacrylamide, 210
    polymethacrylate, 211
    polystyrene, 204, 205
    polyvinyl pyridine, 211
  as sorbents
    electron acceptors, 11
    electron donors, 11
  synthetic, 11, 204–212
Prostaglandins, 39, 41, 63, 64, 71, 72, 77, 156, 177–179
Purines, 40, 68, 137, 138, 140
  oxi, 42